Hal Espen

Caroline Fraser's work has appeared in publications such as *The New Yorker* and *The New York Review of Books*. Her first book, *God's Perfect Child*, was a *New York Times Book Review* Notable Book and a *Los Angeles Times Book Review* Best Book.

ALSO BY CAROLINE FRASER

God's Perfect Child

REWILDING
THE WORLD

REWILDING THE WORLD

DISPATCHES FROM THE
CONSERVATION REVOLUTION

CAROLINE FRASER

Picador

———

A Metropolitan Book
Henry Holt and Company
New York

Designed by Meryl Sussman Levavi
Maps by Mapping Specialists®

The Library of Congress has cataloged the Henry Holt edition as follows:

Fraser, Caroline.
   Rewilding the world : dispatches from the conservation revolution / Caroline Fraser.—1st ed.
       p. cm.
   Includes bibliographical references and index.
   ISBN 978-0-8050-7826-8
   1. Biodiversity conservation. 2. Endangered species. 3. Restoration ecology. I. Title.

QH75.F738 2009
333.95'16—dc22

2009032989

Picador ISBN 978-0-312-65541-9

First published in the United States by Henry Holt and Company

P1

For Hal

Only connect! . . . Live in fragments no longer.

—E. M. FORSTER

CONTENTS

REWILDING
THE WORLD

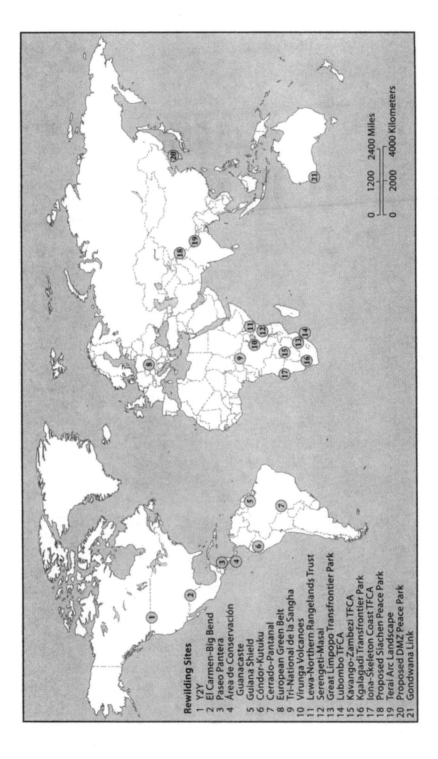

Rewilding Sites

1 Y2Y
2 El Carmen–Big Bend
3 Paseo Pantera
4 Area de Conservación
   Guanacaste
5 Guiana Shield
6 Cóndor-Kutuku
7 Cerrado-Pantanal
8 European Green Belt
9 Tri-National de la Sangha
10 Virunga Volcanoes
11 Lewa–Northern Rangelands Trust
12 Serengeti–Masai
13 Great Limpopo Transfrontier Park
14 Lubombo TFCA
15 Kavango–Zambezi TFCA
16 Kgalagadi Transfrontier Park
17 Iona–Skeleton Coast TFCA
18 Proposed Siachen Peace Park
19 Terai Arc Landscape
20 Proposed DMZ Peace Park
21 Gondwana Link

0   1200   2400 Miles

0   2000   4000 Kilometers

THE PREDICTA MOTH

OVER THE YEARS, COYOTES ATE MANY OF MICHAEL SOULÉ'S cats. For most people, this might have been the end of the story, a nasty reminder of nature's darker proclivities. But Michael Soulé is not most people.

Soulé is a biologist. At the time, he was a professor at the University of California at San Diego, living in the chaparral canyons outside the city. He had grown up in the canyons, poking around in the leaf litter, catching lizards. When the boy became a biologist, he recognized that the chaparral was a unique ecosystem, with its own suite of interdependent plants and animals, the coastal sage scrub home to fox and bobcats, wrentits and spotted towhees, cactus mouse and California quail. But to real estate developers, the canyons were empty wasteland, waiting to be turned into homes. As he watched the progressive paving of the canyons, Soulé found himself even more distressed about the big picture, the loss of the ecosystem, than about the cats. Recent breakthroughs in biology had suggested that fragmentation of habitat inevitably threatened species. As developers carved

Global Map: A sampling of hundreds of rewilding projects around the world involving every large-scale method from transboundary and community conservation to ecological restoration.

the canyons into suburban lots, leaving behind islands of isolated brush, Soulé was alarmed enough to investigate that theory, and he sent students to compile data on the disappearance of birds from thirty-seven forlorn chaparral islands. He also had them collect data on local carnivores, to see if predation was a factor. After two years, as expected, data showed that the number of birds and other species in each patch was diminishing.[1]

But the data revealed something else, something counterintuitive. In canyons with coyotes, a greater diversity of birds survived. Canyons without coyotes supported fewer species. Having seen ample evidence that coyotes were responsible for his disappearing pets—cats flying through the cat door as if "chased by the devil"—Soulé had a theory: more coyotes meant fewer cats.[2] Fewer cats meant more birds. Coyotes were eating not only cats but also other midsized predators, such as foxes. Coyotes were acting as a control. Without that control, the midsized carnivores ran wild in an orgy of predation that Soulé termed "mesopredator release."[3] Another study confirmed it: one in five coyote scats contained domestic cat.

Before long, scientists were realizing that much of the country was suffering from a bad case of mesopredator release. The artificial absence of wolves and other large predators gave cats, dogs, raccoons, and foxes license to grow fat on wild birds from the beaches to the mountainsides. Soulé had observed just one manifestation of a crucial new scientific discovery: predators do not merely control prey. They control other predators, and by doing so, they regulate species with which they never directly interact. They regulate biological processes down the food chain. As scientists study the unbalanced and fragmented systems humans create as they alter the environment, they are realizing how interdependent species are. In a way, all of us are now living in a scientific experiment similar to that which San Diego developers created by carving up the canyons. We have unleashed forces we are still struggling to comprehend.

This global experiment is comparable to the one Americans unwittingly set in motion in the 1950s by sowing the land with toxic pesticides. In the fable that opens *Silent Spring*, Rachel Carson described a "strange blight" settling over a town.[4] Birds fell silent, bees vanished. There was no pollination, no fruit. "Everywhere," she wrote, "was a shadow of death."[5] Carson helped keep that pall from settling over the whole country by inciting a national debate that led to the banning of DDT.

But the shadow has fallen again. This time the problem is neither as concentrated nor as easily tackled as that of pesticides or pollution. This time the problem is the disappearance of nature itself.

The blight we are now experiencing, termed the "demographic winter," is a result of human population growth, rampant development, and the destruction of ecosystems.[6] Together, these forces are acting to reduce the extraordinary diversity of plants and animals to a select, hardy few. Scientists estimate that this period of loss may last a millennium before tapering off and may bring about the elimination of millions of species of plants and animals. For biologists, the worst implication of the demographic winter is that by shrinking and isolating habitat, cutting wildlife populations to the bone, we may be erasing the process of evolution itself. No one can imagine the consequences of that. It seems unthinkable. But field biologists around the world are forced to think it, confronted with the evidence of empty forests and coral reefs reduced to slime. "It's not death I mind," Soulé once said. "It's the end of life that bothers me."[7]

Biodiversity loss is now lining up to be the greatest man-made crisis the world has ever known. Biologists call it the Sixth Great Extinction, or the Holocene extinction event, after our current geologic time period. (The five previous extinction events all came before the evolution of *Homo sapiens*, apparently triggered by a cataclysmic event or combination of events, such as a fall in sea level, an asteroid impact, volcanic activity.) Mass extinctions are different in kind from what specialists term "background" extinctions, the rare but regular loss of between one to ten species per decade. Two hundred and fifty million years ago, the most catastrophic, "the Great Dying" of the

Permian age, wiped out over 90 percent of all species in the oceans and 70 percent on land. It took tens of millions of years for life to recover.

The current extinction rates are alarming enough. Preeminent biologist E. O. Wilson believes we stand to lose half of all species by the end of this century.[8] Of the 45,000 species evaluated in the 2008 Red List, issued by the International Union for Conservation of Nature, 17,000, or nearly forty percent, may vanish.[9] Conservative estimates suggest that the extinction rate in the modern era has reached a hundred to a thousand times normal.

Climate change further exacerbates biodiversity loss, and each of these crises magnifies and intensifies the effects of the other. As the planet warms and dries in some areas, species are pushed out of niches they currently occupy. Some of them, including the emperor penguin in Antarctica and the polar bear in the Arctic, have nowhere to go. Worldwide, as water temperatures rise and ponds dry, exposing amphibians and their eggs to ultraviolet radiation and disease, a third of those species are threatened with extinction. As people burn forests for agriculture and grazing, as they replace native vegetation with monoculture crops that discourage cloud formation, they alter the dynamic relationship between the earth's surface and the atmosphere, initiating further drying and warming, and further species loss.

Why do species matter? Why worry if some go missing? Part of the answer lies in the relationships coming to light between creatures like the canyon coyotes and the chaparral birds. After the nineteenth century's great age of biological collecting, when collectors filled museums to bursting with stuffed birds and pinned beetles, the twentieth and twenty-first centuries have proved to be an age of *connecting*. Biologists have begun to understand that nature is a chain of dominoes: If you pull one piece out, the whole thing falls down. Lose the animals, lose the ecosystems. Lose the ecosystems, game over.

Put another way, in this era of connection, we have learned that everything is interdependent. There are no spare parts. Predators regulate a constellation of other predators and prey; grazing animals regulate grasslands; grasslands and forests regulate climate. The physical world is like a big organic machine, an old car, for example, composed of interconnecting moving parts. Eco*systems*. You can lose inessential cosmetic elements, the bumper, the hubcaps, and it will still run, for a while. But eventually, if you lose enough critical parts, the machine will fail. When parts of an ecosystem are lost—predators, grazers, pollinators—the machine starts stalling, stuttering, failing. The processes of life grind to a halt.

This was the essential insight of conservation biology, a new scientific field launched with the determination to identify threats to ecosystems and to design the methods to deal with them. E. O. Wilson has called it "a discipline with a deadline."[10] The Society for Conservation Biology, founded in 1985, became one of the fastest-growing scientific organizations of its time, bringing together diverse specialties from ecology and population genetics to sustainable agriculture and forestry, revolutionizing the once sleepy field of natural history. Conservation biology and its epiphanies constitute a latter-day Copernican revolution. Just as Copernicus revealed that the earth was not the fixed center of the universe, so conservation biologists have found that *Homo sapiens* is not the independent actor he has imagined himself to be. With agriculture and livestock, we may have stepped outside the local habitats that rule other creatures' lives, making ourselves at home in places too cold or nutritionally marginal for any other primate. But we are still subordinate to natural forces. No species survives in a vacuum, including our own.

The tremendous variety of species held in wilderness areas, particularly the tropics, is our bank and lifeline, our agricultural and medical insurance policy. Three-quarters of the world's food supply comes from twelve plant species, but those species are dependent on thousands of others: pollinators (insects, bats, birds), soil microbes, nitrogen-fixing bacteria, and fungi. The tropical rain forests contain a pool of genetic diversity for important food crops, a source for vital new strains that can be hybridized to fight pests and diseases. Botanists are combing the planet for wild ancestors of soybeans, tomatoes, hard wheat, and grapes, believed to contain valuable genes for drought tolerance and other characteristics, but much diversity has already been lost. Genetic engineering alone cannot replace what hundreds of millions of years of evolution have given us. Wild replacements for pineapples, pomegranates, olives, coffee, and other crops lie in biodiversity-rich areas that must be saved.

In terms of medicine, our most important modern pharmaceuticals, including quinine, morphine, aspirin, penicillin, and many other antibiotics, are derived from microbes, plants, and animals found in tropical and marine environments. The first comprehensive scientific treatise on our reliance on other species, *Sustaining Life: How Human Health Depends on Biodiversity*, published in 2008, confirmed the importance of genetic variety, describing groups of threatened organisms crucial to agriculture and human medicine. Predictably, our close relatives, primates, make up a key group. Contributing to work on smallpox, polio, and vaccine development, primates allow research on potential treatments for hepatitis C and B, Ebola and Marburg viruses, and HIV/AIDS.[11]

The list of threatened plants and animals we rely on is weird and varied, including amphibians, bears, gymnosperms (the family of plants that includes pine trees), cone snails, sharks, and horseshoe crabs.[12] Cone snails, a large genus of endangered marine mollusks, inject their prey with paralyzing toxins that are prized in medical research for their use in developing pain medications for cancer and AIDS patients who are unresponsive to opiates. The blood of the horseshoe crab, which carries antimicrobial peptides that kill bacteria, is being tested in treatments for HIV, leukemia, prostate cancer, breast cancer, and rheumatoid arthritis; it also yields cells crucial in developing tests to detect bacteria in medical devices, and its eyes have allowed Nobel Prize–winning researchers to unravel the complexities of human vision.

Cone snails and horseshoe crabs are exactly the kinds of species that people tend to dismiss, seeing no utility in them, no connection to human need. This was the attitude expressed in 1990 by Manuel Lujan Jr., secretary of the interior during the George H. W. Bush administration, who asked in exasperation, "Do we have to save every subspecies?"[13] It was the attitude expressed in 2008 by presidential candidate John McCain, who repeatedly declared his opposition to the funding of research on grizzly bear DNA. He got a cheap laugh whenever he said, "I don't know if that was a paternity issue or a criminal issue."[14] Medical researchers were not laughing: bears, too, are essential to human medicine. Bear bile yields ursodeoxycholic acid, now used in treating complications during pregnancy, gallstones, and severe liver disease. Denning bears enter a period of lethargy during the winter and recycle body wastes in a process unique in mammals; this process is studied for insights in treating osteoporosis, renal disease, diabetes, and obesity.

If species are crucial to medicine, ecosystems are indispensable to the health of the planet. Ecosystems provide the most basic provisioning services—food, firewood, and medicines—along with the so-called regulating services of a fully functional environment, which include cleaning the air, purifying water, controlling floods and erosion, storing carbon, and detoxifying pollutants in soils. When ecosystems are lost, as they have been through felling of forests and conversion of landscape to agriculture on a vast scale, havoc ensues, triggering human and natural catastrophe on an unprecedented scale. When Americans plowed under the grasslands of the Great Plains, they set the stage for the Dust Bowl, violently exacerbating the effects of cyclical drought. The loss of that ecosystem plunged millions of people into misery and poverty, deepening the Great Depression. Australians sparked their own post–World War II dust bowl by burning millions of acres of native bush and planting wheat. Without eucalyptus and native bush cover to keep rainfall from entering aquifers below, without birds and native wildlife to propagate the bush, water saturated dry subsoil, mobilizing ancient salt deposits. Since then, 10 percent of the wheat belt in western Australia has turned to salt, with 40 percent more poised to follow in the next few decades. In nearby towns, salt erodes buildings, literally exploding bricks in slow motion from within. While the cracks in North America's natural systems are slower to appear than in Australia, they are nonetheless appearing. In the southwestern United States, the water table has been driven to such historic lows that geologists are beginning to see visible subsidence: waves and cracks in the earth.

Gradually, we are realizing that the environment *is* the economy. The planet's rain forests currently function as a "giant 'utility,'" according to Andrew Mitchell, director of the Global Canopy Program.[15] He points out that the Amazon alone releases twenty billion tons of water into the atmosphere daily, providing free air-conditioning, free irrigation, free hydropower. With more tracts of rain forest lost every day, what will it cost to provide artificially the services we currently get for free? A recent study commissioned by the European Union calculates that those lost services, along with the massive release of carbon as forests go up in smoke (accounting for 20 percent of carbon emissions worldwide), add up to 7 percent of global GDP, or two to five *trillion* dollars annually—the equivalent of the total cost of the Iraq War every year.[16]

Signs of these costs are showing up everywhere, because ecosystem services are beginning to fail. After several record-breaking drought years, the Amazon is approaching a tipping point when its "rain machine" may malfunction, putting South American agriculture in peril. Australia's rice and wheat crops are failing due to drought, sending worldwide wheat prices to a ten-year high. Emerging diseases are being unleashed as forests are cut. Honeybee colonies are vanishing, leaving commercial crops from avocados to oranges unpollinated. Water shortages, food riots, failed crops, the worldwide collapse of pollinators: this is only the beginning. Welcome to the demographic winter, the new hot season in hell.

But what if we had a choice? What if we could slow or even stop the Sixth Great Extinction? What if conservation could save us from the demographic winter?

It can. Conservation biologists have developed a number of methods for restoring the balance between ourselves and nature, for saving biodiversity. The most exciting and promising of these methods is re-wilding.[17] Proposing conservation and ecological restoration on a scale previously unimagined, rewilding has become a principal method for designing, connecting, and restoring protected areas—the ultimate weapon in the fight against fragmentation.

Michael Soulé and a colleague, Reed Noss, formulated the essence of the new discipline in a 1998 paper, "Rewilding and Biodiversity: Complementary Goals for Continental Conservation."[18] In it, they boiled the requirements down to three words: "Cores, Corridors, and Carnivores."[19] Core protected areas had long been a feature of conservation design, but Soulé and Noss described national parks and wildlife refuges as only the beginning, the kernels from which larger, mightier protected areas must grow. Cores, they argued, must be continental in scale, preserving entire ecosystems: mountain forests, grasslands, tundra, savannah. Corridors were necessary to reestablish links between cores, because isolation and fragmentation of wilderness areas erode biodiversity: They enabled wildlife to migrate and disperse. And carnivores were crucial to maintaining the regulatory mechanisms keeping ecosystems healthy, harking back to those chaparral canyons. Because large carnivores regulate other predators and prey, exercising an influence on the ecosystem far out of proportion to their numbers, their protection and reintroduction is crucial. Because predators need large areas to survive, "they justify bigness."[20]

Over time, the definition has been broadened and refined by a host of experts. Cores are to be expanded and strictly protected, and their natural fire and flood regimes restored, wherever possible. Corridors are only one type of connectivity, which may take forms other than simple linear strips of land, including patches, or stepping stones, of habitat. Both cores and corridors may require restoration of degraded habitat to achieve large-scale connectivity; carnivores may need to be reintroduced. The category of carnivores has now been joined by "keystone species," creatures that interact so strongly with the environment that they wield an outsized influence. The damming of beavers—altering the course of streams, opening meadows within forests, and creating pond ecosystems—elevates them to a keystone species. The grazing and browsing of elephants, who act as forest engineers, pushing over trees and keeping vast grasslands like the Serengeti open, makes them keystones.

The original proponents of rewilding were careful to propose it as a "complementary" method to those being implemented by nongovernmental organizations like the World Wildlife Fund. Some of those methods are similar to rewilding in their focus on large-scale conservation. "Representation," for example, is one large-scale strategy, focused on preserving representative areas of every identifiable ecosystem, such as savannah, tropical moist forest, tundra, desert, coral reef. The WWF's "ecoregions" program favors representation.[21] Another model, "hotspots," is designed to save unique areas of high endemism, places like the Galápagos Islands, where many species of plants and animals found nowhere else in the world have evolved.[22] The large-scale continental reserves envisioned by rewilding might neglect island hotspots like Madagascar or Java. Likewise, a single-minded focus on hotspots might shortchange areas like the African savannah, which is low in endemic species but enables mass migration.

But rewilding's unique triple focus on protecting and restoring cores, connectivity, and carnivores (or keystones) sets it apart from other large-scale conservation methods and projects. The goals of reintroducing carnivores where extirpated and restoring connectivity—even if it means replanting or regrowing bushland or forest between reserves—make rewilding more ambitious than even the most visionary conservation plans of the past.

Over the last two decades, extraordinary progress has been made. Many landscape-scale rewilding projects have been launched, aimed at restoring "megalinkages" throughout and between continents: Yellowstone to Yukon, Algonquin to Adirondack, and Baja to Bering in North America; Paseo Pantera in Central America; the Terai Arc in Asia; Gondwana Link in southwest Australia; the transboundary peace parks in Africa.

In line with these goals, many countries have moved to place more land under protection. Only a thousand protected areas existed in 1962, representing 3 percent of the earth's surface. Now there are over a hundred thousand protected areas worldwide, expanding conservation to over 12 percent. According to the United Nations' World Conservation Monitoring Center, protected areas now represent "one of the most significant forms of land use on the planet." While not all protected areas are devoted to rewilding, some of the largest reserves in the world, including the transboundary protected areas in Africa, are. Ecological restoration has worked wonders in Nepal's Terai Arc, where monsoonal lands are recovering from intensive human use as people are persuaded to manage forests for conservation and supplement their income with ecotourism and sustainable native crops. In northern Kenya, a privately owned rhino reserve is guiding communities that are rewilding former grazing lands at the same time as it fosters lucrative tourism facilities in a region once devastated by poaching. Native forests and bushlands are being painstakingly regrown in Costa Rica and Australia.

To be sure, daunting challenges loom. Many parks around the world are still "paper parks," without adequate funding or protection. Issues that threaten to stall or derail rewilding have included everything from poaching to the opposition of people living in or around protected areas, which were often proposed or planned without their input. Restoration itself is a flash point for organizations that fear it might undermine protection and encourage environmental depredation. While transboundary parks—stretching across national borders between neighboring countries—seem thrillingly idealistic on paper, implementing them has proven to be fraught, as planners pick apart political and legal knots while local people grow impatient.

As rewilding has entered the mainstream, increasingly accepted by international organizations, it has had to negotiate an uneasy expansion from a scientifically based conservation method into an ambitious social program. The institutions that fund major conservation and rewilding projects, including the United Nations and the Global Environment Facility, have pressed for greater sensitivity and attention to human rights, insisting that projects with a strict focus on biological conservation be expanded to encompass human aid, in the form of so-called community conservation projects. This has been controversial for biologists; scientists have reluctantly found themselves acting as social engineers, trying to design new economic opportunities for traditional pastoralists, changing the way people live on the land. Conservation organizations have rapidly evolved into groups practicing poverty relief on the side, installing biogas facilities in villages in Nepal or providing seed money for microfinance loans to poor women. In consequence, poverty relief has become bitterly disputed in conservation, with some biologists insisting that it is an ineffective and even counterproductive means of habitat protection.

Controversial or not, the unflinching message of conservation biology is that rewilding is not only a scientific necessity but also an ethical responsibility. Biologists no longer shrink from the overtly moral argument that humanity has an obligation to protect and restore wilderness. That responsibility, they contend, goes beyond any utilitarian argument. Noteworthy biologists from E. O. Wilson to Jared Diamond have pushed their colleagues to enter the political arena in the fashion of groups like International Physicians for the Prevention of Nuclear War, while religious leaders have encouraged their flocks to consider the moral implications of destroying creation. Biologists and clergy are being radicalized by the same shared belief, a sign that we find ourselves on the brink of committing irrevocable acts.

Conservation, like the ecosystems it seeks to restore, has also suffered from isolation. Too often dismissed, too often relegated by governments, media, and decision makers to the bottom of priority lists, beneath scores of other pressing social and economic crises, conservation has only recently begun to be perceived as a critical component of any effort to restore a nation's health and revitalize its agriculture, natural resources, and economy. A wider social recognition of our global ecological interdependence has lagged behind. As a result, the response to the Sixth Great Extinction has been slow in coming, weak in urgency, and disorganized in focus. To remedy that, rewilding—the great project of conservation in recent years—needs sustained consideration and scrutiny. Scientists and activists on virtually every continent have been working for decades to advance the science and promote the agenda, pressuring governments to accept the need for it. As science has been transformed into actual projects, there have been mistakes, trials and errors, grandiose claims, and dead ends: that is, of course, how science works. My purpose is to examine rewilding as a central, pivotal enterprise and to find, amid those successes and failures, a sense of its progress.

Far from being a quixotic, utopian quest for a lost Edenic wilderness of the past, rewilding is a necessity, economic and existential. Along with alternative energy, the emerging professions of ecological restoration and management will help to drive the economy in the future. Already thousands of jobs in developing countries have been created in ecotourism, law enforcement, and agroforestry. From design to implementation, rewilding projects create jobs for a host of specialities—soil assessment, land system mapping, wildlife surveys and management, fire management—and for people in the construction and landscaping fields. Already projects are being designed to store carbon over decades in newly planted native vegetation, to restore connectivity and biodiversity in large-scale protected areas, and to train workers in restoring and maintaining wetlands and removing invasive species. Such projects could constitute the centerpiece of a global jobs program in developing and developed nations alike, training workers in environmental science and carbon sequestration.

Most important, rewilding can play a crucial role in addressing climate change. While it cannot stop the crisis by itself—only systemic changes in government policies and corporate practices can accomplish that—expanded core reserves and restored connectivity between them can stabilize and revitalize forests, protect biodiversity, reestablish the crucial balance between predator and prey, and restore the health of coastlines, prairies, deserts, oceans, and river systems. In this sense, rewilding holds the potential to stabilize far more than natural areas in peril: it can enhance and protect national security by sequestering carbon and safeguarding fresh water, fertile soils, cleaner air. It is a Marshall Plan for the planet.

Famously, Charles Darwin predicted that an extraordinary white night-blooming orchid found only in Madagascar—an orchid with a nectary so deep that only an impossibly long implement could penetrate to the pollen within—necessitated the existence of a night-flying pollinator, probably a moth equipped with a nearly foot-long proboscis.[23] No one had ever seen such a creature. Forty years later, a night-flying subspecies of moth with a nearly foot-long proboscis was found feeding from the flower. It was named *Xanthopan morganii praedicta*, the Predicta moth. Nature itself is like that orchid— unique, fragile, locked in a relationship with a transient being on which it is utterly dependent: ourselves. We flew away for a while, thinking we could leave the natural world behind. But our destinies evolved together. We will survive only in a world as complex and biodiverse and interdependent as the one that created us.

PART I

CORES, CORRIDORS, AND CARNIVORES

CHAPTER I

REWILDING NORTH AMERICA

Pluie

THE PROOF WAS A WOLF. IN JUNE OF 1991, A FIVE-YEAR-OLD alpha female, perhaps searching for new territory, embarked on a two-year foray, roaming over five hundred miles through the Rockies, an area of around 40,000 square miles, twelve times larger than Yellowstone National Park. As scientists were learning, this is what wolves do.[1]

She was wearing a radio collar fitted with a satellite transmitter, and Paul Paquet, a Canadian zoologist studying wolf movement for the World Wildlife Fund, was watching her. Paquet found and collared her in the rain, so he named her Pluie. He and a colleague, biologist Diane Boyd, tracked her with growing amazement as she inscribed an enormous circle, starting near Banff National Park in Alberta, Canada, heading south into Montana, skirting the southern boundary of Glacier National Park, swinging past Coeur d'Alene, Idaho, and Spokane, Washington, and then trekking north into British Columbia, making another pass at Banff. Paquet told a reporter, "We thought she was on a pickup truck for a while, she was moving so fast."[2]

Pluie's journey provided key evidence substantiating the theories on which rewilding is based. In the early 1990s, rewilding as a conservation method was still in its formative stages, but Pluie helped move it from a collection of hypotheses to a specific set of recommendations. A perfect illustration of the cores-corridors-carnivores idea, Pluie's traverse showed a top predator traveling from one core area to another, using wilderness corridors to do it. Her journey made national parks look minuscule, highlighting their inadequate size, isolation, and fragmentation. Although Pluie utilized parks such as Banff and Glacier, she clearly needed a space many times their size, a space a dozen times the size of Yellowstone.

The corridors she used were passageways of remaining forest linking large wilderness areas. Some—particularly in the Bow Valley—were bisected by major highways and railroads, forming dangerous bottlenecks or breaks. Pluie's travels let scientists identify the movement corridors, which, once located, could potentially be protected or restored by fencing off highways and providing safe overpasses for wildlife.

Pluie's journey, Paquet told me later, "made very clear what kind of geographic scale we should be thinking of. We were able to show pretty definitively that these theoretical corridors that we imagined were *actually* there and being used."[3] Now that scientists could watch elusive species like wolves traveling across the landscape, they could begin planning to manage, maintain, and even restore the necessary cores and corridors.

Pluie also demonstrated the contribution of "umbrella species," as biologists called wide-ranging animals: protecting the vast spaces they required could provide shelter for countless additional species.[4] If conservationists were able to put in place a series of big enough protected areas linked by corridors, they would be protecting not only wolves but everything else under that umbrella.

In 1993, Pluie lost her collar, which was found with a bullet hole in it. The wolf herself was shot dead two years later, along with her mate and several pups. Her fate matched that of most wolves, bears, and other large animals in today's West: shot or hit by cars, trucks, or freight trains. She was lucky to have survived as long as she did. One of Paquet's graduate students enumerated the toll that traffic on the Trans-Canada Highway and the Canadian Pacific Railway near Banff National Park had taken on wolves:

> In the last 15 years or so, 27 percent of the known wolf deaths have been from the railway, and 60 percent were on the highway. Just 5 percent were natural. . . . The Bow Valley used to have three packs. Now it has one. In 1996, three of the four pups born to this pack were lost to the highway. The next year, none of the five pups born survived, and we know at least one was hit on the railway. During 1998, the pack had no pups and was down to three members.[5]

Such figures—the rapidity with which an entire population of wolves can become roadkill—suggested the urgency of the need to find a way around these bottlenecks. Almost as soon as Pluie had run her course, the data gathered about the places she went and the routes she took were pressed into service to design a wilderness network. Most importantly, Pluie's movements inspired the first major rewilding project designed around cores, corridors, and carnivores, the Yellowstone to Yukon Conservation Initiative. Pluie's story, Paquet later said, "was the founding story of Y2Y. Really, the whole idea evolved out of it."[6]

But we get ahead of ourselves. Before the projects, before Pluie, before the proof, there was a theory. Like Newton's falling apple, Pluie acted as inspiration and demonstration, but scientific journeys begin with questions, not answers. The origins of rewilding—the conservation method designed to slow a wave of human-caused extinctions— are rooted in the most important ecological theory of the twentieth century, a theory that examines the forces governing extinction, the theory of island biogeography.

The Trouble with Islands

Nature is not a closed system. Since 1930, when a British botanist coined the term *ecosystem* to define the complex interrelationships between plants, animals, and microorganisms and the physical elements they interact with—rocks, soil, water, air—scientists began to recognize that wilderness cannot be preserved by sealing it off. To seal off is to interrupt processes that make life possible: natural selection, predation, competition. Because ecosystems contain such an extraordinary diversity of interactive species and processes, because they are not static, they have proved notoriously difficult to classify and study. Wrenched by larger environmental events—climate change, storms, fires, floods—they are capable of shifting, changing, evolving, and disappearing. Only within the past century have we begun to understand the laws that govern the evolution and transformation of ecosystems.

A momentous advance in understanding such systems came in 1967, when Robert H. MacArthur and Edward O. Wilson published *The Theory of Island Biogeography*. Wilson was thirty-seven when this epochal work appeared and would eventually become the most impassioned elder statesman of conservation, writing Pulitzer Prize–winning volumes about natural history and the need to protect biodiversity; indeed, his books popularized the term, a compression of "biological diversity." But in his early thirties, he was still a young Harvard biologist, albeit the world's greatest taxonomic expert on several subfamilies of ants in Melanesia, having spent years collecting in New Guinea and the islands of Fiji, as well as in Australia and South America. Over the years, he had noticed a pattern: the number of different ant species on any given island seemed to correlate to its size.[7]

As told in Wilson's autobiography, *Naturalist*, and David Quammen's history *The Song of the Dodo: Island Biogeography in an Age of Extinctions*, Wilson discussed this pattern and other issues in the emerging, genetics-driven field of population biology with MacArthur, a young University of Pennsylvania mathematician and ecologist legendary for his ability to discern patterns from masses of data and to construct sophisticated mathematical models illustrating general principles. The theory Wilson and MacArthur shaped is an explanation of how natural forces act to control the number of species populating a given area. Paradoxically, the theory that launched a worldwide movement to protect enormous continental areas was inspired by the smallest of land units, the island.

As MacArthur and Wilson noted at the outset, an island "is certainly an intrinsically appealing study object. It is simpler than a continent or an ocean, a visibly discrete object that can be labelled with a name and its resident populations identified thereby. In the science of biogeography, the island is the first unit that the mind can pick out and begin to comprehend."[8] But as they intuited, islands also provided useful information about the conditions that humanity was creating everywhere, even within continents. Islands were not only land masses surrounded by water; they were also isolated habitats surrounded by development. The principles of insularity—reduction and fragmentation—were going to apply "to an accelerating extent in the future" as habitats were "broken up by the encroachment of civilization."[9]

To understand why size and distance were related to the number of species populating an area, the authors examined two crucial ways in which species rise or fall on an island: immigration and extinction. Using data on ants and other species, they worked out that, as new species arrived, a similar number of established species was becoming locally extinct, in a process of turnover. They set forth a mathematical model illustrating how an island's area and its distance from other islands or mainlands regulated the balance between arrivals and displacement. Their model allowed calculation of a number representing equilibrium—predictable and stable—that was based on those factors of area and distance. If the key factors changed, so did the equilibrium. According to the mathematical model, altering the factors of area and/or distance would cause the number representing equilibrium to reset. If a theoretical island grew smaller or more distant, the number reset downward; if it grew larger or closer to other land masses, the number reset upward. Although I have drastically simplified this explanation, leaving out discussion of additional factors (climate, location relative to ocean currents, initial species composition, etc.), the essence is this: the smaller the island and the more distant from other places, the fewer species it supports. As a rule, a 90 percent decrease in the area of an island results in a 50 percent decrease in diversity.

Equilibrium itself, and the species it represented, could vanish when an island or fragment became so small and isolated that immigration stopped occurring. This was known as "ecosystem decay" and could be seen in one illustrative example: MacArthur and Wilson reproduced a series of maps showing the reduction and fragmentation of a woodland in Wisconsin between 1831 and 1950; the maps clearly showed the progressive diminution of wooded remnants to tiny scraps that could support little variety of flora or fauna.[10] As evidence supporting their theory, the authors looked at the famous volcanic eruption of Krakatau Island in 1883, which snuffed out all life under a sterilizing layer of searing pumice and ash. Although there were no comprehensive data on the flora and fauna prior to the eruption, there were for the subsequent "recolonization episode," in which insects, birds, and mammals returned to an island that had lost two-thirds its total area. MacArthur and Wilson calculated that the number representing equilibrium for bird species, based on its new area and distance from other islands, should have settled at roughly thirty species within forty years. The historical data on recolonization seemed to confirm their calculation.[11]

The last chapter of *The Theory of Island Biogeography* described how further testing might be done by reproducing "miniature 'Krakataus'"—eliminating all species or all of a particular class of organisms (insects, fish) from a series of small islands or lakes, either "manually or by poisoning," and monitoring their return.[12] Wilson and one of his graduate students, Daniel Simberloff, re-created the sterile island experiment, performing an exacting census of all species of insects on several tiny mangrove islands in the Florida Keys, then hiring exterminators to tent and fumigate them. Over the following year, Simberloff monitored their return. "To my absolute delight," Wilson later told Quammen, "we watched the numbers of species rise to what was obviously equilibrium within about a year."[13] The experiment confirmed the theory as it related not only to area but also to distance: the most remote of the experimental islands was the slowest to return to equilibrium, and with the lowest number of species.

The havoc that equilibrium would play in small, remote protected areas was immediately obvious to biologists. By extrapolation, the smaller and more isolated an area is, the farther it is from the nearest wild area—the more *islandlike* it is—the more likely it will exhibit the characteristics of an island, including reduced diversity of species and a higher rate of extinction. Quammen distilled the issue of scale to a single, unforgettable metaphor. What do you get when you take a beautiful Persian carpet, he asked, and cut it into thirty-six pieces? Thirty-six separate carpets? Or thirty-six worthless, fraying scraps? Substitute ecosystems for carpet, he suggested, and you begin to see the problem.[14]

In a 1972 paper inspired by the equilibrium theory, the ornithologist Jared Diamond, who had done extensive fieldwork in New Guinea, directly addressed how equilibrium would affect protected areas that were too small and too islandlike. He observed that the government of that country was in the process of setting aside small rain forest "tracts" as preserves. While the intentions were good, Diamond wrote, the outcome was likely to be the opposite of what they planned. The governments of New Guinea and other tropical countries were creating islands within islands, surrounded by "a 'sea' of open country in which forest species cannot live."[15] Diamond argued that the diminution and fragmentation would cause a "relaxation to equilibrium." The size of the preserves would inevitably trim the number of species they protected to a lower number than the forests initially held, undercutting their very purpose. In yet another paper, Diamond set forth initial suggestions for the size and design of nature reserves, the first to be based on the equilibrium theory. Large is better than small, he argued, and reserves grouped together or, better yet, connected to one another would support more species.

Given the evidence already available, many biologists were quick to agree that when it comes to preserving ecosystems, large is better than small, connected is better than isolated, and whole is better than fragmented. But some were resistant, arguing against a rush to judgment, suggesting that protected areas in the real world might prove vastly more complex, each with unique characteristics that might affect the outcome. The intellectual debate over the subject became so vituperatively colorful that David Quammen made it a central focus of *The Song of the Dodo*. The arguments were heated, he explained, because what was at stake was nothing less than saving the planet: "At a time when humanity was cutting forests and plowing savannas at a rapid pace, when habitat everywhere was becoming fragmented and insularized, the equilibrium theory embodied minatory truths. It was not just an interesting set of ideas—it was goddamned important. If heeded and applied, it might help save species from extinction."[16]

The most prolonged and violent debate to come out of island biogeography, a veritable "pissing match," as one biologist put it, was the debate over "SLOSS": "single large or several small."[17] Are single large protected areas better than several small ones? Jared Diamond strongly defended the idea that a single large reserve would tend to preserve more species. Large reserves, he argued, would preserve large carnivores, which need more space; they would provide more protection in the event of extreme climate change. Others thought that the theory had yet to be proved and, ironically, the most adamant critic of the "single large" camp was Daniel Simberloff. He pointed to cases in which officials in Costa Rica and Israel, in a position to make decisions about conservation, nearly threw out plans for reserves that seemed too small and therefore—or so they thought—useless. Small parks might target hotspots of endemism; the conviction that big is better, Simberloff said, is "a cocktail-party idea" with the "trappings of science."[18]

The SLOSS debate eventually wound down, as more and more scientists and conservationists moved toward the big-is-better-than-small hypothesis. Looking around, they could see that national parks, protected areas, and reserves in the United States and around the world were small, fragmented, isolated. As anyone who has driven to Yosemite or Yellowstone knows, parks have indeed become islands of protected land in a sea of development—motels, shops, restaurants, malls, homes, roads—that washes right up to the entrance gates. In addition, many were poorly placed to preserve biological diversity. "National parks," conservation biologists argued, "are essentially square. Few conform to watersheds, mountain ranges, or other . . . features that define natural regions. Most parks are too small."[19] Moreover, those dating back to the nineteenth century revealed the tastes of their creators in their emphasis on aesthetic grandeur—Yosemite's cliffs or Glacier's peaks—a criterion now derided by biologists as "rocks and ice," habitat notably short on biodiversity.

But the theory that large protected areas would preserve more species over a longer period of time than smaller ones was not properly tested until 1983, when William Newmark, a graduate student at the University of Michigan, set out in search of data for his doctoral thesis. He drove around the West, sleeping in the back of his Toyota, visiting a couple of dozen national parks in the United States and Canada, researching archival lists of mammal sightings for each one. He compared the record of mammal species present when the parks were founded to their current complement. His study, published in *Nature* in 1987, was a bombshell, the kind of logical observation that seems obvious only in retrospect.

Newmark found that most national parks had lost species since their founding. Mink and black-tailed jackrabbit disappeared from California's Yosemite. Gray wolf, fisher, striped skunk, and lynx disappeared from Washington's Mount Rainier. Caribou disappeared from the Waterton-Glacier parks that lie on the border between Montana and Alberta. The smaller the park, the more likely it was to be affected: river otter, ermine, mink, and spotted skunk vanished from Oregon's

Crater Lake (248 square miles); Nuttall's cottontail, fisher, river otter, striped skunk, ringtail, and pronghorn from California's Lassen Volcanic Park (165 square miles); white-tailed jackrabbit, red fox, and spotted skunk from Utah's Bryce Canyon (56 square miles). Only a single protected area—the largest—maintained its original complement of species: the combined area of Kootenay, Banff, Jasper, and Yoho national parks in Canada, at over 7,700 square miles. "Without active intervention," Newmark wrote in *Nature*, "it is quite likely that a loss of mammalian species will continue as western North American national parks become increasingly insularized."[20]

Newmark's study established that our national park system had essentially proved the equilibrium theory of island biogeography. Some early naturalists had feared as much: in 1949, in *A Sand County Almanac*, Aldo Leopold had written that "the National Parks do not suffice as a means of perpetuating the larger carnivores."[21] Now the fear was based on fact: remote from one another and undersized, small parks could not protect species over time. The press took notice. In February 1987, in a major article in the *New York Times*—"Species Vanishing from Many Parks"—Michael Soulé warned that "the scale is beyond anything that people had appreciated or feared."[22] Jared Diamond pointed out that one of the messages of Newmark's study was that large predators are particularly vulnerable to undersized, isolated protected areas: "Certain species are prone to extinction—if they go extinct in park A they may also go extinct in parks B, C and D. Big predators like grizzly bears are a prime example. If you find that they've already gone extinct in five national parks, you've got to expect they're in danger of going extinct in the remaining seven."[23]

Conservation biology is a small world: Michael Soulé sat on the committee at the University of Michigan that supervised Newmark's dissertation. The study percolated in Soulé's mind as he went to his next job, at UC Santa Cruz. Sitting in his kitchen one day, Soulé was talking to his friend Arne Naess, the Norwegian philosopher who founded "deep ecology," a branch of ecological philosophy claiming that all species share an equal right to survive and thrive. He recalled looking out the window and being suddenly gripped by a sense that "we know what the problems are." Having diagnosed the disease—protected areas were too small and too far apart—he knew it was time to prescribe the remedy: restoring connectivity. "Reconnect the isolated patches of wildlands throughout North America," he said to Naess.[24] He began saying it to everyone who would listen.

Reed Noss, a young doctoral candidate in wildlife ecology at the University of Florida in Gainesville, was already convinced; he began working on a reserve design that would institute large core reserves and corridors between them for the Florida panther, a critically endangered subspecies, and the Florida black bear.[25] Noss's design, incorporating federally owned land, envisioned this densely populated and developed state setting aside 60 percent of its land for a reserve network. The design proved so influential that, in 1990, the Florida state legislature appropriated $3.2 billion for a rewilding program to buy land to restore wildlife linkages. In 1994, Noss and a colleague, Allen Cooperrider, published *Saving Nature's Legacy*, laying a foundation for how to site and design networks of reserves with core protected areas surrounded by buffer zones, where human use and development would be limited or controlled, linked by corridors or other forms of connectivity.

Connectivity tied everything together, and it took over the conservation world the way the iPod took over music. In comparison, cores and buffer zones were relatively old news: core reserves first appeared in the 1970s as a central feature of UNESCO's "Man and the Biosphere" program, which identified a number of existing parks representing the world's ecosystems as core "biosphere reserves" and urged their protection.* MAB also promoted the idea of bolstering core reserves by enveloping them within buffer zones where human activity

---

* UNESCO stands for United Nations Educational, Scientific, and Cultural Organization. Its system of World Heritage Sites, established by treaty in 1972, designates areas of great cultural and historic significance; the designation is sought after as a means to protect such sites and encourage tourism. Likewise, the Ramsar Convention on Wetlands, from the same era, provided a legal framework for protection of coastal wetlands and freshwater ecosystems.

would be limited to protect and maintain populations of endangered species.

As connectivity became the hot topic of conservation biology, international agencies took notice. Soon corridors appeared in planning conferences and additional treaties and were lauded at the 1992 Rio de Janeiro "Earth Summit." They influenced the resulting Convention on Biological Diversity, a binding treaty requiring ratifying governments to conserve biodiversity through sustainable development. Even the Man and the Biosphere program was revised in 1995 to encourage the planning of corridors between core reserves.

Experts increasingly refined the roles that core reserves, buffer zones, connectivity, and ecological restoration would play in rewilding. Cores were defined as "strictly protected" areas where human access, activities, and particularly roads were limited. In the United States, cores included national parks, wilderness areas on federal lands, some state or provincial parks, and reserves managed by private conservation groups, such as the Nature Conservancy. While cores were familiar from the Man and the Biosphere model, emphasis was now placed on expanding their size substantially.

What would corridors look like? In *Continental Conservation*, Soulé and biologist John Terborgh put it simply: "One size does not fit all."[26] While corridors implied "linear habitats," the definition was expanded to include not just narrow pathways but also wide swaths of habitat permitting daily and seasonal movement, stepping stones, matrixes, mosaics of habitat, or ecological networks combining many forms of connectivity. Any one type could be tweaked to address the spatial scale—big carnivores need more space than rodents—or stretched north or south, or spread into a variety of altitudes to allow species sensitive to climate change to find safe haven.

Corridors linking core areas would allow not only seasonal migration but also dispersal. As animals mature, they leave their home range, looking for their own territory. Mother grizzlies drive off older cubs before mating again; young beavers launch themselves downriver or across forests, seeking places to build new dams and lodges; the young of most species must seek new territories. Dispersal maintains the genetic health of a population, as individuals find mates from distant groups. Confined populations, by contrast, become inbred, susceptible to reduced fertility, genetic disorders, and immune system problems. But so much territory has been altered or destroyed that there is less and less opportunity for dispersal. Corridors would help, permitting movement away from areas under development or degraded by climate change. They would allow recolonization of newly restored habitat. Corridors might be "stopover" mountain meadows where migrating rufous hummingbirds could refuel on their way to Mexico or hedgerows alongside organic strawberry crops in California, providing food and shelter for birds and small mammals. They might be windbreaks of trees in North Dakota, breeding sites for tree-nesting birds. They might be as simple as a repurposed underpass allowing frogs to hop under a highway or as major as a newly fence-free Africa, permitting elephants to resume their stately continental migrations. But all would, as Newmark put it bluntly, "reduce the risk of extinction for many species."[27]

Unlike past environmental campaigns, many of which were based on opposition to dams, construction, or oil drilling, the call for corridors was positive and appealing. Corridors made sense not only to scientists, who could point to a growing body of empirical observations supporting the notion, but to activists, environmentalists, or anyone else concerned about the degradation of local landscapes and parks. Even people without a background in biology or animal behavior could empathize with the need to get from one place to another.

And if connectivity was appealing, restoration was as visionary as science could be, holding out the hope and promise that humanity could heal the environmental damage that had already been done. It was both a radically new conservation tool and an idea with emotional force. In earlier decades, restoration consisted largely of superficial, cosmetic methods employed by mining or logging companies to backfill pits or replant despoiled areas with a scattered remnant of what had been destroyed or removed. Those kinds of rehabilitation rarely rose above disguising aesthetic disasters or forestalling erosion. Now scientists were experimenting with methods of restoring ecosystems on a far more complex and meaningful scale, reintroducing communities of plants and animals, regrowing coral reefs, even replanting tropical forest. They were seeing promising, sometimes astonishing results. Scientists recognized that these methods would be indispensable to continental-scale conservation. So much damage had been done to areas between parks and protected wilderness that it was impossible to imagine enlarging or connecting core reserves without restoring land between them. "Without restoration at large spatial scales," *Continental Conservation* advised, "the goal of protecting all species and ecosystems cannot be achieved."[28]

But restoration was also controversial. For years, the Nature Conservancy, World Wildlife Fund, and other international environmental organizations had based major fund-raising campaigns around the notion that nature, once lost, was lost forever. Ecological restoration contradicted that. Maybe nature wasn't lost forever; maybe ecosystems could be recovered. If so, NGOs wondered, would the public become confused; would donors lose the motivation to give? Would government and industry, real estate developers, oil companies, and agribusiness be emboldened to step up their already voracious consumption of land, arguing that it could be restored later? But the biologists felt that these caveats represented timidity at a time when bold, urgent action was required.

Rewilding was bold, urgent, new, radical. If existing links between parks and reserves could be exploited, fine. But if corridors were not available, they would have to be re-created. If that meant restoring damaged or destroyed forests or grasslands between them or reintroducing carnivores, then that is what had to be done.

With the publication of *Continental Conservation* and a raft of other books and papers explicating connectivity and restoration, the conservation biology community in North America was reaching a consensus, offering a comprehensive show of support for rewilding as a solution, a remedy, a plan. But no one had ever restored a continent before. Where should cores, corridors, and restoration be planned? How should these large-scale networks of protected areas be designed? Who would carry out these projects; how would they be managed? By whom and for how long? Where would the money come from? Rewilding represented an almost infinite number of consequential decisions and tasks. It was a prospect of staggering, overwhelming dimensions. In terms of time, money, scientific expertise, and technological prowess, rewilding made NASA's job of sending a man to the moon look finite and manageable. Every decision made would have real-world consequences affecting land use, water, and wildlife management on a scale never before contemplated. Every decision would also affect an unpredictable and volatile constituency: human communities.

Rewilding in the Real World

In the 1990s, the American environmental landscape presented a discouraging prospect for the new wave of conservation biologists and rewilding activists. Major private conservation organizations, including the Sierra Club, World Wildlife Fund, Nature Conservancy, and Wildlife Conservation Society, had invested decades and millions of dollars in established programs organized around their own priorities, from endangered species protection to buying tracts of private land. Many programs included elements of what would become rewilding, but none were devoted to the promotion of cores, corridors, and carnivores together. It would be difficult to convince such entities to back new, untested, expensive schemes.

Public agencies overseeing environmental issues and land use—the Department of the Interior (including the U.S. Fish and Wildlife Service, the National Park Service, and the Bureau of Land Management) and the Department of Agriculture (which includes the U.S. Forest Service)—were even less promising. Their priorities, altered by successive presidential administrations and appointments, had been highly politicized in recent disputes over the Endangered

Species Act and the disposition of logging and grazing rights in the West. For decades they had been heavily influenced by ranching, mining, and logging interests.

So rewilding advocates in North America struck out on their own, creating new groups, essentially seeking to reshape the environmental movement. In 1991, the same year that research on Pluie began, an eclectic group of scientists, activists, and businessmen—including Doug Tompkins, founder of the Esprit and North Face clothing companies; Dave Foreman, a founder of Earth First!; Michael Soulé; Reed Noss; and David Johns, an attorney and environmentalist—banded together to form the Wildlands Project, a group devoted to moving conservation biology from the realm of theory to action. In 1992, the project published a special issue of its magazine, *Wild Earth*, on "Plotting a North American Wilderness Recovery Strategy." The issue was filled with ambitious maps outlining a proposed Adirondack wilderness reserve system, a southern Rockies ecosystem, a southern Appalachian wildlands network sweeping up the Great Smoky Mountains National Park in a skein of protected areas between Georgia and Virginia. There was also a map of "Paseo Pantera," a plan to create a "Path of the Panther" through the narrow Central American isthmus.[29]

With money from Tompkins, the project distributed seventy-five thousand copies, and Johns remembers seeing it in far-flung locales all over the globe, in the labs and homes of biologists from Europe to Siberia. Throughout the 1990s, the idea of connectivity flew around the world, infiltrating scientific conferences, papers, graduate theses, environmental courses, grassroots meetings. By 1993, conservationists and biologists were debating the steps that should be taken to maintain and restore connectivity in what came to be called Yellowstone to Yukon.

The Yellowstone to Yukon Conservation Initiative, formally launched in 1997, promoted a conservation vision for a vast corridor

Yellowstone to Yukon: The Yellowstone to Yukon Initiative (Y2Y) aims to forge a single wildlife corridor by connecting isolated parks, national forests, and some of the largest roadless areas left on the continent.

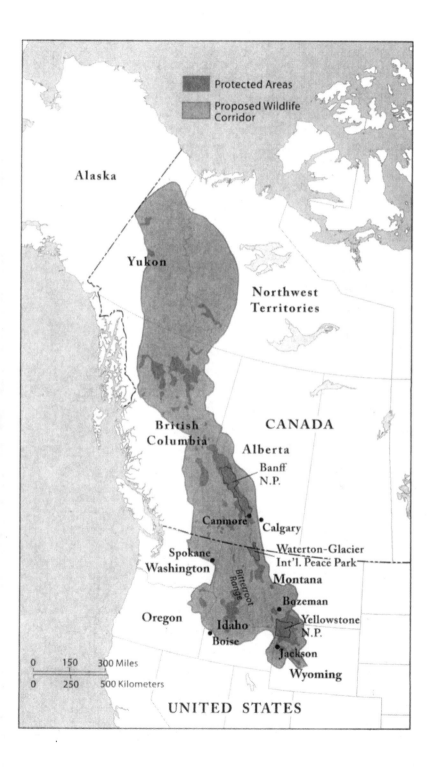

Alaska

Yukon

Northwest Territories

CANADA

British Columbia

Alberta

Banff N.P.

Canmore

Calgary

Waterton-Glacier Int'l. Peace Park

Spokane

Washington

Montana

Bitterroot Range

Bozeman

Oregon

Yellowstone N.P.

Idaho

Boise

Jackson

0   150   300 Miles

0   250   500 Kilometers

Wyoming

UNITED STATES

of the northern Rockies, from Canada's Yukon down through the Waterton-Glacier national parks and south into the system of parks, national forests, and wilderness areas that make up what biologists define as the Greater Yellowstone ecoregion. A Canadian environmentalist named Harvey Locke, an early board member of the Wildlands Project who had been captivated by the accounts of Pluie's journey, had an immediate sense of the potential of the Yukon and Yellowstone "myth," as he called it, arising from the romantic wilderness mystique of Robert Service and Jack London.[30] Locke would later say of Y2Y, "I chose those words, 'Yellowstone to Yukon,' because they're deep symbols in people's brains. If I say those words in Stuttgart, Germany, in Toronto, in New York, or in Tokyo, everybody knows what I'm talking about."[31]

Y2Y quickly mobilized environmentalists, mountaineers, and activists in Montana and Wyoming. Rick Bass, Doug Peacock, and Rick Ridgeway wrote essays and gave interviews publicizing it. The Craighead Environmental Research Institute in Bozeman, founded by charismatic twin brothers and grizzly bear biologists Frank Craighead Jr. and John Craighead, whose work on bear territories and behavior predated Paquet's with wolves, began coordinating research projects in the region. Gradually Y2Y seeped out of conservation circles and into the popular culture, championed in catalogs put out by Yvon Chouinard's outdoor clothing company, Patagonia, and discussed on the television drama *The West Wing*. By 2007, there were eight hundred member organizations.

Y2Y was an initiative and a vision. It was not an implementing project and would not carry out or fund corridor or restoration projects on its own. Instead, the initiative would lobby, educate, promote. Much like the Wildlands Project (now Wildlands Network), Y2Y served as a clearinghouse and organizational network, setting up conferences and workshops for joint rewilding projects between conservationists from Canada and the United States. The group facilitated dialogue between the conservation community and members of Native American and First Nation groups, made small grants to research projects, published a sophisticated map of the region that identified priority areas for conservation and potential threats to connectivity, and excelled at promoting the visionary potential of connectivity, maintaining a high profile in features in the *New York Times*, the *Washington Post*, and metropolitan papers in the United States and Canada.[32] While the group was quick to deny that it intended to turn the Y2Y region into a vast park excluding human use, its promoters intended to influence land management throughout the area by warding off road projects and developments that could interfere with the habitat and movement of wolves, grizzly bears, and their prey.

The most immediate example of Y2Y's influence was its successful lobbying for highway crossing structures for wildlife in the Banff area, which Pluie had so perilously navigated. Widely adopted in Europe for decades, crossing structures are popular not only with conservationists but also with insurance underwriters who value their potential to reduce accidents. In the United States, for example, deer-vehicle collisions alone occur up to one and a half million times each year, costing some two hundred lives and $8.8 billion annually; collisions also imperil the survival of twenty-one endangered and threatened species.

As a result of Y2Y pressure, Canada in the midnineties allocated $3.3 million to build two wildlife bridges and twenty-two underpasses providing passage across the Trans-Canada Highway, the lethal road that cuts the Bow Valley in half. Wolves and grizzlies initially proved reluctant to try the 160-foot-wide bridges, which had been covered in rock, dirt, native trees, and vegetation to replicate a natural crossing. "Not all wolves use them," Paquet told me. "Wolves are like humans. Some are willing, some are not."[33] For the past decade, the Banff Wildlife Crossings Project, a long-running research project supported in part by Y2Y, has monitored that willingness, finding that animal use increased steadily over the years. Grizzly bears utilized the crossings seven times in 1996; in 2006, over a hundred times.[34] Scientists have begun to theorize that grizzly mothers are teaching their cubs to find and use the structures.

To expand their data, Banff Project scientists strung barbed wire hair snares across the underpasses, a low-cost method of identification for bears, wolves, and other big creatures passing through. Footage from cameras mounted on the underpasses shows bears and mountain lions approaching the wire cautiously, sniffing and peering around. Then most of them burst over or under the wires, galloping off, leaving samples behind. Using these hair snares, camera footage, and beds raked to reveal tracks, scientists have found that eleven species of large mammals have used the crossings over 104,000 times since 1996. Wildlife collisions have been reduced over 80 percent, and those involving elk and deer by 96 percent. Given that people involved in wildlife collisions end up spending an average of two thousand dollars per accident for repairs, and that many are injured or even killed in such accidents, the Banff crossing structures may have already paid for themselves.

Transportation departments in the United States watched the Banff experiment closely, and highway mitigation projects are now planned throughout the country. In 2008, the Washington State Department of Transportation approved the construction of fourteen wildlife crossing structures on Interstate 90 through the Cascade Mountains: narrow bridges and culverts will be replaced with wider ones to allow wildlife to cross under the highway.[35] Exclusion fences will funnel animals to the crossings and keep them off the road. Two vegetated overpasses will be built. In Montana, forty wildlife crossings were added to a highway reconstruction project on the Flathead Indian Reservation; in Florida, eight alligator underpasses were built on Highway 1 through the Florida Keys. In Arizona, desert bighorn sheep were fitted with radiotelemetry collars that yielded eighty thousand "waypoints," GPS readings indicating where sheep were trying to cross Highway 93 in the Black Mountains near Hoover Dam.[36] The data revealed that bighorns shun underpasses, fearing predators, and prefer high crossings, so state and federal transportation and wildlife officials designed three overpasses that should reunite two groups of the Black Mountains bighorn long separated by the road. Construction began in 2008, and more sheep overpasses are being planned around the country.

Impressed by the success of Y2Y in improving wildlife passages across highways, regional groups took up the mission. In Southern California, the South Coast Wildlands Project launched a campaign to identify threats to connectivity, honing in on Coal Canyon, a critical corridor for mountain lions and bobcats between Anaheim and Riverside. In 2000, the group successfully lobbied the California State Parks agency to purchase a 700-acre tract slated for development and to close and convert an underpass into a wildlife-friendly passage for big cats. "What do you call a freeway interchange ripped up to create a wildlife underpass?" the South Coast Wildlands Project exulted on its Web site. "A good start."[37]

Y2Y led to an explosion of corridor planning across North America and around the world. The Sky Island Alliance scanned the desert southwest of Arizona and New Mexico, focusing on weak or broken connections between mountain ranges. The Algonquin to Adirondack Conservation Initiative (A2A) tackled the corridors of the Great Lakes region compromised by extensive development, researching ways to bolster connectivity in a region anchored by Algonquin Provincial Park in Ontario and Adirondack State Park in upstate New York. Networks of marine reserves were sketched out, including a blueprint of twenty-eight "Marine Priority Conservation Areas" in the coastal Pacific from Baja, California, to the Bering Sea (B2B).[38] As with Y2Y, these cross-border initiatives aimed to engage North American nations in joint conservation planning, to identify species of common concern, and to begin managing and monitoring transportation issues and fisheries collectively.

The North American initiatives were matched and even exceeded by the scale of rewilding projects being planned internationally. With corridors entrenched in international agreements on conservation, large-scale protected areas and transboundary projects proliferated.

This was perhaps the greatest effect of Y2Y—altering the scale on which conservation was planned. Y2Y inspired activists to be visionary, ambitious, even utopian. The new thinking represented a paradigm shift: conservationists were now considering continents and planning for a millenium. One participant said, "We look at things that are too small and we think too short term. . . . Y2Y has completely changed that whole thinking."[39]

"A Corridor in People's Minds"

But there was just one problem. Many biologists felt that activists had begun advocating for corridors before knowing where to put them. Bill Newmark, for example, believed corridors were a biological necessity. "The issue of whether or not corridors are a wise conservation strategy, I don't think people question that anymore," he told me. But he also thought that there were dozens of questions to be answered before planners could begin to figure out how and where to design potential corridors. When it came to Y2Y, he was dismissive of something that existed, as yet, largely on paper. "It's a corridor in people's minds," he said dryly. "Whether or not animals really use it is another question."[40]

A number of scientists and organizations set out to answer that question and a host of others. Joel Berger, a senior scientist for the Wildlife Conservation Society (WCS) working out of a Teton, Wyoming, field office, focused on the Great Plains ecosystem, leading a team designing a migratory corridor for pronghorn antelope, one of the most important ungulates in North America, with half a million roaming Wyoming alone.

The fastest land animal in the New World, perhaps the fastest long-distance runner on earth, pronghorn migrate astonishing distances, traversing mountain passes and steep canyons, for reasons that are still not completely understood. A population of around two hundred to four hundred pronghorn run an annual marathon from their summer range in Grand Teton National Park, south of

Yellowstone, to a winter range to the southeast, in Wyoming's Green River Basin, near the town of Rock Springs, returning in the spring. Sometimes reaching speeds of fifty-five miles an hour, the pronghorn run thirty miles a day without stopping to feed, the longest and most dramatic land migration in the lower forty-eight states. Archaeological evidence shows that the route has been used for some six thousand years, but it has become lethal, blocked and constricted by highways, development, and rampant gas drilling unleashed by the second Bush adminstration. Berger and his team have seen up to six animals killed at once, in a single vehicle collision. Looking at the historical record, they came to believe that many previous migratory pronghorn routes— one of which went through what is now downtown Jackson Hole— had already been choked off by development. Thus it was all the more important to preserve this last intact route: while the species itself was not technically endangered, although reduced by 95 percent since the 1800s, major impacts to pronghorn migration might spell eventual genetic fragmentation and doom.[41]

Fitting ten pronghorn with radio collars in 2003, the team mapped their journey from the summer to the winter range. The resulting computerized map identified several bottlenecks, including one funnel in midroute only 120 meters wide, which housing developments and parking lots threatened to close completely. Using the data, they designed a plan for the first National Migration Corridor, ninety miles long and a mile wide, virtually all of it on public land managed by the Bureau of Land Management, a federal agency charged with "balancing" energy needs with a responsibility to protect the environment. While the corridor has gained popular support to the north in Jackson Hole and other towns around the Grand Tetons, gas-drilling communities in southern Wyoming oppose it. The state itself lacks statutory authority to manage federal lands, so the nation's first migration corridor is still awaiting a moment when the political and economic stars align in its favor.

In 2006, Jon Beckmann, one of Berger's WCS team, filled another data gap, this one in the understanding of carnivore movement. He ran a pack of scat-detector dogs through the Centennial Mountains, near the Idaho-Montana border. The dogs were trained to detect scats of black bear, grizzly, mountain lion, and wolf. Laboring for months, covering two hundred and thirty-two square miles of terrain, the painstaking canines found 660 scats and thirty hair samples. DNA testing showed they achieved 98 percent accuracy in identifying the samples.[42]

The data provided a gold mine of information to determine exactly where carnivore superhighways were located and what kinds of human activities were causing gridlock. The findings persuaded the Bureau of Land Management to close 40 percent of the roads in the western Centennials and helped stop a 1,200-home, eighteen-hole golf course development that had been scheduled to be built right in the fast lane of this busy corridor.

Bill Newmark, too, was gathering information on how to restore connectivity. To identify principles of prey behavior, Newmark used teams of volunteers provided by Earthwatch, an international NGO in the forefront of rewilding research, to study wildlife trails in Idaho's northern mountains. He wanted to know how the landscape influenced the movement of moose, elk, and deer, critical to the survival of carnivores. In Trail Gulch, a historic area on the east side of the north fork of the Salmon River, where Lewis and Clark crossed the Bitterroots and where grazing was permanently prohibited, Newmark deployed his volunteers to complete a vegetation study, and I tagged along. Data on food availability, combined with information on the steepness of slopes and other landscape features, would all contribute to a detailed picture of where ungulates were going and why.

Walking in pairs along transects spaced seventy meters apart, we followed straight lines drawn across a map in order to provide an accurate sampling of vegetation. This meant struggling up and down steep hills and across perpendicular fields of sliding scree. Every ten meters, the volunteers dropped a quadrat, a meter-square measuring device made of PVC pipe, and identified the percentage of food plants (western wheat grass, clover alfalfa, lupine) within it. Newmark himself, a lean, grizzled figure in his fifties, was always first to finish, indifferent to hail, fog, or lightning strikes nearby. Below, vultures and small planes rode the thermals.

One day, he remarked that this place nearly killed Lewis and Clark, whose men hated "these most terrible mountains . . . as steep as the roof of a house."[43] Even in an age of four-wheel drive, the mountains remain intimidating, the skyline pierced by jagged peaks. After finishing the transects, it routinely took hours to descend, as the team took the hills at an angle, trying not to fall headfirst.

At another site, equipped with handheld GPS units, the team mapped actual wildlife trails made by elk and deer. Some were barely visible tracks; others looked like highways for the cloven-hoofed. Newmark spent the better part of a day running down one of these major migratory routes with a sophisticated backpack GPS unit, accurate to less than a meter. As he ran, the unit recorded the data, to be downloaded onto a topographical map of the area, yielding a sophisticated visual picture of elk movement. "They're using the same rules we would use," he said later, "taking these steep saddles down to the stream." He was wondering about elk motivation: was saving energy more important during long migrations or was predator avoidance the priority? Understanding their behavior could affect the potential design of functional corridors, which would have to meet the needs of the animals that would actually use them.

Hiking through Trail Gulch, from the trailhead into the mountains and back again, we passed through natural choke points, narrow creek beds where countless deer, elk, and moose had met their end. Predator scat showed that wolves and mountain lions had lain in wait and found their opportunities. The defiles were littered with a chaotic tangle of fallen trees, stumps, skulls, and bones. These corridors, at least, were fully functional and as efficient as an abbatoir.

They were also rare. Such glimpses of a healthy balance between predator and prey can be seen only in northern pockets of the West where the wolf is making a comeback. Asked how large predators survived in Africa when many were driven extinct in the Americas, after the arrival of *Homo sapiens*, Newmark stated a simple evolutionary fact. "They evolved with humans," he said, "and they learned, *that's* a nasty predator."

Biologists, for all their carefully calibrated qualifications, excel at this kind of startling bestial insight, stepping outside our species to see how we affect others. And the picture was not a pretty one. Even when Newmark, Berger, and other biologists finished filling in the gaps, they knew they would face resistance from the top predator in this and every ecosystem, one loath to give up any ground.

CHAPTER 2

THE PROBLEM WITH PREDATORS

The Green Fire

IN ALL OF NORTHERN IDAHO, BILL NEWMARK MAY HAVE BEEN the only person who cared about the big scientific picture. In the tiny, struggling rural towns of Salmon, North Fork, and Gibbonsville or across the border in Wisdom, Montana, there was little interest in a complex analysis of elk behavior or the purpose of wolves. Hunting, fishing, and recreation were good. Conservation was bad. Wolves were competitors; we were better off without them. Y2Y's message had failed to penetrate. On a bulletin board in the Antlers Bar in Wisdom, a tacked-up snapshot showed a live wolf with one leg caught in a leg-hold trap. Someone had written on it, "It's a start." Nearby, a bumper sticker: "Save 100 Elk, Kill a Wolf."

Rewilding was being compared to a Marshall Plan for the environment, at least by scientists aware of the gravity of the biodiversity crisis. But if biologists felt the urgency, the rest of society did not. While Newmark was gathering data, the governor of Idaho announced that he wanted to kill 550 of the state's wolves when they were removed from the endangered species list. He would be happy, he crowed, to be the first to shoot one.[1] He crowed again in March 2009, when the Obama administration's interior secretary, Ken Salazar, a fifth-generation Colorado rancher, announced the removal would stand.

Hostility to predators is atavistic, so hardwired in our instincts and behavior that it has taken us thousands of years to comprehend a simple scientific fact: the environment *needs* predators. They regulate ecosystems in ways we can never re-create artificially.

The first to understand this was Aldo Leopold. If Michael Soulé was the Martin Luther of conservation, reforming a tired discipline with rigorous new rules, then Aldo Leopold was preaching the gospel in *A Sand County Almanac*. Called the "urtext for the ecological restoration movement," the *Almanac*, published in 1949, a year after Leopold's death, has become the secular bible of rewilding, beloved by biologists and activists alike.[2]

Ahead of his time, Leopold intuited the importance of predators as a vital regulatory control in the ecosystem and articulated a "land ethic" based on the assumption that "land is not merely soil."[3] He urged conservationists to keep "all the parts" of the ecosystem in order to rebuild it: he envisioned and experimented with restoration ecology, planting forty-eight thousand native pines on his patch of the unique "sand counties" of Wisconsin.[4] Modern conservationists from Paul Ehrlich to E. O. Wilson have recognized themselves in Leopold's lament, quoted in countless scientific papers: "One of the penalties of an ecological education is that one lives alone in a world of wounds."[5]

Leopold shot one of the last wolves in the Southwest during the 1920s, as a fieldworker for the U.S. Forest Service, and his recollection of the event, written years later, is unsparing and painful to read, with its retrospective pangs of regret. Eating lunch on a rimrock above a river, he and his colleagues watched a wolf swim the current and joyously greet a half dozen pups waiting for her on the bank:

> In those days we had never heard of passing up a chance
> to kill a wolf. In a second we were pumping lead into the
> pack, but with more excitement than accuracy: how to
> aim a steep downhill shot is always confusing. When our

rifles were empty, the old wolf was down, and a pup was dragging a leg into impassable slide-rocks.

We reached the old wolf in time to watch a fierce green fire dying in her eyes. I realized then, and have known ever since, that there was something new to me in those eyes—something known only to her and the mountain. I was young then, and full of trigger-itch; I thought that because fewer wolves meant more deer, that no wolves would mean hunters' paradise. But after seeing the green fire die, I sensed that neither the wolf nor the mountain agreed with such a view.[6]

He went on to retail the horror of watching "many a newly wolfless mountain" as it was browsed "first to anaemic desuetude, and then to death."[7] The environment without carnivores, he wrote, was soon "dead of its own too-much."[8]

Since Leopold's day, studies in virtually every type of ecosystem throughout the world have confirmed the fact that top carnivores, known as "apex predators," regulate the processes of life. Some of the most astonishing evidence came from John Terborgh, a biologist who has spent his career conducting fieldwork in the jungles of Manu, Peru.

In 1988, in response to an E. O. Wilson essay in *Conservation Biology* titled "The Little Things That Run the World"—"it is a common misconception that vertebrates are the movers and shakers"—Terborgh fired back in "The Big Things That Run the World."[9] He talked about prey mammals taking over Barro Colorado Island, created by the flooding of the Panama Canal and emptied of large predators like jaguar and puma. Densities of prey there were, in some cases, over ten times greater than normal, which meant lots of little creatures eating seeds. He theorized that this would alter seedling and tree communities and ultimately the forest itself. He called for "a major research program" to study these effects, not realizing that he would end up running it.[10]

In 1990, Terborgh stumbled on an opportunity to test his Barro Colorado assumptions: In Venezuela, a new hydroelectric dam had flooded a valley, leaving stranded hilltop "islands" of forest floating in a newly created lake, Lago Guri. As the lake filled, predators fled, seeking higher ground, leaving the newly created islands populated only by the smaller creatures left behind. On his first visit, Terborgh was astounded at the weird fallout, calling the islands "a god-forsaken place." Another writer described them this way: "Some islands looked as though they'd been plowed by phantom bulldozers. Others had apparently been cloaked in thickets of thorns by some satanic gardener."[11] After a team studied the islands, the data painted a horrific picture. Safe from predators, howler monkeys proliferated on some islands, but they were not enjoying their freedom from fear. Normally social animals, they were living alone, attacking one another, and killing their own infants. By denuding trees, they caused surviving plants to protect themselves with toxins, so meals provoked vomiting. Many plants are capable of deploying extraordinary chemical defenses against herbivory—predation by herbivores—by inducing a rapid rise in levels of toxins that can repel or kill those feeding on them. Other species, such as *Acacia collinsii*, the bullhorn acacia of Central America, marshal armies of ants that live in the tree's thorns, feed on a substance produced by its leaves, and attack anything or anyone unwise enough to approach. On islands with howler monkeys, the instability caused by the absence of predators and superabundance of herbivores set off a vicious chain reaction. When water levels dropped, the monkeys fled to a more felicitous environment.

On other islands, predators of leaf-cutter ants (armadillos and army ants) were absent, and the ants ran amok, carrying everything green off to their underground nests, leaving a *Sleeping Beauty*–style thicket of impenetrable thorny vines, destroying all remaining life, plant and animal. Far from a predator-free paradise, the islands of Lago Guri were "dead of [their] own too much." In a 2001 *Science* paper, "Ecological Meltdown in Predator-Free Forest Fragments," Terborgh and colleagues reported that after a few years almost 75 percent of vertebrate species had been lost from the smaller islands without jaguars or pumas.[12] Leopold was right.

Closer to home, the same scenario has played out. When wolves and bears were exterminated in most of the lower forty-eight states, our trigger-happy forefathers unwittingly set in motion a biological experiment in ecosystem impoverishment. In Yellowstone National Park, the last wolf was killed in 1926. With their chief predator gone, elk herds grew, delighting human hunters and becoming such heavy browsers that the Park Service was forced to cull them. In 1995, wolves were reintroduced after an absence of nearly seventy years. Their favorite prey in Yellowstone was elk. Soon, the results of the experiment—what happens when a top predator is subtracted from the system, what happens when it returns?—became evident.

In 1997, William Ripple, an Oregon State University forester, came to Yellowstone to figure out what was happening to aspen. Widely found throughout North America, quaking aspen once represented an important plant biomass throughout the park, covering an estimated 4–6 percent of the northern range of Yellowstone, but its occurrence had fallen to 1 percent by the time Ripple arrived.[13] He and a colleague conducted a study of growth rings revealing that aspen regenerated continually from the late 1700s to the 1930s, but after 1930 regeneration ceased. They found similar patterns in cottonwoods and willows. They theorized that wolves had an indirect effect on aspen growth: without predators, elk had browsed trees to death for decades, becoming comfortable in dense thickets and rivers, two of the most dangerous places for them when wolves are present. The scientists launched a long-term study, this time of aspen stands inside and outside of established wolf pack territories, comparing sample plots of aspen, counting elk pellets in the plots, and measuring the height of aspen suckers. Initial results showed that aspen were growing to a greater height in areas of high wolf use in certain habitats.

While the study continues, certain implications seemed clear: During the wolf-free era, heavy browsing spelled doom not only for the trees themselves but for a host of other creatures dependent on them. Beavers love aspen, and their population in Yellowstone crashed after wolves were removed. Without beavers to create ponds, wetland ecosystems—aquatic plants, amphibians, birds—were devastated. When wolves returned, grazers and browsers resumed normal patterns of behavior, preferring safer, open areas over the dense cover and streamsides that provided excellent lurking opportunities for carnivores. Keeping elk wary and on the move, wolves gave aspen and other young trees the opportunity to grow and become reestablished; beavers bounced back; a healthy, resilient biodiversity was restored.

Wolves accomplished what biologists called "landscape-level change."[14] When browsers had the run of the place, feasting on tender vegetation that shaded rivers, they eliminated shade and raised the water temperature. Trout are exquisitely sensitive to changes in temperature; they need cool, well-oxygenated water to reproduce and avoid stress. Warm water contains less oxygen. The return of wolves moved elk away from rivers; trees and shrubs grew over the water, shading it; the water temperature dropped; trout benefited; fly fishermen rejoiced. Studies showed that, in addition to effects on elk, sixteen species of vertebrates were affected by wolf predation. Populations of coyote dropped in density. Grizzly bears, capable of driving wolves off a kill, feasted on carrion. Other scavengers benefited as well, including bald and golden eagles, ravens, and magpies. Songbirds and amphibians also began to return. "Soon after we bring wolves back, plants are flourishing again," Ripple told the *Seattle Times* in 2005.[15] What he termed "the ecology of fear" proved more consequential than anyone had foreseen.[16]

The studies of what Terborgh called the "pathological ecology" at Lago Guri led scientists to understand that predators stabilize the environment, exerting "top-down" control.[17] Plant reproduction, Terborgh wrote, "is particularly sensitive to distortions in the animal community."[18] Subtract predators, and plants go haywire. "As soon as we get rid of wolves, plants stop flourishing," Ripple noted.[19] The technical term for these downward cause-and-effect spirals is "trophic cascades"—trophic meaning nutritional—and the forces exerted by predators and other keystones are "top-down" forces or "top-down regulation."[20] In the same way that white-tailed deer populations explode in the absence of predators, with disastrous consequences throughout North America, a large, competitive mussel species, *Mytilus californianus*, in the Pacific Northwest can proliferate out of control, taking over the seafloor, if its predator, the intertidal sea star, goes missing, eliminating a more complex array of algae and other species. As oceans are rapidly emptied of their biodiversity, entire marine ecosystems are failing, causing catastrophic effects on world food supplies.

Top predators control other predators: the number of foxes, jackals, coyotes, and smaller carnivores is directly related to predators at the top of the food chain. Take out a major predator, and the whole system feels the repercussions. Nearly exterminated by the nineteenth century, sea otters staged a spectacular comeback after a 1911 treaty banned hunting, but recently their numbers in Alaska and along the Pacific Coast have declined by 70 percent. James Estes, sea otter biologist for the U.S. Geological Survey, now believes that they are being eaten by killer whales since populations of the whales' common prey (seals, sea lions, and other whales) have dwindled, victims of overfishing. As sea otters disappear, sea urchins deforest rich stands of underwater kelp, impoverishing this important marine ecosystem. As kelp forests vanish, so do the fish that live among them, which feature predominantly, along with sea otter pups, in the diet of bald eagles. In areas where sea otters have disappeared, bald eagles are now predating on other sea birds, affecting their populations.[21] "Carnivorous animals are important," Estes has said. "We have to stop thinking of them as passengers on this earth and start thinking of them as drivers."[22]

Inevitably, an ecosystem robbed of its top predators begins a remorseless process of impoverishment. As early as 1955, Robert MacArthur, coauthor of the equilibrium theory of island biogeography, suggested that the more varied and numerous the complement of predators, the more stable and resilient the ecosystem.[23] The wild swings currently playing out in ocean ecosystems are one manifestation of instability caused by removing predators from the system.

What biologists were discovering about trophic cascades in Yellowstone, marine ecosystems, and the tropical forests of South America dramatically emphasized the need for rewilding and the necessity to restore major predators wherever possible, as fast as possible. But as biologists took action, supporting projects reintroducing wolves, lynx, otters, and other species, they ran up against vehement public opposition. The scale of wilderness needed by predators for all the activities of their lives—breeding, competition, dispersal—is enormous, setting them up for conflict with people. While an average wolf pack's territory may measure 200 square miles, dispersals of over 500 have been recorded; as Pluie's epic journey suggests, expanding populations over the long term might require 30,000 to 120,000 square miles as individuals disperse. Similarly, a single male mountain lion might inhabit a home range of several hundred square miles, but according to some estimates, a healthy, viable population may need, over the long term, between 22,500 to 90,000. The state of Colorado is 103,000 square miles. Given the size of the corridors biologists were hoping to protect and the space carnivores need to recover, it was inevitable that the conflicts would grow as fearsome as the predators themselves.

The Problem with Predators

Biologically, the reintroduction of the gray wolf in Yellowstone was an astonishing success. Fourteen wolves were released in 1995; there are now over four hundred in the Greater Yellowstone ecosystem, with a total of around fifteen hundred roaming Montana, Idaho, and Wyoming. But politically, it was a disaster.

Ranchers hate wolves. They recently seized the chance to reenact the historic carnage of the nineteenth century when hunters, ranchers, and federal exterminators shot, trapped, and poisoned their way through the West. The federal government delisted gray wolves as an endangered species in the northern Rockies in March 2008, and a mere three months later, nearly 10 percent of the wolf population was wiped out, over 130 animals. The toll worked out to an average of one wolf per day. Wyoming declared that hunters should consider virtually the entire state a "free-fire zone."[24] The slaughter was so extreme that a federal judge intervened, stopping the hunt in mid-July, ruling it "arbitrary and capricious."[25]

Predators are the stumbling block for rewilding. In response, rewilding organizations have begun a careful cultivation and refinement of the art of compensation. While predators represent a hard conservation sell in the West, they are being sold. Never underestimate the power of a bribe.

Defenders of Wildlife, a Washington, D.C.-based group promoting support and reintroduction of predators, from grizzlies to river otters, has consistently refused to see ranchers as the enemy: ranches, after all, represent significant chunks of land that can themselves act as corridors for wildlife. Aware that ranchers would unfairly bear the financial brunt of the wolf's return, Defenders early on established a widely praised compensation fund to reimburse them for livestock taken by wolves. Since 1987, the Bailey Wildlife Foundation Wolf Compensation Trust, designed "to shift economic responsibility for wolf recovery away from the individual rancher and toward the millions of people who want to see wolf populations restored," has paid out over a million dollars for losses of several thousand sheep and cows.[26] Although only 0.11 percent of livestock losses are caused by wolves, the burden on individual ranchers can be heavy, so in 1998 Defenders launched a second program, to pay for range riders, fencing, alarms, and other innovative methods of protecting livestock.[27] With the help of these programs the wolf has made unlikely friends. Curt Hurless, an Idaho rancher, was adamantly opposed to wolves, insisting that the first wolf pack that attacked his cattle in 2000 be shot. Soon after, he accepted an alarm system from Defenders triggered by radio collars and changed his mind. He began selling his own hand-cast wolf tracks in his store. "I enjoy that we have wolves in Idaho," he told the *Idaho Statesman*.[28]

Defenders was determined to pursue wolf recovery in other suitable habitats, particularly in the Southwest, where there were plans to reintroduce an endangered subspecies. Multiple populations create redundancy, a genetic safety net in case one population is wiped out by disease. More important from the rewilding perspective was the prospect of linking wolf populations through corridors, expanding the ecological integrity that was returning in Yellowstone.

In the Southwest, however, wolf reintroduction came up against obstacles even less tractable than ranchers. Launched three years after Yellowstone's, this project has been a failure, because the federal government has deliberately undercut it at every turn. The *lobo*, or Mexican gray wolf (*Canis lupus baileyi*)—ironically named after Vernon Bailey, a turn-of-the-century Department of Agriculture biologist who actively pursued wolf extermination—was the world's most endangered subspecies of wolf, nearly extinct in 1979. The U.S. Fish and Wildlife Service (USFWS) was required by the Endangered Species Act to restore the wolf; its stated goal was a hundred wolves in the wild by 2007. Zoos were enlisted for a captive-breeding program, and on a January day in 1998, Aldo Leopold's granddaughter watched as Interior Secretary Bruce Babbitt opened a crate in the Blue Mountains near the Arizona-New Mexico border, releasing an alpha female Mexican wolf into an enclosure. A few months later, the first eleven wolves left for the wild. Environmental leaders were there representing Defenders of Wildlife, the Audubon Society, and the Ministry of Environment for Mexico. The director of the USFWS, Jamie Clark, told a news reporter, "Today we are putting the green fire back on the mountain."[29]

By August 1998, the newly released alpha female had been shot dead with a high-powered rifle. Rumors spread during the first year of the program that a Catron County, New Mexico, rancher would pay a bounty of $10,000 for a dead wolf.[30] By the end of that year, all the wolves in the wild were gone—shot, missing, or removed.

The rural community near the release site was exercising something known in the West as "shoot, shovel, and shut up."

It turned out that the government's plan was designed to fail, violating everything that wildlife biologists had learned from Yellowstone about large-scale reserves, wolf behavior, dispersal, and population dynamics. By designating the Mexican wolf as a "nonessential experimental population," the agency executed an end run around the Endangered Species Act, avoiding a key requirement, that of establishing a "critical habitat" for the species. Without taking into account the fact that wolves roam widely to establish new territories, the plan restricted the wolf to a specific area within the Gila and Apache national forests in Arizona and New Mexico. No one from the government explained how the wolves were supposed to know not to leave the area. In addition, any wolf found to have killed livestock three times in a year could be shot or removed from the wild. In the northern Rockies, wolves that scavenged dead livestock carcasses were protected under federal law, but the Mexican wolf was not.

The government was undermining wolves at the behest of the livestock industry, which rallied citizens in Catron County, well known for its commitment to states' rights and private property. At nearly 7,000 square miles, Catron is larger than Connecticut. Its population of 3,543 earns, per capita, $14,000 a year. Michael Robinson, of the Center for Biological Diversity, a Tucson-based nonprofit that has repeatedly sued to improve the wolf program, once called the region "the Appalachia of the West." The last mines and sawmills closed years ago, and low cattle prices have decimated ranching. While the county is undeniably poor, the most vociferous opponents of wolf recovery have been industry lobbyists and absentee ranchers eager to protect their low-cost grazing leases. A ranch hand for Mexican businessman Eloy Vallina, owner of the Adobe-Slash Ranch in New Mexico, recently admitted to branding cows near a wolf den to tempt the wolves to attack, hoping to force their removal.[31]

Unlike the Yellowstone ecoregion, an international tourist destination, Catron County has little experience with tourism. Yet recreation would be a logical source of revenue: in western states, over 43 million people spend $33 billion annually on hunting, fishing, and wildlife watching. Mountain biking, climbing, and adventure sports bring millions of additional revenue to local communities. The reintroduction of wolves in Yellowstone—where over 150,000 people visit every year specifically to see them—is estimated to have brought $35 million in annual revenue to Idaho, Montana, and Wyoming. A 2006 study concluded that successful Mexican wolf recovery could generate similarly substantial regional benefits from ecotourism and the sale of "wolf-friendly" beef.[32]

Such inducements failed to persuade locals or officials. Paul Paquet and other independent biologists repeatedly advised the USFWS to allow wolves to roam outside the confines of the national forests and to require ranchers to remove or apply lime to cattle carcasses so that wolves would not eat them.[33] The agency refused to revise the plan; environmental groups sued to force its hand. Wolves have killed only ten cows a year out of thirty-five thousand in the area, but the USFWS has authorized the shooting of eleven wolves since 1998, including six pups from a single pack and a male described in the scientific review as "genetically irreplaceable."[34] Over the course of the reintroduction program, thirty wolves have been illegally shot, thirty more died due to accidents or mishandling, and thirty-four were permanently removed from the wild. By 2007—when there should have been a population of a hundred in the wild—there were only fifty-two. In a 2008 report, the government finally acknowledged that the recovery plan was outdated but failed to issue a new plan or to revise the old one. Meanwhile, the USFWS—which Michael Robinson has termed a "predatory bureaucracy" in a recent book by that title—continues to excel at what has long been a core competency: killing wolves.[35]

During an era when big business and corporate interests reigned supreme, government was not on the side of rewilding; indeed, government became the declared foe of science, firing biologists from virtually every federal agency with a mandate to protect wildlife, opposing science-based solutions to climate change, trying to cripple the Endangered Species Act, and thwarting responsible attempts to restore the balance of nature. In response to this intransigence, western governors and environmental groups were forced to strike out on their own to address the biodiversity crisis, using state legislatures, private money, and the power of the courts. In 2008, the Western Governors' Association launched a Wildlife Corridors Initiative to coordinate all available data on crucial habitat and connectivity, in an effort to preserve wildlife and the revenue it brings.[36]

But by and large, rewilding was on its own. In the absence of governmental support, the movement would need to come up with its own techniques, resources, and solutions. And in many unique and inspirational ways, it did.

"Buy More Cats"

As the Mexican wolf program was getting under way, soon to descend into chaos, another predator, the jaguar, appeared in an unlikely place and attracted unlikely allies. It was as if a carnivorous version of the Virgin of Guadalupe had appeared in the wilderness: so transformational and almost religious was the euphoria induced in hunters, conservationists, and biologists alike that support for the animal immediately coalesced in plans for a core reserve and corridors designed to protect it. Rewilding suddenly found new support just south of Catron County, among a group of conservative Republican ranchers in the mountains of southeastern Arizona.

On 7 March 1996, a fourth-generation sixty-six-year-old rancher and hunting guide named Warner Glenn, of the Malpai Ranch near the borders of Mexico, Arizona, and New Mexico, was guiding a client on a mountain lion hunt, riding a mule and leading a pack of hunting hounds into the Peloncillo Mountains, when he heard his hounds sound off, chasing a cat. After he caught up with them, he said to himself, "God Almighty! That's a jaguar!"[37] After a lifetime hunting in the region, it was the first wild jaguar he had seen—the first seen in the United States in decades—and he documented the experience by taking ten remarkable photographs, holding his camera blindly over the edge of a chasm where his dogs had cornered the animal. After he snapped the pictures, later published in *Eyes of Fire: Encounter with a Borderlands Jaguar*, he pulled the dogs off, allowing the jaguar to escape.

Six months later, another hunter, Jack Childs, treed a jaguar in the Baboquivari Mountains in Arizona.[38] Like Glenn, he took photographs and then let the jaguar go free. Both Glenn and Childs, lifelong hunters, have become the jaguar's most ardent promoters. The rarity of the cat transfigured it, in their eyes, into something intrinsically valuable, something worth preserving. Already active in conservation, one of the founders of the Malpai Borderlands Group, which had embraced conservation science, putting a million acres under protective easements, Glenn felt that his group was partially responsible for the return of the big cat.

The unexpected jaguar encounters established incontrovertibly that there was a surviving population of desert jaguars, and the discovery galvanized the Malpai ranchers and the conservation community in the American Southwest. While the federal government only reluctantly and under pressure of litigation declared the jaguar endangered, it has so far refused to designate a critical habitat or recovery plan for the species. So several grassroots groups banded together to preserve jaguar habitat that still existed and to restore "safe passage corridors" from northern Mexico to the southern Rockies, including the Sky Island Alliance, based in Tucson; Naturalia, a Mexican conservation organization headquartered in Mexico City; and the Malpai ranchers.

In 2003, members of these groups launched the Northern Jaguar Project, to purchase ranchland in the Mexican state of Sonora, where the northernmost breeding population of jaguars lay hidden in deep canyons inaccessible by road. That year, they bought 10,000 acres; in 2007, they added 35,000 acres of adjacent land, sold to them by an eighty-four-year-old rancher who told the new owners that his habit had been to shoot jaguars whenever he had the chance. Also, Naturalia collaborated with the nearby Yaqui tribe to link the new jaguar reserve with Sierra Bacatete, a Yaqui reserve on the coast of the Gulf of California, where jaguars and California condors have been seen feeding on beached whales.

Peter Warshall, a gray-bearded biologist for the Northern Jaguar Project—who has worked on conservation issues for over thirty years for the United Nations, the U.S. Agency for International Development, and for a number of NGOs—recruited biologists and young Mexican graduate students to survey the new Mexican jaguar reserve. With camera traps, motion-sensitive cameras affixed to trees that snap a photo when an animal crosses in front of a light beam, they documented five species of felids: jaguar, bobcat, ocelot, jaguarundi, and mountain lion. Specialists counted over two hundred species of birds, including military macaws, elegant trogan, and the southernmost nesting population of bald eagles; ninety-two species of butterflies; and rare species of turtles, fish, insects, and plants. Warshall recently described the unique biodiversity of the area, where temperate, cooler climate species overlap with neotropical ones: "There are really peculiar mind-bending combinations of mesquite, oak, and palm trees, and a species of morning glory that has become a tree whose flowers are fueling stations for bats and all kinds of hummingbirds."[39]

In a bid to engage neighboring ranchers in the vision for the jaguar reserve and to convince them to stop shooting big cats, Warshall came up with a new twist on the kinds of compensation schemes that Defenders of Wildlife had pioneered. Instead of paying ranchers for dead cattle, why not pay them for photographing live cats? Warshall disliked the prospect of waiting for livestock carcasses to turn up to be examined for signs that a predator was responsible. "The forensics are grim," he noted of such cases, and so were the logistics: Trying to reach carcasses on remote Mexican ranches would be prohibitively time-consuming. Thus, the Community Photo Survey Contest was born, turning a formerly negative approach on its head.

With funding from Defenders, ranches adjacent to the reserve were equipped with remote cameras. A local *vaquero*, or cowboy, was trained to monitor and maintain them in exchange for a stipend. Ranch owners who agreed not to shoot cats on their property were paid for each photograph of a living cat, $500 for jaguar, $100 for mountain lion. In the first year, the program paid out 66,500 pesos, or $6,500, for photos of lions, ocelots, and bobcats. Warshall recently asked the son of a rancher who had been paid for eight photographs what he and his father were planning to do with the money. He turned to his father and said, "Buy more cats!" Warshall said. "That's the kind of thing that's changing attitudes in these corridors down there."

Warshall was tapping into a new social reality in Mexico. While its government was slow to establish a park system and has never made substantial investments in its protected areas, Mexican citizens had recently begun to embrace their natural history. To aid in the purchase of the first 10,000 acres, Naturalia convinced a Mexican bank to issue a collectible limited-edition series of five-peso coins, each featuring a different endangered species. Wildlife posters that Naturalia hung in a town hall near a new protected grassland in northern Chihuahua were so coveted that people took them off the walls to display in their homes. Craig Miller, who works on jaguar and Mexican wolf issues for Defenders of Wildlife, met with ranchers in Sonora and was invited to a rancher's home after the contest was under way. He was startled to see hanging on the dining room wall a photograph from one of the remote cameras "of the ass end of a cougar."[40] Despite the unflattering view, the rancher was so proud of the photo, for which he received $500, that he had framed it beautifully in local mesquite.

During a recent scouting expedition in Sonora, looking for evidence of how jaguars were traveling north, Sergio Avila, a young Mexican biologist with Sky Island Alliance, showed me what conservationists were up against. Scrambling down a creek bed in the dry Sierra foothills, we picked our way with some difficulty over miles of black PVC pipe littering one rocky canyon, cow shit dissolving in what was left of a seeping stream. Above, on the ridgelines, severely eroded rills snaked downward: overgrazing had destroyed vegetation holding the hill together. Whenever it rained, water washed bare soil down the arroyo, like talons raked through dirt. The bright spot was the ranchers' attitude, which seemed almost comically flexible compared with that of their American counterparts in Catron County. At "Rancho Agua Caliente," an overseer enthusiastically claimed that "*el tigre*" used to cross the road "right on this wash." As we hiked out of one canyon, we were hailed by several men on horseback, including the owner of the land, who introduced himself as Raul, gazing down at us from a big pied horse. Avila explained that he was evaluating the region for conservation.

Raul broke out with a big smile when he heard, "*conservación.*" "Si! Si!" he said, waving a cigarette. "*Yo quiero conservar por el futuro! Yo quiero invitar!*" He welcomed photographers, conservationists, anyone who might want to enjoy the land. He had an entrepreneurial eye for sources of income other than cattle, like his Sonoran neighbors near the jaguar reserve, who are organizing deer hunting trips for gringos. He had an *onça*, he said, mountain cats, maybe an ocelot.

"All right!" Avila exclaimed, as Raul and his *vaqueros* turned their horses and galloped off. "I'm happy and hopeful! You do something with people over here for a year, or half a year, and then the people on the other side of the mountains, they hear about it." This was the kind of strategy that grassroots rewilding groups excel at: creating support by putting a value on predators.

The jaguar continues to inspire reverence on both sides of the border. Jack Childs launched his own nonprofit in 2001, the Borderlands Jaguar Detection Project, setting up camera traps that eventually documented two jaguar males—one of which turned out to be the animal Childs had treed years earlier—patrolling a huge territory of around 525 square miles in a 50-square-mile loop from Sonora to Arizona.[41] Passing the cameras at intervals, the males appear to be trolling for prey and females. In 2006, Warner Glenn and his daughter were hunting mountain lion when they encountered another male jaguar, this one in the Animas Mountains of New Mexico.

In the spring of 2009, "Macho B," one of the male jaguars photographed repeatedly and thought to be resident in the United States, was caught in an Arizona Fish and Game snare baited with female jaguar scent. The animal was fitted with a radio collar and released, but twelve days later he was recaptured, faltering and ill, and subsequently euthanized.[42] After a public outcry and investigations by the *Arizona Daily Star*, it remained unclear why the agency had snared the animal, but it may have something to do with the lawsuit filed by Defenders and other organizations, aimed at forcing the USFWS to revisit its decision not to create a recovery plan for the jaguar. Only a few weeks after Macho B's death, a federal judge ordered the agency to reconsider that very decision, saying that it had not articulated "a rational basis" for its refusal.[43]

The jaguar was making its comeback in remote areas of the Southwest at a time that could not have been more inopportune politically. After a massive wall was constructed across the San Diego–Tijuana section of the U.S.-Mexico border in the 1990s, illegal border crossings had shifted to the desert terrain of Arizona, a shift that resulted in hundreds of migrant deaths as well as environmental damage to fragile habitat caused by tons of abandoned detritus. The Sky Island Alliance and other rewilding groups aggressively fought the expansion of the border wall into Arizona, New Mexico, and Texas, but to no avail: the Department of Homeland Security waived requirements to comply with existing environmental regulations. Nearly seven hundred miles of fifteen-foot-tall steel fencing is being constructed, along with damaging roads, lighting, and infrastructure.

But Warshall remained hopeful, arguing that the fence did not change the goals of the Northern Jaguar Project. In his view, jaguars could breed in the Mexican reserve and eventually, perhaps when the highly unpopular wall comes down, the population would be able to use natural mountain corridors to disperse north. "In twenty, thirty years, after the Berlin Wall falls," he said, "they'll be able to come to the United States." Warshall also predicted that the methods used to convince Mexican ranchers that big cats are worth saving—camera contests and compensation for photographs—would prove effective in the United States, with other species. "I have a feeling you'll start seeing camera traps in wolf country," he said.

And indeed, although there was no sign that wolf opponents in Catron County were beginning to soften their opposition, Defenders of Wildlife had opened a new front in the wolf wars, strengthening support for the Mexican wolf among a critical Arizona constituency, the White Mountain Apache. The tribe has an exemplary conservation record, breeding and reintroducing endangered native Apache trout to streams, establishing its own sport fishery. Since 2000, the White Mountain Apaches have accepted captive-bred wolves and allowed dispersing individuals on their 1.6 million acres of high-quality spruce and ponderosa forests on the Arizona border, adjacent to the wolf recovery area. While tribal livestock managers have suffered some losses, the tribe is planning an ecotourism program to combine wolf tracking with performances by Apache Crown dancers that highlight their cultural relationship to the wolf, creating jobs and revenue. Defenders recently began negotiating with the tribe's livestock association to bring the camera contest to the Apache, documenting the return of the wolf to their land. Why is the tribe succeeding while Catron County fails? Krista Beazley, biologist, tribal member, and manager of its wolf recovery program, sees the answer as simple: empathy. After long persecution, the Apache look at wolves and see themselves. "It's the same thing that happened to us," she says.[44]

As lawsuits over the Mexican wolf program and the jaguar recovery plan wind through the courts, every ounce of such empathy remains crucial. When it comes to predators, we could learn to pull our punches. We could practice a little tolerance. Think of those howler monkeys on their wasted islands in Lago Guri. That's what happens when you're all alone.

CHAPTER 3

CORRIDORS IN CENTRAL AND SOUTH AMERICA

Categories of Concern

THE WORLD WAS FILLING UP. OUTSIDE NORTH AMERICA, CON-
servationists bent on bringing rewilding to every corner of the planet
found themselves contending with issues outside their expertise: poverty, land reform, injustice. Even seemingly inhospitable places—
forests, jungles, volcanoes, mountainsides—turned out to house surprising numbers of people or to provide nearby populations with their
living: fuel, firewood, charcoal, bushmeat, wild plants. Millions of
landless poor were scraping by on land they could never hope to own,
often as squatters. Indigenous people had long-standing and legitimate historical claims to land that were or were not recognized by
laws and governments. In such places, obtaining land for conservation meant displacing people. In its relatively short history, conservation
had already earned an unfortunate colonial reputation, a reputation
that continued to color everything it hoped to achieve.

Historically, as governments or institutions created national parks
or wildlife preserves, they removed people from them. Yellowstone
established the model. After Congress designated it as the world's
first national park in 1872, park management excluded from it a number
of Native American tribes whose members had used the region for

hunting, gathering, and ceremonial activities. Since the post–World War II era, the same pattern has continued to cause conflict and resentment. The very language of the world's largest confederation of conservation organizations, the International Union for Conservation of Nature (IUCN), reflected these inequities.*

Over the years, the union has defined and redefined a classification system for parks and reserves. At present, the highest categories, I and II, cover national parks and nature reserves, protecting their "ecological processes" through "strict" control of human use; category III protects notable landforms or geological formations (such as caverns, caves, or seamounts); and categories IV–VI allow controlled human use of protected landscapes or habitat management areas through "sustainable natural resource management."[1] The most tightly managed of reserves, those described in category I, "where human visitation, use and impacts are strictly controlled and limited to ensure protection of the conservation values," seem to exclude human habitation, although the language is not specific. And to add to the confusion, the IUCN in a separate resolution has recommended that indigenous people not be required to relinquish traditional lands and that local people be consulted during the creation of new protected areas.[2]

Biologists designing rewilding and restoration projects remained committed to "strict" protection of core reserves, which was generally understood to preclude human habitation and use. But increasingly, as the conservation community began working with the aid community over how to negotiate with and compensate people displaced by the growing number of protected areas worldwide, aid workers from other disciplines—sociologists, anthropologists, political consultants—condemned the very idea of strict protection as colonialist, even racist.

---

* This body currently brings together over a thousand conservation organizations and government agencies, including such well-known entities as the World Wide Fund for Nature (the World Wildlife Fund in the United States), the Nature Conservancy, the World Bank, the United Nations, and USAID. The union is the leading voice of the international conservation community, and its regulations govern the selection and management of most of the world's protected areas.

A world of trouble was concealed in the IUCN's definitions, and rewilding projects increasingly collided with the enormous worldwide aid bureaucracies. The first casualty was the project known as Paseo Pantera.

The Path of the Panther

Central America was a potentially ideal location for the world's first major international corridor project. It harbored exceptional biodiversity, with rich marine ecosystems and coral reefs, cloud forests on its volcanic mountain chains, and astounding levels of endemism—huge numbers of species found nowhere else on earth. Its forests teemed with 17,000 species of plants (2,941 endemic) and 440 species of mammals, including charismatic predators (jaguar, mountain lion, and many smaller species of cat), primates (howler monkeys, spider monkeys, squirrel monkeys), and the shy tapir, a rare, strange neotropical mammal, part terrestrial, part aquatic, distantly related to the manatee.[3]

Some of the most spectacular birds in the world were to be found in Central America, over a thousand species, including the resplendent quetzal, a wonderfully gaudy green and red creature with foot-long tail streamers, a native of the cloud forest. The neotropical land isthmus also served as a migratory flyway for over 225 additional birds. Its beaches were crowded with sea turtles, and its forests sported colorful poison dart frogs, crocodiles, and over a thousand other reptiles and amphibians. Despite its relatively small size, representing 0.5 percent of the earth's land surface, Central America loomed large, with an estimated 8 percent of the world's total biodiversity.

But there were also problems that undermined the region's environmental richness. Poverty was so endemic that it was precipitating a potentially calamitous environmental crisis: half the population subsisted below the poverty line, with many lacking access to clean water, basic health care, electricity, and education. Those eking out a life in rural areas were dependent on natural resources, on forests for wood and bushmeat, on rivers and streams for water, and on small plots of land to grow corn and keep livestock. Erosion cost lives in the wake of hurricanes and flooding: over half the forests of the isthmus had already been cut down, and the rate of forest loss remained high, at 1.4 percent a year. A full 90 percent of the primary, or old-growth, forests were gone, and the secondary forests that remained made up a patchwork of small, isolated fragments.

To conservation organizations and the United Nations, the environmental protection of Central America seemed like a means to a gloriously harmonious end: jobs created through ecotourism would address persistent poverty, and natural resources could be secured through sustainable development. In the early 1990s, the Wildlife Conservation Society—inspired by the Florida reserve designed by Reed Noss and his colleagues at the University of Florida—asked a team of mapping experts there to produce a similar map of existing parks, reserves, and extant forests.[4] The result suggested that the best potential corridor lay along the Atlantic Coast. The map did not exactly present a promising picture: most existing parks were small, isolated, and poorly maintained. Connecting a chain of some six hundred protected areas, the corridor would depend on convincing governments to set aside remaining forested remnants. But there were bright spots, including large contiguous blocks of southern Mexico, Belize, and Guatemala that could be utilized. Costa Rica, which had developed its ecotourism market and had a much lower poverty rate than the rest of the region, also held promise.

The WCS planners called the proposed corridor Paseo Pantera, for "the path of the panther," referring to the mountain lion whose range reaches from Canada all the way south to Argentina. This would be the first time that a major corridor project was launched outside North America.

As Paseo Pantera moved forward, it attracted many additional organizations and funders; at times, it seemed like the entire international conservation community was involved. Soon all seven Central American countries and the five southern states of Mexico had pledged to cooperate. But after governments endorsed the project in 1994,

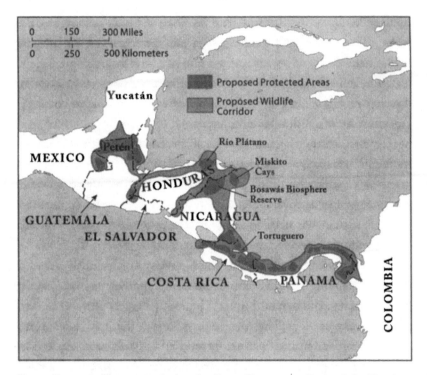

Paseo Pantera: The original plan for Paseo Pantera, or Path of the Panther, envisioned a corridor running from Mexico's Yucatán Peninsula south through Central America. (Based on the original map by Brian Evans in *Wild Earth*, Special Issue, 1992.)

indigenous groups protested, concerned that land traditionally used by their people would be annexed into parks. Some of these groups threatened to ignore new protected area boundaries.

In response, governments, conservation organizations, and funding agencies redefined the project as an "integrated conservation and development project," a type of human aid scheme heavily promoted by the United Nations and other funding agencies and based on the premise that conservation areas would survive only if they addressed human concerns and served human needs. The rise of such programs represented a major turning point in the fortunes of modern-day conservation, and during the 1990s organizations pumped up their budgets by qualifying for large grants to pay for rural development near protected areas. As these programs proliferated, they adopted the language of human aid organizations: all parties involved were "stakeholders"; educational or vocational training was "capacity building."

Proponents of such programs argued that people must come first, that the environment was a lesser issue, that people could be persuaded to respect protected areas only if they could derive some benefit from them. They were joined by revisionist critics who railed against "neoliberalism"; some went so far as to reject the existence of the extinction crisis. Biologists, for their part, believed that many in the aid community did not comprehend the compelling motivation behind conservation biology and its attempt to forestall a wave of extinctions. They watched in dismay as more money was spent on human aid than on conservation. They believed development near parks played a counterproductive role, drawing more people into buffer zones and degrading habitat.

But they were overruled to meet the indigenous opposition. The project was renamed the "Mesoamerican Biological Corridor" and refashioned as a multifaceted environmental and sustainable development project promoting an array of activities—leadership training for indigenous people, organic farming, community forestry, aquaculture, fishing, honey production, handicrafts, ecotourism, and butterfly farming, among them—that would serve local economies in and around the protected areas. A formal agreement was signed in 1997, and a new regional bureaucracy was created to administer the project. Conservation groups were relegated to the status of "stakeholders," just one among many competing voices.

The complexities of wrangling so many governments, constituencies, municipalities, bureaucracies, and dozens of wary indigenous groups soon became overwhelming. At the heart of the problem was land use. The Mesoamerican corridor identified four zones—cores, buffers, corridors, and multiple-use zones—but made little progress in reaching agreements with villages over what the zoning would mean, how it would be enforced, and where hunting and fishing could continue. Workshops and forums held in local communities failed to impress rural people who had seen wars waged for decades over land reform. In interviews, rural people made it plain that they did not want "participatory" consultation; they wanted decision-making power. While they might agree in theory to the establishment of biological corridors, in practice they expected something in return. Instead of fashioning a compensation scheme, the project continued to hold endless conferences, meetings, consultations. In 1998, a coalition of fifty groups representing the indigenous and peasant peoples of Central America presented its own, competing project for an indigenous biological corridor.

By this time, many of the original biologists felt that the project had lost its way. It promised too much to too many and no longer seemed to be based on conservation principles or geared toward restoring connectivity. A 2001 *Science* article acknowledged that much of the vision had been lost. What remained on the maps was a "braided network" of hundreds of narrow strips, a mile or so wide, that might be devised to connect protected areas.[5] What nobody seemed to know was whether these funnels would provide viable habitat and connectivity for wildlife.

Controversy plagued the project, particularly after it became linked in the minds of critics with the Puebla-Panama Plan, a massive $50 billion project funded by the Inter-American Development Bank to develop factories and transportation from Puebla, a Mexican manufacturing hub, to the Panama Canal. That plan called for expanding airports and seaports, hydroelectric dams, pipelines, investment in electricity grids and telecommunications, and a vast new road system. By hiring conservation consultants to produce reports on cultural and ecotourism projects, and by making only minor commitments to conservation, the Puebla plan incited charges of "greenwashing." It also touched off strikes and protests that shut down airports, blocked border crossings and roads, and stopped work on dams throughout southern Mexico, Guatemala, El Salvador, and Nicaragua. While one airport and the hydroelectric dams were eventually dropped from the project, the rest of the development is proceeding.

If conservation organizations did not explicitly support the aims of the Puebla plan, they did not oppose it either. In a region with a long history of corporate exploitation going back to the days of United Fruit, their neutrality incited intense suspicion among civil protest groups. Conservationists also refused to oppose a major trade accord, the Central America Free Trade Agreement, completed in 2005, that promised to bring development—factories, roads, dams, mines—to the same areas that the Mesoamerican corridor project proposed to protect; they feared that such a political stand would cost donations, a significant proportion of which originated from major corporations or political entities.

Thus the world's first major conservation corridor project came to be the target of bitter criticism and opposition led by anthropologists, sociologists, human rights workers, and members of campesino groups, universally denounced as an adjunct to development supported by Shell, ExxonMobil, Monsanto, Dow Chemical, and many other multinational corporations associated with pollution and environmental depredation.[6] There it has languished, distributing grants, paying for research, and by all accounts accomplishing little in terms of setting aside protected areas, taking care of existing parks, or creating corridors between them. Costa Rica, Mexico, and Belize planned to move ahead on their own with plans to create biological corridors, but so far none of these have reached completion. In 2001, a major report acknowledged that, while the original project remained urgently necessary, public support and awareness were "quite limited."[7] Conservationists, it noted, feared that the project's expansion had burdened the project with impossible expectations: "social and economic problems that it cannot solve."[8]

By 2006, hundreds of millions of dollars had been spent but only a single protected area had been created.[9] That didn't stop the Second Mesoamerica Protected Area Congress from celebrating the stalled project or the Wildlife Conservation Society, the organization that launched Paseo Pantera, from agreeing to reexamine it. Alan Rabinowitz, the group's jaguar expert, commended Central American nations and Mexico "for agreeing to such a far-sighted initiative."[10] Recently, in *National Geographic*, the project was rebranded yet again, as "Paseo del Jaguar," without reference to its checkered past.[11] "Far-sighted" was perhaps not a fitting description for an enterprise so diffuse, bureaucratic, opaque, unaccountable, and ineffectual. An enormous amount of money has disappeared into the Mesoamerican corridor, with little to show for it. One day, it may achieve what it set out to do. But in the meantime, it has become a metaphor for everything that can go wrong with corridor projects.

When it comes to rewilding, modesty may prevail. As I would see in Brazil, when projects are sharply delineated and strictly limited, when land-use issues have been clarified by law, when there is less money up for grabs, more gets done.

Fragments of Brazil

In contrast to the debacles of the Mesoamerican Biological Corridor, the plan to restore connectivity between two core areas in Brazil—Emas National Park, a tiny remnant of savannah, or cerrado, in southwest Brazil and the Pantanal, one of the largest wetlands in the world—was well defined and eminently practical. Located in an area of low human density, the Cerrado-Pantanal Ecological Corridors Project has been spared the complex land-tenure issues that plagued the Central American project. With nowhere near the financial support or government attention of that larger plan, this small Brazilian project has made great gains, solidifying an unusual array of corridors for a host of endemic, endangered wildlife and winning the trust and cooperation of local people.

Emas is an extraordinarily biodiverse place, rivaling Carnaval in its strange and colorful characters attired in masks and feathers: jaguars, maned wolves, giant anteaters, macaws, and the rapt Brazilian biologists who follow them in the night. The eminent naturalist George Schaller, who visited the area in 1975, called it "the Brazilian Serengeti," and the vistas of grasses bending in the wind, flocks of ostrich-

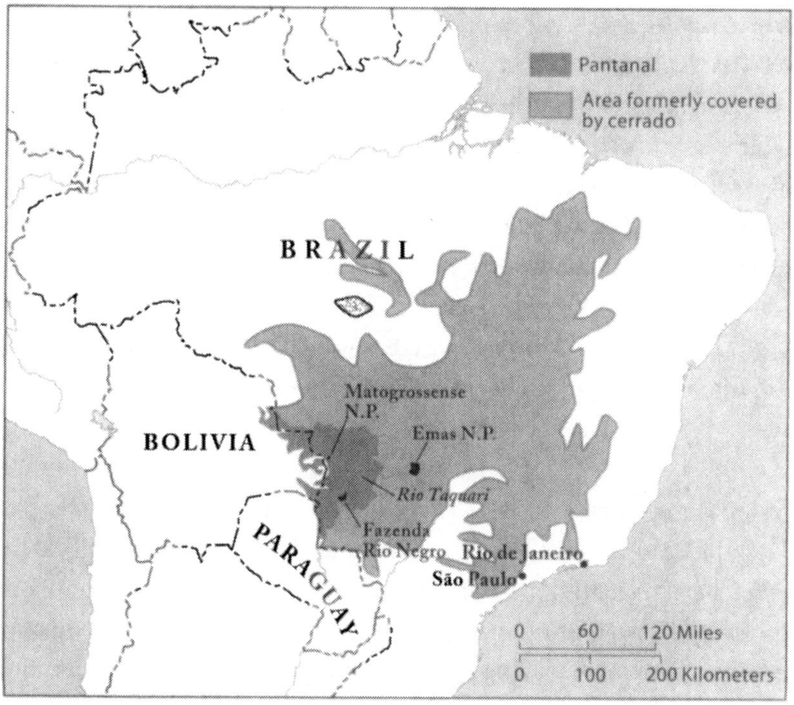

Pantanal

Area formerly covered by cerrado

BRAZIL

BOLIVIA

Matogrossense N.P.

Emas N.P.

Rio Taquari

PARAGUAY

Fazenda Rio Negro Rio de Janeiro

São Paulo

| 0 | 60 | 120 Miles |

| 0 | 100 | 200 Kilometers |

Brazil: With eighty percent of Brazil's cerrado, or tropical savannah, plowed under, biologists hope to reestablish riparian corridors between Emas National Park, a stronghold of the region's exhilarating wildlife, and the Pantanal, one of the largest wetlands in the world.

like rhea, and herds of pampas deer recall the African plains. But Emas is tiny, at only 510 square miles, half the size of Rhode Island, surrounded on all sides by sorghum, soybeans, and cotton. Few Brazilians have ever heard of it, but it is hard to find a biologist who has not: the park is renowned as a UNESCO World Heritage Site and Biosphere Reserve: It is one of the last fragments of South American tropical grassland savannah left on earth.

The cerrado once covered 500 million acres, three times the size of Texas. It was justly famed for its wildlife: 935 species of birds, 10,000 species of vascular plants, hundreds of which were found nowhere else, and almost 300 species of mammal, including the maned wolf, giant anteater, giant armadillo, crab-eating fox, pampas deer, white-lipped peccary, ocelot, puma, and jaguar. But the once richly biodiverse grasslands that stretched thousands of miles across the interior of the continent have been plowed under to create one of the widest stretches of intensively cultivated monoculture in the world; 80 percent has been converted to soybeans, corn, cotton, and other crops. While Brazil's huge investment in agriculture has yielded great gains for the economy, it threatens to alter the country's climate and rob it of some of its most valuable plants and animals. While Emas and a few other protected savannah fragments are a wonderful representation of a once mighty ecosystem, they are too small to protect their species far into the future.

The Pantanal, too, is isolated. A seasonal floodplain of 74,000 square miles, home to jaguars, caimans, and giant river otters, the Pantanal is also a UNESCO and Ramsar site, occupying two Brazilian states and parts of Paraguay and Bolivia. While a small area is protected by Matogrossense National Park, 99 percent of the Pantanal is privately owned, playing host to a unique frontier cow culture: the area has long been a stronghold of *pantaneiros*, local ranchers who maneuver some eight million cattle through the rainy season, when 80 percent of the land is flooded. Surrounded by agriculture and deforested highlands, the Pantanal is increasingly under siege from pollution and development.

Brazilian biologists Leandro Silveira and his wife, Anah Tereza de Almeida Jácomo, who have been living and researching in Emas since 1994, have made it their goal to patch together a mosaic of undeveloped land and riparian habitat between the core areas of the Pantanal and Emas using another evolution of the corridor, the stepping stone model. To accomplish this, they must convince property owners in the region to set aside the right land in the right places, creating a matrix of vegetative cover and riverine habitat that will stretch from the floodplains to the savannah, allowing genetic exchange and stabilizing both ecosystems. If either the Pantanal or Emas were to remain isolated, some of the species and populations within them would eventually wink out, in the same way that species in national parks in the United States have begun to vanish. Given its limited size, Emas may not be large enough to maintain the genetic viability of predators like the jaguar over time. If the jaguar were to disappear, that could trigger the unforeseen consequences that eroded biodiversity in the Yellowstone ecosystem in the absence of wolves. Without jaguars, other predators might come to the fore or prey numbers could explode, with deleterious effects on many organisms, fauna and flora.

To produce the best possible design and management plan for such corridors, the Brazilian biologists are intent on gathering the basic knowledge that will be needed. Learning as much as possible about the species in Emas and the Pantanal will be crucial to designing a matrix of natural patches and restoring connectivity. Many of the species in both regions have never been studied; too little is known about everything from the size of territories to breeding behavior. While jaguars are found elsewhere, those in Emas behave in ways unique to the region, developing behavioral adaptations to their habitat. Silveira, Jácomo, and their team have put in exhausting hours, tracking animals by night, recording results by day. They employ several veterinarians who shepherd teams of volunteers, graduate students, and researchers from around the world who pitch in daily to help them check live traps laid throughout the park on the edges of its dirt roads. Their home and a nearby office outside the park have become a research hub where prodigious amounts of data are constantly being gathered, gone over, and entered in computers.

I recently spent a couple of weeks riding around the park with Silveira, Jácomo, and their veterinarians and volunteers, checking live traps, processing radio-collared animals and photographs from camera traps, and conducting an ongoing fauna census throughout the park, which is not open to visitors except by special arrangement. Driving past blackened firebreaks, burned regularly along the roadsides to prevent catastrophic wildfires like the one in 1994 that swept through most of the park, we constantly scanned for and tallied numbers of pampas deer and flocks of rhea, which run through Emas flapping their flightless wings like petticoats behind them. The horizon was studded with millions of reddish termite mounds, home to owls, snakes, ants, and nine-banded armadillos. Each day, we checked two loops of live traps baited with pigeons, often finding a maned wolf or crab-eating fox inside. The maned wolves—not true wolves but ancient relatives of the dog—looked like long-legged coyotes with russet fur, their feet dipped in charcoal.

Animals that had never been captured were fitted with radio collars after their blood was drawn and they were measured and weighed. Recaptures—collared animals caught more than once, so accustomed to a free meal that they become "trap happy"—were beginning to provide answers to questions about life expectancy and population saturation. Long-term studies such as these yield essential information about the carrying capacity of the ecosystem for different species. Patterns of die-off tell biologists when an area has reached its capacity for a particular species and thereby aid in projecting optimal sizes for a protected area. If an area is too small to support the number of mountain lions it already has, then the area will need to be enlarged—or a corridor in and out provided—to prevent the population's becoming inbred and genetically weak.

Already, Silveira told me, the study of radio-collared jaguars within the park has helped reveal their movement preferences. In South America, the jaguar is a jungle animal, and the jaguars of Emas have never ventured out of the dense vegetation of the park and its riparian gallery forests to make their way through open agricultural fields. Puma and maned wolf have—one day we saw a maned wolf standing in a field of dried stalks, peering around above the crop—but the jaguar will not set foot in open fields. That means the jaguar needs wild, junglelike corridors in and out of Emas and other fragments of the cerrado. Silveira received valuable confirmation of this after he invited University of Washington biologist Sam Wasser and his graduate student Carly Vynne to Emas. Wasser has pioneered the art of training dogs to do what dogs do best: find scat, nuggets of information on diet, health, DNA. Wasser and his students have trained dogs to locate species-specific scat: grizzly bear, spotted owl, jaguar, even whale. "I'm always thinking about how you can get more poop out of the woods," he once said.[12] Working with Silveira's dogs—he has a kennel of hunting hounds—Vynne recently found twenty-four samples of jaguar scat, yielding data on diet and movement that helped to confirm what the radiotelemetry was already telling them: the jaguars of Emas need dense cover.

With this information, Silveira and Jácomo are mapping out their matrix of intact riverine forest and wild lands between Emas and the Pantanal. Silveira hopes to take advantage of Brazilian law, which stipulates that owners in the cerrado are required to preserve all riparian areas and 20 percent of their land from development, deforestation, or agricultural use, a policy enforced more effectively here than in the Amazon. He sees cultivating relationships with local landowners as the key to the process. Before he arrived, no one believed that jaguars lived in Emas. When he found them, locals accused him of releasing the animals himself. Now he and Jácomo have made it their business to develop close partnerships with all thirty-nine farms surrounding Emas. They often discuss with landowners which parts of their land should be preserved, and how. "There are jaguars in Emas, and jaguars in the Pantanal," he said. "We need animals coming from the Pantanal. If every thirty years, one male comes from the Pantanal and mates with a female, that's enough. We don't need to have jaguars in the corridor, but if a male comes dispersing, he needs to be able to get to the other end."[13]

Silveira and Jácomo have been luckier than their counterparts in Central America: population density in the huge monoculture areas is low, and in the Pantanal relatively few people own and manage large spreads. To improve relations with ranchers in the Pantanal, they have established a Defenders-style "Jaguar Conservation Fund" to compensate ranchers who lose cattle to jaguars. They also run a program providing education and free medical and dental care to workers on eleven ranches in the Pantanal. If a rancher or his workers kill a jaguar, they're out of the program. So far, all have complied, and Silveira claims others are eager to join: "The *pantaneiros* are very traditional, they don't like outsiders at all. We were the first to come and say to them, we want to be partners. We want to work on a management policy to manage wildlife." Silveira and Jácomo have mastered the art of the conservation quid pro quo: luring ranchers to protect wildlife with a deal they can't refuse.

One night out in the field, driving with the biologists, I was manning the spotlight in the back of the pickup when Silveira slammed on the brakes, leapt out of the driver's seat, and raced off into a dense tangle of grass and shrubs, shouting for me to follow him with the spotlight. After an intense struggle, he and his wife, wielding a huge net, dragged their booty back to the truck: a giant armadillo, sedated, its enormous five-inch-long digging claw waving feebly. The rare subterranean species is practically unknown to science and in grave danger of extinction owing to habitat loss. This was the first female ever caught. The next day, Jácomo drew blood, took measurements, and gingerly attached a radio collar to the animal's carapace, essentially a giant fingernail. She was to be released the following day, after the anesthesia had worn off. But during the night, she used her claw to lift the heavy wooden door of the cage, jumped out of the back of the pickup, and took off. The signal from her radio collar showed her heading back into Emas.

So far, Silveira's team has captured and released 755 carnivores, 121 of which have been radio-collared and tracked. The process has provided insight into where and how to design the network of 250-mile-long corridors linking the cerrado and the Pantanal. Those corridors are now under way: in 1999, Brazil established the Headwaters of the Taquari State Park, which lies near Emas and whose waters flow into the Pantanal. In 2003, Conservation International, using Silveira's data as a guide, negotiated the creation of a 271,700-acre biodiversity protection area composed of several private reserves and the Rio Negro State Park.[14] These reserves will anchor the main corridors at each end.

Silveira and Jácomo are also mining their data—which prove that roadkill has escalated an alarming 33 percent in the cerrado—to propose changes to the highway system bisecting potential Pantanal-cerrado corridors. Publishing eight years of roadkill data, the pair have demonstrated that highways present a devastating ecological barrier. They have called for the same kinds of wildlife-friendly over- and underpasses, bridges, and culverts that have worked in Europe and Canada.[15]

Funding is lean, so the team is remarkably frugal. Many a day was spent tracking down a radio collar that was sending a "mortality signal," a special tone given out when there has been no movement for over two hours, indicating a dead battery, lost collar, or dead animal. It was important to know if an animal had perished, but it was equally important to locate a lost—or even consumed—collar. Radio collars are expensive, costing several hundred dollars apiece; those with GPS or satellite-tracking capability cost two or three times that. Silveira and Jácomo once wrote and published in the *Herpetological Review* a field report recounting how one of their radio collars, worn by a crab-eating fox, ended up in the stomach of a green anaconda that measured ten feet, four inches long and weighed thirty-three pounds (sans fox).[16] A month after their equipment went missing, they tracked the anaconda, still sending a signal, to the shore of the Rio Formosa, where it was basking comfortably. They followed it into a marsh and captured it, corralling it inside a temporary fence where the snake regurgitated the collar and the remains of the fox. (Snakes expend a great deal of energy in digesting their food, and the process can take a long time.) Once they had recovered their property, the snake was allowed to reingest its meal and go on its way.

A hundred and fifty miles away, at the Fazenda Rio Negro, a former cattle ranch owned and managed by Conservation International, where Leandro Silveira holds medical and dental clinics for members of the Jaguar Conservation Fund, I saw the famously dense forests of the Pantanal and the profusion of wildlife they host. Caimans basked by the score beside the Rio Negro, watched over gravely by herds of capybara, the largest rodents in the world, their distinctive square faces turning in the same direction at once as they listened for the suspicious furtive sounds of predators. A giant anteater strolled in its peculiar rocking gait beside one of the *baias*, the shrinking pools stranded by the retreat of the water during the dry season. In pastures in front of the *fazenda*, a few humped white cattle were grazing— they keep down the grasses, which cause fires during the dry season—as pairs of hyacinth macaws perched above them, perhaps the most beautifully colored birds imaginable, a deep violet blue, with rings of yellow around their eyes and beak. The Pantanal boasts the largest jaguars in the world, fattened on capybara, and one afternoon, upriver, I saw on the beach the tracks of a jaguar that had emerged from heavy overhanging foliage nearby to walk briefly along the sand. It was entirely possible to picture the future riparian corridors between Emas and the Pantanal: they will look just like this.

Every day, every night, Silveira and Jácomo sent their team out the door, exhorting them to "maximize the effort!" In these remote fragments of Brazil, I never heard any aid jargon about "capacity building" or "stakeholders." I saw mutual respect between cattlemen and biologists; I saw a committed corps of volunteers helping to compile fauna census data, radio-tracking data, data on the health parameters and behavior and territories of rare, endemic wildlife. I saw passion and conviction and the will to make rewilding happen. I saw what no one has yet seen in the Mesoamerican Biological Corridor. I saw results.

CHAPTER 4

RECONNECTING THE OLD WORLD

The European Green Belt

THE FORTUNES OF REWILDING VARY ENORMOUSLY ACROSS EUR-
asia. In Russia, the great steppes and boreal forests remain largely
intact, and conservationists focus on preservation, trying to safeguard
populations of brown bear, Siberian tiger, and migratory cranes. China
was deforested centuries ago and struggles to hold on to slivers of
wilderness where the World Wildlife Fund hopes to link patches of
Sichuan forest, the last habitat of giant panda and umbrella bamboo.
In the Middle East, wetlands in Israel and Iraq continue to serve as
ancient stopovers for migratory birds as they cross the globe from the
Arctic to Africa. A newly formed conservation group, Nature Iraq,
published the country's first Arabic-language *Field Guide to the Birds
of Iraq* in 2007, and the Mesopotamian marshes, among the largest
and most biodiverse wetlands of the region, are being restored. But
the most innovative corridor project in the Old World lies in the heart
of densely populated Western Europe, a region that has, in Aldo Leopold's words, "a resistant biota . . . tough, elastic, resistant to strain."[1]

Despite American pretensions to the title, modern conservation
is largely a European invention. It was a German, Ernst Haeckel, who
coined the word *ecology*. The International Union for Conservation of

European Green Belt: The "death zone" of Europe's Iron Curtain, bristling with barbed wire and land mines, divided the continent for nearly half a century, but conservationists now see this narrow strip as the backbone of a new system of nature reserves.

Nature was founded in Europe during the late 1940s by, among others, Julian Huxley, grandson of Darwin's friend and defender, the naturalist Thomas Henry Huxley. To raise money for that organization, in 1961, Huxley rallied other British nature enthusiasts, including Peter Scott, son of explorer Robert Falcon Scott, to launch the World Wildlife Fund, envisioned as "Nature's Red Cross" and the IUCN's chief fund-raising body, its appeals led by England's Prince Philip and Prince Bernhard of the Netherlands.

But in recent years, European conservation has seemed more smothered than inspired by bureaucracies. A signatory to the Convention on Biological Diversity, the 1992 UN treaty that provides goals and incentives to save biodiversity, the European Community has an oceanic appetite for proliferating secretariats, policy pronouncements, and programs, including Countdown 2010, aimed at pushing governments to reverse biodiversity loss by that year; the "Habitats Directive," requiring members to maintain and restore threatened habitats and species; and Natura 2000, a program to preserve core habitats across Europe.* While laudable in intent, the evidence suggests that these byzantine plans are failing: the 2008 Convention on Biological Diversity conference in Bonn, Germany, glumly acknowledged that the rate of species loss was accelerating, not falling.[2]

In contradiction to this melancholy record stands the longest, strangest, most creative rewilding initiative in the world, the European Green Belt. Fashioned out of the former Iron Curtain, the European Green Belt is a continent-wide experiment in ecological restoration, ecotourism, and environmental tolerance. A test case in transforming former military or industrial installations, so-called brownfields, into recovering ecosystems, the Green Belt is a model for converting war-torn wastelands into wilderness havens.

The European Green Belt will eventually run over five thousand miles, through twenty-two countries, from the Barents Sea in the north to the Black Sea in the south. The project is divided into three regions: the Fennoscandian Green Belt (Norway, Finland, Russia), the Central Green Belt (Germany, the Czech Republic, Austria, Slovakia, Hungary, Slovenia, and Italy), and the Balkan Green Belt (along the former border separating Croatia, Serbia and Montenegro, Macedonia, Romania, Bulgaria, Albania, Greece, and Turkey). It will link core protected areas—transborder parks like Germany's Bavarian Forest and the Czech Republic's Sumava National Park—and major riparian areas such as the Danube-March floodplains of Austria and Slovakia, a Ramsar site and the largest floodplain in Central

---

* President Clinton signed the Convention on Biological Diversity in 1993, but it was never ratified by the U.S. Congress.

Europe. Ambitions for the Green Belt are grand: its proponents hope that the network will spark ecotourism and sustainable farming practices in an expanding band on either side of the former border, promoting regional wildlife management and transborder cooperation.

The statistics of the Iron Curtain, out of which the Green Belt was born, are staggering. The German-German border alone ran for 870 miles, with forty-eight thousand border guards—one for every forty yards—patrolling in East Germany and over twenty-two thousand border police and customs officers in the West. Sandwiched between barbed wire and metal fences, the so-called Death Zone bristled with defenses, containing a narrow minefield; a patrol road for military vehicles (the *"Kolonnenweg"*), two parallel tracks created by buried concrete blocks; a strip where herbicides were applied to leave the ground bare so as to detect footprints; and a ditch lined with concrete or metal plates, to prevent vehicles from trying to blast through the barriers. In East Germany, thousands of people were forcibly moved away from a three-mile-wide area adjacent to the Iron Curtain. Altogether, there were 794 miles of twelve-to-fifteen-foot-high metal fences, 144 miles of minefields, and 517 miles of antivehicle trenches, as well as hundreds of watchtowers, electric alarms, and automatic spring guns. After reunification, 818 watchtowers and over a thousand land mines had to be removed.

Socially and economically, the Iron Curtain had a devastating effect on the areas it divided, blocking transportation, splitting communities, cutting off farmers from local markets. Environmentally, however, the border enforced a strange respite on the land, taking it out of intensive use. The Iron Curtain became its own ecosystem. Ironically, once the mines and fences were gone, the former Death Zone, generally between fifty and two hundred meters wide, was revealed as one of the most undisturbed natural areas on the continent.

For wildlife—particularly birds, reptiles, mammals, and plants that do not adapt readily to intensive agriculture—the Death Zone was anything but. Not even herbicides ruined it. Liana Geidezis, a biologist for the German Green Belt, has described it: "The curious thing . . . is that it is less polluted than the surrounding land. Herbicide was used to keep the strip clear of foliage . . . but this was less than the fertilisers and other chemicals which were used on either side of the border by farmers."[3] What was even better, from the conservationist point of view, was that the Green Belt cut through virtually every distinct ecological biome in Germany, creating an environmental cross-section of 109 key habitats, from streams, bogs, and riparian areas to coniferous and alluvial forests, heaths, and grasslands.

Germany, where the border between East and West was the most militarized in Europe and the most heavily fortified on earth, has taken the lead in launching the Green Belt. Even before the Iron Curtain fell, German naturalists, birders, and conservationists had their eye on this strip. Staff members of the BUND—the German federation of Friends of the Earth—conducted a census of bird species from the West German side as early as 1979. A decade later, in December 1989, with the Berlin Wall crumbling, the BUND organized a meeting of over four hundred conservationists from West and East Germany in the town of Hof, near where the German-German border met Czechoslovakia. Here the Green Belt project was launched. The meeting produced an exuberant welter of mixed metaphors: the former Death Zone could be a "lifeline, " a "string of pearls," an "important backbone with ribs," a chain of habitats linking nature reserves and crucial riparian areas across Germany, taking advantage of the fact that rural areas of East Germany had remained undeveloped and undisturbed, a haven for wildlife.[4] During the BUND's "Biodiversity Day" in 2003, five hundred experts fanned out along the German Green Belt. In twenty-four hours, they found over fifty-two hundred different species of plants and animals.[5]

The German vision of the Green Belt as a "green backbone" was soon adopted by other countries seeking to meet the Habitats Directive on saving biodiversity, and in 2004, at an international conference in Fertö-Hanság National Park—a Hungarian-Austrian transborder park on the Green Belt—the International Union for Conservation of Nature adopted the notion of extending it throughout Europe.

Participating groups and countries began buying or repurposing land, adding onto the Green Belt and strengthening the backbone in ways large and small: Croatia declared that it would protect a major riparian corridor along the Drava and Mura rivers; Bulgaria began planning a park on its mountainous border; and in Germany, the Protestant Church invited the BUND to restore habitat and protect threatened storks and orchids on eight hundred of its properties in Bavaria and Thuringia. The greening of the belt had begun.

On a hazy, hot summer afternoon near the town of Mitwitz, I met Stefan Beyer, a tall, stooped biologist working on the project. Mitwitz lies only a few miles to the west of the former border, and Beyer took me to the Green Belt, walking me along the former *Kolonnenweg*, blocks where soldiers once passed in jeeps. Pushing through thickets of reeds, grass, and skunk cabbage, Beyer led me to a footbridge over the Steinach River, once rent in two by barbed wire, and showed me how ditches lined with concrete plates—originally intended to keep East Germans from driving across the stream to freedom—had been converted into ponds dense with native plants and amphibians.

Standing in this former death zone, where we would once have been shot, all we could hear was a deafening chorus of birdsong, crickets, and frogs. This skein of streams, the Steinach and the Foritz, feeds into one of the most richly biodiverse areas in the country. In the distance, we could see forested hills, the low ranges of Thuringia and Upper Franconia: the Green Belt provides connectivity from rivers and wetlands into those mountains for predators like the European lynx and the *Wildkatze*, or wild cat, an animal that looks startlingly like a domestic tabby but for its blunt, thick tail. Moths and butterflies were staggering through the air in their erratic journeys, and shrubs and grasses grew high overhead. Many threatened or endangered species that are of particular concern to the Green Belt project find a haven here, including the *Eisvogel*, or European kingfisher, a brilliant azure blue, with a rust-colored breast and cheek patches and a rapierlike black beak. According to folktales, the kingfisher was one of the passengers on Noah's ark, originally drab in color, but the sun burned its cheeks and its back absorbed the blue of the sky; it is now a poster bird for the Green Belt. Also thriving in the area are the bluethroat (a robin with a blue bib), woodlark, red-backed shrike, winchat (a tiny, sparrowlike bird that sits on local fences, flicking its tail), marsh harrier, and the *Laubfrosch*, or common tree frog (now endangered in Thuringia and Bavaria), along with a plethora of grasshoppers, orchids, butterflies, wildflowers, and dragonflies.

Rewilding here, Beyer said, means buying the most valuable conservation properties near the Green Belt and expanding habitat by reconstructing ponds and "wet meadows" for these species. The project converts former wheat fields into wild meadows, taking them out of cultivation and preventing the growth of trees season by season in order to preserve habitat for frogs and other wetland creatures. It persuades farmers to err on the side of biodiverse exuberance and to adopt less intensive, wildlife-friendly methods. It encourages national farming associations to alter expectations about how much of the country must be given over to agriculture. Many farmers try to squeeze in two harvests, one early in the summer and one in the autumn, but sacrificing the extra harvest and allowing fields to mature for an extra month or two before mowing can mean reproductive salvation for many species of lepidoptera. Like Y2Y, the Green Belt also tries to discourage the building of new roads across the belt: Germany already has the densest network of roads in Europe, and keeping the belt intact is an important way to discourage further development near the corridor.

These have not been easy arguments to make, even in Germany, where the Green Belt remains popular. Standing on the Green Belt itself, which looks like an unruly overgrown hedgerow running through well-groomed fields, Beyer pointed out an area where a farmer had mown land bought by the Green Belt, following a common practice of cultivating strips of land that appear unused. Beyer went to great lengths to track down the farmer who had mown the area after the Green Belt bought it, in order to insist that the land was to be left alone. The man was not pleased, arguing that empty, unused land should be put to use and not left fallow. "He was angry," Beyer said dryly. "I was very angry too."[6]

Melanie Kreutz, a young landscape planner for the Green Belt project office in Nuremberg, agreed that roads and agriculture constitute the biggest threat to the project. Reestablishing economic connections between East and West remains "a really important theme in Germany," she said. "So it's hard to say no if people need another highway."[7] The eastern state of Thuringia, for example, has resisted the establishment of the Green Belt, wanting to expand roads and factories. As in the United States, the federal government's policies on land use have also become an obstacle. In 1996, a "border land law" gave former landowners along the border the opportunity to buy back their land. The Green Belt is now 20 percent privately owned and 65 percent federally owned; the rest is in the hands of municipalities, public authorities, or conservation groups, chiefly the BUND. For the past few years, the BUND has led a popular movement pressing the federal government to donate its border properties—which it had planned to sell on the open market—to the states or NGOs specifically for nature conservation, but some federal ministries have opposed the land transfer, frustrating further progress.[8]

While Beyer and his Green Belt partners seemed well along in rewilding small sections of the corridor, converting the grim apparatus of death into a place seething with life, there were other problems. Cautiously embracing the Green Belt, Europeans have nonetheless rejected the carnivores that go along with it. In France, sheepherders riot against wolves. Norway and Switzerland, with wolf populations in the single digits, have opened hunting seasons. And in Germany, a bear named Bruno inspired the kind of hysteria normally reserved for serial killers.

A Problem Bear

At the height of World Cup frenzy in June 2006, a wild brown bear wandered into Germany for the first time in 170 years. For a few weeks, the animal, dubbed "Bruno" by the press, lived a charmed life, adored by fans following his feats in the media, giving the slip to pursuers who hoped to relocate him away from Bavarian tourist towns. Described as "brazen," "a rogue," and "a problem bear," Bruno wandered north through the Italian Alps and into the mountains south of Munich, breaking into beehives, living large on sheep and goats. His most notorious exploit occurred in the lakeside resort of Kochel am See, where he snatched a little girl's pet rabbit and guinea pig, sat down in front of the local police station, and ate them.[9]

Historically, the three largest predators on the European continent were bears, wolves, and lynx. Wolves once roamed all of Eurasia and the British Isles. The brown bear was native to all continental temperate regions, and lynx were found across Europe. By 1800, the wolf was gone from England and low-lying coastal areas on the continent. By the nineteenth century, only small, isolated populations survived outside the Carpathians, Russia, and Scandinavia. The bear had a similar fate—extirpated in England by the Middle Ages and hunted down to isolated pockets in Spain, France, Italy, and the Balkans. Lynx survived only in major mountain ranges and in Scandinavia and Russia.

But as European populations moved to cities and livestock production was replaced by imports, forest cover increased by over 70 percent between 1960 and 2000. Wolves reappeared across the Alps. Lynx, solitary animals with a low dispersal rate, were less successful, but as biologists noted in a recent assessment, "By the late 1960s the tide had turned, and today most populations [in Europe] are increasing or stable."[10] The Swiss Lynx Project reintroduced the species in the Alps in the 1970s and 1980s, and the WWF's "Link the Lynx" project aims to reconnect fragmented populations. Bears, too, began to make a comeback, reintroduced in France, Austria, and Italy and aided by stronger legal protection.

Bruno—also known as JJ1, a two-year-old born in the Adamello-Brenta National Park in northern Italy as part of a five-year effort to reintroduce the species—was, in fact, doing exactly what the European Union wanted him to do, working his way toward the Green Belt core area of the Bavarian Forest. Committed to protecting endangered species in member countries, the EU had funded a 2003 project to restore bear populations in the Austrian Alps, Italy, and Slovenia. Had Bruno found a mate—perhaps in Austria, with a population of twenty to thirty bears—he would have played an important genetic role in his species' recovery. But on 26 June 2006, a hunter shot him, with the blessing of the Bavarian environmental ministry. It was an incident that exposed how frail the support for predators was and how deep the ignorance about them, even in Germany, where environmentalism is mainstream.

Such was Bruno's popularity that his killing raised an international outcry. The environment minister who gave the kill order received death threats. Italy, with only twenty bears of its own, lodged an official complaint, demanding the return of Bruno's body. "Killing animals that belong to a protected species is a barbaric act [that] destroyed years of work," said the president of WWF Italia.[11] Bavaria decided to keep the corpse, stuff it, and put it on display in Munich's Museum of Man and Nature, next to the remains of the last brown bear killed in Germany, in 1836.

In Passau, near Germany's Bavarian Forest National Park, I spoke to Helmut Steininger, who has long been involved in endangered species protection. In the 1970s, he traveled to Poland to track down some of the last *Wisent*, the European race of bison, along with wolves, lynx, and fish otter to repopulate the German park. The Bavarian Forest is now a transborder park, managed in concert with the Czech Republic's Sumava, just over the border. Together they form the largest contiguous forested area in Western Europe and a major protected core reserve crucial both to the Green Belt and to large-scale network design plans.

At first, Steininger said, it was hard for people accustomed to well-groomed parks to get used to the chaotic natural succession of the forest, after trees were downed by wind storms or destroyed by a recent invasion of bark beetles. As in the United States, when the parks department persuaded a reluctant public to accept a charred

Yellowstone after the 1988 fires, Steininger and the Bavarian forest management argued against cutting down dead trees, which would eventually fall and provide a basis for new growth. The bark beetles only exposed an existing weakness in the forest, regrown from monoculture spruce seedlings after the war; if nature was left to take its course, the new forest would be more diverse, more stable, and more resilient to insects or storms. Hiking around the park, where bark beetles had opened up meadows for wildflowers, punctuated with dead standing trees that provided good nesting sites for birds, I could see the mountains of the Czech Republic. "We want to see what nature will be doing," Steininger said patiently, and the sentiment was echoed by placards posted at trailheads, asking visitors to accept the seemingly dead parts of the forest as "A Catastrophe with a Future," a sign of renewal to come.[12]

I saw fish darting in dark streams in the surviving spruce woods, but otherwise the Bavarian forest seemed strangely empty. The only area where I saw wildlife was in the "Nature Park" section of the forest, where the animals Steininger had brought to the park in the 1970s had been installed. The Nature Park was large and beautifully designed, with paths integrated through naturalistic exhibits, allowing visitors to stroll next to *Wildschwein*, or wild boars, feasting on piles of bread and corn. But however cunning its design, it was unmistakably a zoo. While a number of birds and other small species have been successfully reintroduced in the wilderness areas of the park—ural owl, raven, hazel grouse, and three-toed woodpecker— the predators that Steininger had brought back from other, wilder countries had never been released into the wild: Germany, it was felt, was just not ready to deal with them. That fact alone could pose a far more stubborn obstacle to the Green Belt than any road.

The Rebirth of the Neusiedler See

In Austria, a lake along the border with Hungary once famous as the site of dramatic, often tragic attempts by Eastern Europeans to escape the Iron Curtain by wading through vast beds of reeds, has become a model for future restoration along the Green Belt. Intensively used over centuries, its surrounding land cultivated in tiny family plots over generations, its waters drained and funneled into canals, the lake and wetlands of the Neusiedler See-Seewinkel and Fertö Hanság National Parks are in the process of being restored to a mosaic of different habitats. So far, the process has yielded greater protection for the area's unusual biodiversity and economic opportunities for farmers that seem to meet the ever-elusive criteria of being truly sustainable.

The second-largest steppe lake in Central Europe, now a trans-boundary park jointly managed by Austria and Hungary, the shallow Neusiedler See lies only 120 meters above sea level, surrounded by reeds and saltwater pools formed a couple of thousand years ago. The lake has no outlet to the sea and is fed only by rain, and repeated evaporation of the water has concentrated different types of sodium in the soils. Sea aster and salt cress—typically coastal plants—cluster around the saltwater pools, and the pools' filling and drying plays a role in moderating the warmer climate of the area, which has a far longer growing season than surrounding areas, encouraging a uniquely varied flora and fauna. Some depressions are too saline for anything to grow; others, covered by layers of gravel and sand, become intensely fertile pockets. Two saltwater birds, the pied avocet and the Kentish plover, nest here, while dense reedbeds provide nesting, feeding, and resting areas for freshwater birds and small mammals, crustaceans, snails, newts, frogs, and toads. For decades, the meadowlands surrounding the lake had been whittled away, compressed into an ever-smaller area by encroaching agriculture and development. By the simple method of creating buffer zones—much of them off-limits to agriculture—the meadowlands and reedbeds around the lake have been restored to a biodiverse wonderland of orchids, dragonflies, butterflies, and songbirds.

Alois Lang, the Green Belt coordinator for Europe, had worked on the transformation of the Neusiedler See and gave me a tour of the new park at the lake, explaining its importance to conservation and describing how an intensively developed area was being rewilded. On the way, we stopped at Eckartsau Castle, a former hunting retreat of the Hapsburgs, now headquarters of the Danube Floodplain National Park. Hunting trophies of Franz Ferdinand and his wife, Sophie—foxes, roe deer, birds of prey, waterfowl, a ragged spoonbill hanging from the ceiling—adorned one macabre staircase, testifying to, among other things, the incredible potential of the ecosystem being restored. While Neusiedler See is not directly linked to the Danube, its restored wetlands make a natural stepping stone at the base of the eastern Alps, linking protected areas throughout Austria and north to the Czech Republic and Poland.

As we walked near the lake, watching flocks of geese stretch into the distance, Lang explained that after Neusiedler See was declared a park in 1994, hundreds of acres of buffer zones were established around it. Land was taken out of cultivation, and farmers were paid an agricultural subsidy (equal to that for growing certain crops) from the park's conservation budget. Eventually 60 percent of the fields within the buffer zone were left unused. Much of the remainder has been turned into small vineyards, and selling wine has become a common sideline in local villages.

Another program in the buffer zones that proved popular among farmers—particularly those who worked under military supervision in Hungary before the fall of the Soviet Union—was a return to traditional, nonindustrial farming practices, including pasturing Hungarian gray cattle, a rare and ancient breed beloved among local people that helps create the rich shortgrass meadows that wild plant and bird species rely on. Organic produce and the meat of the Hungarian cattle are sold at a premium in nearby restaurants and markets. Farmers are also allowed to rent grasslands to equestrian businesses, and around the park itself, a favorite summer destination for vacationers from Vienna, many locals have developed lodging, restaurants, and touring facilities that offer canoeing or birding.

A quiet, determined man, Lang is a landscape planner, not a biologist, with a sophisticated grasp of local history, customs, and the ways in which borders have frustrated and wounded people over centuries. His knowledge allowed him to weave Green Belt elements together: conservation planning, ecotourism, and what he calls "valorization," enhancing the value of the Green Belt's economic and cultural raw materials. He was partial to methods of persuading people to come to the Danube park or Neusiedler See to stay for a few days and absorb the local landscape and culture through birding, hiking on a new system of paths, horseback riding, wine tasting. This, he felt, was infinitely preferable to just passing through. The longer the stay, the more money spent and the more interest taken in conservation. He was broodingly critical of a plan promoted by a mountain biking enthusiast from Berlin, Michael Cramer, to turn the former military vehicle lane into the "Iron Curtain Trail." "If the aim is to encourage people to cycle through as fast as possible, where's the valorization?" he said.[13]

Lang feared that outsiders like bikers from Berlin might arouse suspicion or heighten resistance. Each country and community on the belt should be free, he felt, to develop its own unique economic opportunities, things that appealed to local people, like the Hungarian cattle. Imposing or importing activities foreign to the culture could alienate people who had already suffered considerable impositions in the past. "You cannot market an 8,000-kilometer trail," he said. "You *can* make the Green Belt a well-known initiative and enable local markets to take advantage of it." Just as Neusiedler See park management had encouraged farmers to develop a new market for organic produce, local people farther down the belt could develop and sell their own handicrafts and foods. "We are dealing with nature conservation in very poor areas," he said. "These things have to be done on the local level. If someone comes in from outside and says, 'I will bring you ecotourism,' that's always dangerous. You need to support local people in their own activities."

Lang had me climb one of the former Iron Curtain guard towers beside the Neusiedler See, now a birding tower strategically placed for viewing shorebirds, marsh harriers, and Europe's famed migrants from Africa, the great bustard, the hoopoe, the bee-eater. "This area has the highest diversity of birds in Central Europe," he said proudly. The drying wind of the Neusiedler See whipped past us, and we could see grass and reeds bending before it. The reeds, too, are a sustainable crop: select areas are rented to local reed-harvesting companies, and many surrounding homes and businesses boast traditional thatched roofs, often topped with platforms for nesting cranes.

From one of the towers, Lang pointed out a herd of Hungarian gray-necked cattle. At lunch, we sampled wafer-thin slices of their intensely flavored prosciutto-like meat on a *Sommersalat* of local greens and herbs. Pasturage of these rare breeds of cattle, water buffalo, Przewalski's horses (wild Asiatic horses once native to Mongolia that prefer grazing off saline soils), and white donkeys has been encouraged in the park's buffer zone to maintain the meadows, as has cultivation of organic methods to produce hay for the stock, all benefiting local farmers. Lang was insistent that conservation and sustainable economic development could never again be considered separate activities. If environmental and economic activities were developed together, everyone would see the benefit, everyone would support both. People who might not be conservationists themselves—farmers, winegrowers, lodge owners—could develop a vested interest in the success of the transborder park and its biodiversity, simply by virtue of the fact that tourists and birders were bringing in revenue. Lang said, "The farmer who sells his wine to a bird-watcher, what's important for him? He knows that nature conservation allows him to earn money. The landowner whose wife is renting out five rooms as a bed-and-breakfast, who is a member of a local hunting association, he knows what sustainable use is, without even using the term."

Beyer's ponds and Lang's Neusiedler See are baby steps, pilot projects that suggest what the large-scale corridor might one day achieve. As with Leandro Silveira's staged rewilding in Brazil, the trick to gaining traction on large-scale projects may lie in starting small. Projects work better on a human scale. While the big plans seem doomed to creep along for decades, these nodes along the Green Belt

activists something to build on, amassing small successes
biodiversity and economic support as they go. And if they
...ight even persuade people to tolerate predators.

Reclaiming Romania

Perched on a bench in a hut in the forest, I watched as a game warden
in a green woodsman's cap flung corn into "feeding stations," rusting
metal bins. He was looking over his shoulder. As soon as he was done,
he hustled back under the hut, built on stilts, and scampered up the
stairs, pulling the door shut behind him. There were bears out there.

Left over from an evil age, the hut was one of Nicolae Ceausescu's
notorious bear-hunting hides, now repurposed for ecotourism like
Alois Lang's Iron Curtain guard towers. It was dim in the falling
dusk. Tourists sat in anxious silence, peering through a narrow window, as mice rustled overhead. Then a bear materialized, a big brown
shape in the fading light, and then another, jumping on the feeding
station, flinging the cover aside, hoovering up corn. It was a halluci-
natory Eastern European fairy tale scene: the cigarette-smoking
warden in a feather-trimmed cap, the capering mice, the bears clank-
ing around under the hide like trolls under a bridge. In Romania, this
is nothing special.

The forests of Eastern Europe—Poland, the Czech Republic,
Romania—constitute the largest core areas in proximity to Western
Europe and the best hope of repopulating extirpated wildlife in that
part of the continent. Romania, with its Carpathian Mountains largely
intact, holds the largest wolf and bear populations west of Russia,
despite suffering from some of the most grievous pollution on the
continent. But important as these cores are, many are not protected.
Conservationists are intent on putting as many of them as possible
under some form of protection and on forging connectivity between
these forests and protected areas in the West. The fear of Helmut
Steininger and other conservationists who work in the borderlands
between East and West is that large parts of these countries' forests

could be subject to development or logging unless their status is better defined.

Both the pollution and the haven are down to Ceauşescu. While disfiguring the landscape with mines and noxious factories, the dictator loved bear hunting, and the animals were reserved for him alone. The fruit of that autocracy is 5,000 bears, 3,000 wolves, 2,000 lynx, and a sheepherding culture largely inured to living with predators. Romanian shepherds, who spend months of the year on remote hilltops eating homemade cheese and guarding 4.5 million sheep and 1.5 million cattle, make American cowboys look like crybabies.

Although Ceausescu's policies inadvertently saved predators, it was Romania's ancient farming and sheepherding methods—free of pesticides, fertilizers, and mechanized equipment—that truly saved the country's biodiversity. Preserving those traditions, which offer a fascinating draw for tourists in themselves, may simultaneously boost the rural economy and sustain the wilderness. In the Carpathians, I met Christoph and Barbara Promberger, biologists from Germany and Austria who worked for a decade compiling data on Europe's large carnivores, radio-tracking wolves, and developing proven methods—trained guard dogs and electric fencing—of reducing wolf predation on sheep, methods they promoted through rural meetings and local television.[14] Since then, the Prombergers have pursued a double approach, urging the small Romanian town where they live, Sinca Noua, to retain its sustainable wildlife-friendly agriculture while strengthening its economy through ecotourism. They also hope to win some form of designated wilderness protection for much of the valley in which Sinca Noua and the bear hide lie. The townspeople have welcomed the biologists' support for their traditional methods, which are largely in line with EU standards for organic agriculture, and the town has become Romania's first declared "ecovillage," with plans for an organic butchery, bakery, and cheese factory.

The pastoral way of life in such villages as Sinca Noua is a revelation to anyone accustomed to modern monoculture and industrial-scale farming. In fields during autumn, men and women are scything hay by hand, loading it high onto carts or stacking it in an ancient conical form of haystack still seen throughout Eastern Europe and Asia. Most households keep a few cows and pigs in small barnyards attached to their homes. Cattle and water buffalo are grazed on communal fields in the hills above town, and in the evenings a shepherd goes to fetch the livestock. When the animals reach the narrow cobblestone streets, they trot to their individual barns of their own accord. If the gates are not open, they kick them impatiently until the family lets them in. Although cars and buses pass through town on the main road, the preferred form of local transportation is still the donkey or horse cart.

The relationship between farmers and the wilderness that surrounds them is a model of coexistence. Christoph Promberger pointed out the separation between the communal haying fields and the grazing areas, which are universally respected. As I hiked around the countryside, it became immediately apparent how the old small-scale methods of agriculture in Romania have preserved biodiversity. Monoculture fields, bathed in chemicals, have a cold, empty, unpopulated quality. In the Romanian countryside, within minutes, I saw dozens, perhaps hundreds of species, flourishing in and around cultivated land: bees, butterflies, moths, dragonflies, birds, and beetles. Frogs popped up from the mud puddles. Raptors patrolled overhead. Less than a mile from town, returning from a hike, I saw that a brown bear's tracks were imprinted over my earlier footprints.

The Prombergers hope that the town's certification for organic agriculture will help bring greater revenue for products produced there and convince new generations to continue these methods of farming. In recent years, many young people have left for jobs in the cities and overseas; I met one woman whose children had decamped for Canada. Many of those still farming in the old way were themselves old, and the hard labor and simple lifestyle seemed unlikely to appeal to many. I camped one night in the orchard of an elderly man whose entire life had been spent on a beautiful, remote hillside. His tiny white farmhouse had a view that stretched for miles but none of the amenities of modern life, no indoor plumbing, no electricity. He made do with an outhouse, an outdoor tap, and old-fashioned lanterns. To pass the time, he played Romanian folk tunes on a recorder. The Prombergers wish to preserve enough enclaves of traditional farming in the valley to support a sturdy rural economy based on cultural tourism, ecotourism, and sustainable agriculture. But in a world increasingly driven by technology, where urban areas hold most jobs and opportunities, it seems unlikely that farmers like my host will be replaced.

And there are other threats, other intrusions, to this peaceable way of life. On the drive to the bear hide, Barbara Promberger said that even protected forests are constantly encroached by illegal building, as people from the cities slap up weekend homes without permits, a symptom of the corruption that still plagues the country. She pointed out one rogue building, saying, "This is going on all over Romania. So many beautiful valleys have been ruined like this."[15] Nor is there much government support for the country's forests. I asked the game warden about the hardships of the job. He pointed to a bike, his only transportation. During the alpine winter, he has no vehicle or cold weather gear to patrol an area of hundreds of square kilometers. His answer to a query about poaching was an angry nod.

The Prombergers have rallied experts on insects, birds, snakes, and amphibians to perform a comprehensive fauna census of the area. Seeing private investment and land ownership as one way to secure the future of the Carpathians, they are launching an ambitious campaign to raise money to buy land in the core of Piatra Craiului National Park, much of it privatized after the fall of Ceaușescu, some already logged. Their goal is to put together the largest wilderness area in the EU, to keep Romania's predators safe for perpetuity.

I recognized the Leandro Silveira touch: go-it-alone, bootstrapping conservation heavily reliant on biologists living where they work, developing their own programs, relying on their own relationships with the place and its people. There was no bureaucracy, no fancy office, and precious little money, except what they raised themselves. Like Silveira, they were not waiting for government or for NGOs for approval. They were just doing it.

The Accidental Corridor

All rewilding projects seek to bring together the same elements: cores, corridors, and carnivores. Many of the ambitious early initiatives— Yellowstone to Yukon and the European Green Belt—were biological in design but inspirational in nature, offering a vision of re-creating connectivity across the landscape while promising people a new emotional connection to the land. Activists worked doggedly, steadily, practicing the arts of persuasion, from educational entreaties to compensation schemes to soft bribery. They effectively established, beyond dispute, the importance of corridors. Yet some of the most ambitious projects have had only partial success and have been achingly slow to catch on. If there is one defining characteristic of conservation, it is urgency.

There is one perfect corridor—not a conservation project at all but a strip of no-man's-land—that suggests a metaphor for the difficulties and opportunities of such voluntary, grassroots initiatives. In a region where there are no large-scale core protected areas left, it is a fully restored corridor for wilderness, where tigers and leopards hunt in freedom, a mysterious space known only by the beasts and birds who flourish there in solitude.

In July 1953, the Korean Demilitarized Zone was scorched earth. Now it is Eden. Eden with a million land mines. The 38th parallel, a border 155 miles long and 2½ miles wide, guarded by two million North and South Korean soldiers, is believed to be the best-preserved piece of land on earth. It is also the most dangerous. No human being has set foot in it in fifty-five years.

Quietly, without human intervention, the DMZ rewilded itself. Surrounding it, the Korean Peninsula is among the most densely populated places on earth. South Korea's population is approaching

50 million; North Korea can barely feed 23 million. The peninsula has been intensively developed virtually in its entirety, its forests felled, land converted to agriculture or industry. While six-party talks between North and South Korea and their intermediaries have stalled, the DMZ still shines as a coveted prize. Conservationists want it preserved. Developers want it for factories. Farmers from North Korea have already planted a strip of it along a highway between the countries. The DMZ holds all the peninsula's remaining biodiversity; it is the last stronghold of Korea's wildlife, with "near-primary forest vegetation."[16] Partial surveys conducted with satellite imagery and spotting scopes reveal 2,700 species of plants and animals, including 1,100 plants, 50 mammals, and 80 species of fish. Its wetlands, plains, and forests are home to 67 endangered species, including Asiatic black bears, lynx, the endangered Amur goral mountain goat, and one of the rarest cats in the world, the Amur leopard. Siberian tigers have been spotted. Reports of occasional explosions testify to the fact that occasionally one of the heavier mammals triggers a land mine. But overall, lack of human interference has encouraged an ecological rebirth. Hundreds of species of birds migrate in and out of the DMZ, including half of the world's last 2,000 red-crowned cranes, white-naped cranes, and black-faced spoonbills. There are only 1,000 black-faced spoonbills left on earth; 90 percent of them breed in the DMZ.

An unintentional triumph, the DMZ is more fully functional, less developed or disrupted or disturbed than Y2Y or the Mesoamerican or the Green Belt. It only took a million land mines to do it.

Land mines are a scourge, but those in the DMZ are saying something about conservation, something mutely eloquent about what it takes to regain an Eden. Explosive devices are surely not the best way to keep tigers, leopards, and cranes safe from humanity. But without land mines, or the conflict that placed them there, the DMZ would certainly have been converted to factories and rice paddies long ago. Land and wildlife recover when we leave them alone, but the question remains: Can we find the will to restrain ourselves without the threat of annihilation? Can we do it in time?

Belts:
1
2
3

Org:
1.) Bund

Images

Familiarity / Place
Europe & Maine Rivers

↳ What was going on
before you?
- We Need people with History
↳ To work on these org farms.

Green Belt as Bridge

PART II

AN AFRICA WITHOUT FENCES

CHAPTER 5

PEACE PARKS AND PAPER PARKS

Corridors with Leverage

"The land that war protected": that was how E. O. Wilson and a leading Korean conservationist described the accidental corridor created by the DMZ.[1] In a *New York Times* op-ed article, they suggested that the DMZ be transformed into an international transboundary peace park, a place where visitors could reflect on the long conflict and celebrate the reunification of the Korean people. A peace park, they believed, could anchor a new cooperative binational system of nature reserves, protected landscapes, maritime zones, and "nature villages," where farmers from both countries could grow organic produce. They envisioned it marking an end to the environmental degradation of the peninsula and harmonizing the poisonous relationship between the countries.

Within the last decade, outsized hopes and expectations like these have shown up in plans for peace parks in virtually every major war zone and disputed border area in the world. In the realm of political science, peace parks are rewilding amplified, a potential panacea for conflict, rooted in conservation.

The world's first peace park was a safe bet. In 1931, Rotary Clubs in Montana and Alberta, Canada, seeking to promote "peace and goodwill" among nations, voted to endorse the merger of Waterton Lakes National Park in Canada, created in 1895, and the adjoining Glacier National Park in the United States, established in 1910.[2] The U.S. Congress and the Canadian Parliament approved the measure in 1932, creating the Waterton-Glacier International Peace Park. The designation was largely symbolic: there had never been a fence between the parks; the border crossed a rugged, mountainous, and heavily glaciated area, unlikely to invite trespassing; the parks would continue to be managed and financed by their respective countries' parks departments. The move chiefly celebrated the friendship between two nations that share the longest undefended border in the world.

The relationship has proven fruitful. The parks have cooperated on management issues, including search and rescue missions, wildlife management, and transboundary firefighting. The peace park has allowed collaborative studies on everything from climate change and glacier retreat to the effects of woodpeckers on pine beetle infestations. Since the 1970s, the parks have coordinated research on invasive species, sharing seed and greenhouse facilities for the propagation and restoration of native plants. While there have apparently been tensions caused by increased border security in recent years, the Waterton-Glacier Peace Park has been a beneficial, noncontroversial, and unqualified success.[3]

Despite that example, for years, no one moved to emulate Waterton-Glacier, although the term *peace park* was resurrected following World War II in urban memorials in Japan commemorating victims of the atomic bombings. But it was not until decades later that the peace park idea was revived in its original sense, as an instrument to attain both conservation and political cooperation.

In reports issued during the 1980s, the United Nations World Commission on Environment and Development acknowledged the "accelerating deterioration" of the global environment and natural resources, calling on nations to protect at least 10 percent of their land, later revised to 12 percent.* Although both figures were criticized by conservation biologists as too low, the recommendations powerfully influenced international land planning, particularly after the World Bank created the Global Environment Facility in 1991, which offered nations financial incentives to put larger areas under protection.

The United Nations, at its chartered University for Peace in Costa Rica, posited peace parks as a method by which countries could raise their percentage of protected land while garnering other benefits, including improved relations with neighboring states and coordinated natural resource management. Peace parks were also promoted in academic forums on "peace-building" and international relations at the Woodrow Wilson International Center for Scholars in Washington, D.C. Soon enough, the academic concepts began filtering out into the real world.

In 1994, Jordan and Israel were negotiating a peace treaty. The countries share twenty-five miles of shoreline at the northern end of the Gulf of Aqaba, a narrow finger of the Red Sea, and as part of the treaty, they agreed to develop a binational Red Sea Marine Peace Park that would bring together Jordan's Aqaba and Israel's Coral Reef Reserve near Eilat. While the reserves, located on opposite shorelines, were not contiguous, cooperative management promised to address conservation problems, including oil spills, nutrient runoff from land, ship traffic, and damage to reefs caused by fishing nets and tourists. Both countries were committed to improving the condition of their parks, since both attracted significant tourist revenue from divers and snorklers.

The United States was part of the trilateral team negotiating the treaty and brought in the National Oceanic and Atmospheric Administration to work up a detailed study of human impacts in the gulf,

---

* These reports were the so-called Bali Action Plan (1982) and the Brundtland Commission (1987), named for its chairman, Gro Harlem Brundtland, a Norwegian diplomat. The Brundtland Commission convened in 1983 and issued a report, "Our Common Future," published by Oxford University Press in 1987.

which allowed Jordan and Israel to develop a joint plan to regulate environmentally sustainable tourism and to develop appropriate off-shore mooring. Park staff began work on restoring damaged coral reefs in both areas. The collaboration is credited with improving the health and ecological resilience of both parks. The peace park showed what might easily be done with shared marine environments, but it also exposed the limitations: improved conservation management would not necessarily reduce hostilities in the region.

But the next peace park succeeded on both fronts. Ecuador and Peru had disputed the border between them since 1941, with tensions escalating into a three-week armed conflict in 1995. Conservation International (CI), an American group, had been working intensively in the disputed area, known as the Cordillera del Cóndor.⁴ Little explored, the region was crossed by the Cóndor mountain range and massive rivers; like much of the rest of the tropics, it was a trove of biodiversity. Its unique forests, composed of dwarf palms, bromeliads, and orchids, sheltered over four thousand plants, many endemic, as well as Andean bear, mountain tapir, jaguar, manatee, and macaw. CI fieldworkers were in the practice of negotiating safe passage with the Ecuadorian army, and they engaged officers in discussions about conservation and its potential in resolving the border dispute. They also learned that the local indigenous group known as the Shuar wanted the military conflict resolved in order to secure the conservation of their territory. Discussions with the military and the Shuar, along with assurances by several countries guaranteeing withdrawal of forces, led to a revival of the peace park scheme.

By the end of 1995, a peace agreement was signed and troops withdrew from the area. Three years later, after establishing a permanent border, the neighboring countries signed a treaty designating the area for conservation. Since then, two small independent national parks have been created on either side of the border, and plans are in the works for a transborder Cóndor-Kutuku Conservation Corridor, which aims to link national parks and reserves that include areas for both the Shuar and biodiversity protection.

The treaty signed between Ecuador and Peru catapulted the notion of peace parks as an instrument of conservation and peace-keeping to the forefront of the conservation agenda. Both political scientists and conservationists began studying the kinds of treaties that might be fashioned to create transborder parks, inspired in part by the example of global collaboration set by the 1959 Antarctic Treaty. Owned by no sovereign nation, with no native human population, Antarctica was so important to science that even nations still laying claim to pieces of the polar pie signed a document to ensure international management and strict conservation. With the Antarctic Treaty as a model, peace parks and transboundary areas were established through bilateral or multilateral negotiations leading to a "memorandum of understanding," a form of gentleman's agreement outlining the physical boundaries of the area and creating committees tasked with management and implementation.

While Ecuador and Peru were working out their differences, the International Union for Conservation of Nature began codifying a systematic approach to peace parks and transboundary conservation. In 2001, it published a key document, *Transboundary Protected Areas for Peace and Co-operation*, establishing definitions for "parks for peace" and "transboundary protected areas." Parks for peace were transboundary parks dedicated to protecting and maintaining biological diversity and to "the promotion of peace and cooperation." Transboundary protected areas straddled one or more borders, were dedicated to protecting and maintaining biodiversity, and were managed cooperatively.* They were not parks. By definition, parks fell into the most restrictive IUCN categories, with little human use or habitation, whereas transboundary protected areas allowed more flexibility, as in the protected area shared by Ecuador and Peru, part of which remained

---

* The definitions were later expanded to include networks or clusters of conservation areas, some of which might lie far from an international border. An example is the proposed Central Asian Transboundary Biodiversity Project involving a collaborative management plan for four discontiguous protected areas in a transborder region shared by the Kyrgyz Republic, Kazakhstan, and Uzbekistan.

the home of the Shuar people. Unlike peace parks, they had no political prerequisites.

The biological, or conservation, goals for both peace parks and transboundary protected areas were the same as for earlier corridor projects: putting large areas under some form of protection; safeguarding cores, establishing buffer zones, and restoring corridors; maintaining biodiversity and keystone species. But while the cores, corridors, and carnivores model was based on biological design criteria, peace parks and transboundary protected areas were political animals: they sought to address the worst effects of arbitrary borders, drawn on maps, that crossed, blocked, and interrupted ecosystem processes; they set up a legal framework that would leverage and institutionalize cooperation between nations on issues involving natural resources in shared ecosystems. The kind of international bureaucratic quagmire into which the Mesoamerican Biological Corridor had sunk was exactly what the peace park model sought to avoid.

Many corridor projects could have been redefined as peace parks or transboundary areas: the European Green Belt, for example, crossed numerous borders and included some transborder parks. But the belt itself had not been established by means of international treaties; it was, instead, an initiative undertaken by local environmental groups. While corridor projects overwhelmingly emphasized conservation, peace parks were apt to focus on harmonizing relations between countries, development, and poverty alleviation, with conservation as one among many objectives. And though the Green Belt, too, sought to encourage transboundary management of neighboring parks, its pilot projects were most often headed by biologists working on a conservation agenda, not by politicians or aid specialists. As peace parks and transboundary projects took off, this distinction became ever more critical: The politically based model favored the constellation of goals embodied by "peace" more than those suggested by "park."

Like the corridor model, peace park and transboundary projects were largely untested, except for the safe, stolid example of Waterton-Glacier, which had, after all, come into being without political or social challenges—or peace issues, for that matter. Some biologists were wary on the grounds that nations might base their creation on false criteria, choosing to place them in border areas with less than optimal biodiversity. Protection of the most endangered and critically important areas should be the priority, they felt, not well-meaning but misguided attempts to rope together conservation and peace. Anthropologists and sociologists, on the other hand, were quick to point out that peace parks were often "top-down" projects, imposing conservation on people by government fiat, a practice bound to fail since it neglected all-important local support.

But most conservationists and biologists saw peace parks and transboundary areas as corridors with leverage, a natural evolution of the corridor model backed by the political authority that could move rewilding from grassroots initiatives to the forefront of governmental land-use agendas.

As the United Nations expanded its vision of the importance of sustainable development, conservation organizations and agencies experienced explosive growth, and once modest wildlife protection outfits found themselves elevated to the position of advising or assisting governments. Budgets, staff, and institutional power increased with funding from private foundations and corporations, as well as bilateral and multilateral support. Launched by a small group of ecologists in the 1940s, the Nature Conservancy had previously confined itself to the United States but now expanded globally, becoming the largest conservation organization in the world, with assets of nearly $5 billion in 2008. Managing a hundred million acres of land worldwide, the group works in over thirty countries. The World Wildlife Fund in the United States, a branch of the international World Wide Fund for Nature, has a million and a half American members and boasts its own international program. Conservation International, founded in 1987 by breakaway employees of the Nature Conservancy, brought in substantial funding from the Gordon and Betty Moore Foundation, including the largest single contribution ever made to conservation, $261.2 million.[5] Together, the five largest conservation groups spend over $1 billion a year on conservation and their budgets are bolstered by hundreds of corporate sponsors.

While such spending hardly seems excessive in light of the billions expended on wars and other priorities, the growing budgets, corporate donations, and grand ambitions quickly attracted scrutiny. Former employees criticized high executive compensation for "raconteurs of mass extinction."[6] Steven Sanderson, president and CEO of the Wildlife Conservation Society, took home over $825,000 in 2005. A trifling sum compared with Wall Street compensation, but in the perennially cash-strapped world of conservation, high salaries appeared extravagant. Questions were raised about conservation's corporate ties to mining, logging, cement, and oil companies; many sponsors were notorious polluters, including DuPont, General Electric, ExxonMobil, and Dow Chemical. In 2003, the *Washington Post* ran a series of articles exposing corporate favoritism at the Nature Conservancy that led to an IRS audit and a congressional investigation.[7] All major groups came in for criticism by indigenous-rights advocates, who argued that conservation was creating "refugees" by the million.[8]

But while the money, corporate sponsors, and questions were multiplying, the peace park and transboundary movement had achieved an extraordinary momentum. In 1988, only fifty-nine parks or protected areas shared a national border. By 2001, the number had grown to 169, involving over 650 individual protected areas; by 2005, over a hundred countries were involved in planning 188 transboundary complexes, incorporating 818 protected areas.[9] More were under discussion. The United Nations had asked, and it had received. But it remained to be seen whether this model would accomplish conservation goals.

As this major shift of resources and long-range planning occurred, peace parks were largely subsumed by the broader category of transboundary protected areas. Since peace parks were dependent on the resolution of wars or conflicts, many proposals fell into a state of semipermanent uncertainty, tantalizingly out of reach until governments could resolve their differences. Others, peace parks in name only, united areas where conflicts had already been put to rest.

The DMZ Peace Park, for example, waits for a permanent peace between North and South Korea. A proposal for a Siachen Peace Park in the volatile Kashmir border region disputed by India and Pakistan, the highest-altitude battlefield on earth, is similarly indefinite.[10] The plan seeks to establish some measure of protection for the extraordinary Siachen Glacier, the longest in the world and a part of the critically endangered and fragile Himalayan ecosystem. Climatologists and scientists have warned that glaciers in the region are retreating due to climate change and that villages downriver are in immediate danger from flooding meltwater, the breakdown of ice dams, and catastrophic landslides, a situation that an enlarged protected area and the relocation of villagers to safer ground could prevent. But the Siachen Glacier and everyone living beneath it remain hostage to the military standoff between hostile powers.

Odds are also against peace parks in the Middle East and Central Asia. Conservation organizations have examined the possibility of a peace park in the Wakhan Corridor, the thin strip of Afghanistan reaching to China through Tajikistan and Pakistan, a high-altitude valley traversed by wolf, lynx, Marco Polo sheep, and snow leopard. To date, realities on the ground—which include not only the war in Afghanistan but also China's sensitivity about Tibet and its northwest region—have stymied the proposal.[11]

In the catchment for the Tigris-Euphrates river system, thought to be the original Garden of Eden, a Hawizeh-Azim Peace Park has been proposed for the marshy region between Iran and Iraq.[12] Marsh Arabs have lived sustainably in the area for centuries, sharing habitat with endemic and endangered species. While the marshes in Iraq are currently being restored, the greater marshland area between Iraq and Iran has been virtually uninhabited for hundreds of years. The ecological degradation of the area has also been the work of centuries, as the marshlands were drained and damaged by dams and canals upstream. Saddam Hussein's attack on the Shiite Marsh Arabs after the first Gulf

War poisoned much of the Iraqi section, draining the rest. Now, however, surviving Marsh Arabs are returning to their ancestral wetlands, restored by the reinundation of water, with reeds and fish populations returning. The vision has been gaining support from the Iraqi government, which is considering applying to list the marshes as a Ramsar and World Heritage Site. But the plan is dependent on water continuing to flow. Currently, hydroelectric dam building projects are under way in both Iran and Iraq; Turkey and Syria also plan dams. Without water—and cooperative management—the marshes will dry up again.

There were even thoughts of a potential peace park on the vexed border between the United States and Mexico, originally proposed in the 1930s by the Rotarians after their success with Waterton-Glacier. The idea gained traction during the Clinton administration, when Interior Secretary Bruce Babbitt and his Mexican counterpart signed a "letter of intent" that might have led to a Sonoran Desert Peace Park.[13] The process was undone, however, by hysteria over immigration and security during the Bush administration's "war on terror" and its construction of the massive border wall and accompanying infrastructure. At the same time, Conservation International was working to create a transboundary protected area between Big Bend National Park in Texas and private property across the border in Mexico bought for conservation purposes by Cemex, an international cement company. The transboundary plans for El Carmen-Big Bend are progressing, but the peace park remains on indefinite hold.

As the Big Bend example indicates, utopian plans have often stalled while transboundary protected areas—not dependent on cessation of hostilities—have forged ahead. But distinctions between peace parks and transboundary protected areas would evaporate in a place where peace had long been in short supply. The one continent where both models have been tried widely—where there were high expectations, intensive financial investment, and elaborate conceptual fantasies about what peace parks could achieve—was the continent where conservation has perenially been crippled by civil war, disintegrating governments, and poverty: Africa.

From Penitent Butchers to Paper Parks

Westerners find it difficult to grasp the enormity of Africa. The largest desert in the world, the Sahara is roughly the size of China. The United States, Japan, and Europe fit handily into an outline of the continent, with room to spare. Maps using the Mercator projection distort the size of Africa to fit the planet's curves on a flat surface, making it appear the size of Greenland. Africa is fourteen times that size, with more than 20 percent of the earth's land mass, second only to Asia.

Geographically, climatic patterns govern the life of every living thing: the ocher sands of the Sahara annually press farther south, with drought and desertification acting as major forces motivating wars in the Sudan, Chad, and the Central African Republic. A great green belt of tropical rain forest occupies the midcontinent, leaving sweeping stretches of savannah in the east and the south. Animals in Africa evolved in these spectacularly large tropical grasslands, characterized by periods of drought and temperature extremes. Elephant cover vast territories, and wildebeest follow mass migrations over long distances: survival hinges on the ability to find water.

Equally difficult to grasp is the overwhelming profusion of Africa's wildlife. The continent holds the greatest aggregations of large terrestrial mammals on the planet. There are dozens of species of hoofed mammal alone. They range from the tiny dik-dik, duiker, and suni—antelope scarcely bigger than rabbits—to the large roan antelope, kudu, nyala, bushbuck, waterbuck, and eland. There are antelopes that live in swamps (sitatunga) and forests (okapi, bongo) and deserts (oryx). There are species of stunning beauty, the gemsbok or southern oryx, with a mask of black on its face and horns tapering to an elegant V. Or the sable antelope, deep black with a white underbelly and enormous curving horns. There are leapers and jumpers—gazelle, impala, springbok—and rock climbers like the klipspringer. There are giraffe, zebra, buffalo, rhinoceros (white and black), hippopotamus, pygmy hippopotamus, aardvark, wildebeest, wild boar, red river hog, giant hog, common warthog. There is a deer that looks like a zebra, the zebra duiker. There are five species of rock hyrax, a crazy-looking shrewish creature with teeth like a prairie dog and toes like an elephant. And tipping the gargantuan end of the ungulate scale is the majestic African elephant. Those are just the ungulates.[14] Multiply them by the evolutionary extravagance of primates, cats, wild dogs, hares, bats, snakes, mongooses, crocodiles, tortoises, frogs, birds, fish, and insects. Africa is a cornucopia.

There is no need to rehearse again the appalling slaughter that the colonial powers visited upon the continent. But it is worth noting that the great white hunters of the Victorian age treated Africa's wildlife with the same brutality as its people. There was once a time when the springbok—now the national animal of South Africa—migrated through the Cape in herds so vast they took eight days to pass and measured miles across. But by the late nineteenth century, wildlife in South Africa had been nearly extirpated through indiscriminate big game hunting, slaughter of elephants for ivory, and systematic conversion of land to agriculture.

The havoc wrought by unregulated hunting was the inspiration behind conservation efforts in Africa. Postwar, both UNESCO and the World Wildlife Fund were created in response to fears that African wildlife was on the verge of being wiped out. But the initial preservationist campaigns were based on a sense of colonial entitlement. In 1887, Theodore Roosevelt, among the worst offenders, founded the Boone and Crockett Club, dedicated to "ethical" hunting. In 1903, the Fauna Preservation Society was started in Britain by so-called penitent butchers, former big game hunters. They weren't that penitent: Having collected their share of trophies, they wanted to ensure that every white man so inclined would be able to continue the hunt; their chief goal was to deny Africans the same right.

Colonial governments also founded Africa's first national parks, beginning in the Belgian Congo in the 1920s. During the 1940s, Kenya followed up with a number of other parks and reserves, including Amboseli, Tsavo, and Nairobi national parks. All were modeled on the American national park system, and governments did not hesitate to evict Africans from the land. In South Africa, people were driven from what became Kruger; in Kenya, the Maasai were excluded from areas where they had traditionally grazed their cattle. Africans saw little difference between land seized for conservation and that taken by whites for their own use: the finest agricultural lands in the east and south of the continent, generally free of malaria and tsetse flies, ended up in the hands of Europeans. David Western, who grew up in Tanganyika (now Tanzania) during the 1950s and served briefly as head of the Kenya Wildlife Service, recalled hearing black Africans angrily condemning the parks: "First they took our animals, then our land."[15]

With the end of the colonial era, as nascent governments struggled and failed, many parks were abandoned. Dictators like Mobutu Sese Seko in Zaire (formerly the Belgian Congo) exploited institutions established for conservation, siphoning off money meant for parks. With vanishing budgets, few paid rangers, and weak protection, organized poaching rapidly grew into an international trade in wildlife products—elephant ivory, rhino horn, and other coveted bones, shells, and skins—that now rivals the drug trade as the most lucrative form of illegal trafficking in the world.

Parks in name only, or "paper parks," protected areas across the continent were ravaged by poaching, illegal logging, road building, and encroachment. Governments began indiscriminately "hiving off" land, regazetting reserve boundaries in exchange for payoffs from logging companies, allowing settlers inside to graze, burn, and plant crops. As the continent's population soared, people flocked to parks to strip them of the very resources—water, wildlife, timber, edible plants—the institutions were meant to protect. Paper parks led to the phenomenon known as "empty forest syndrome," a term coined by a conservation biologist, Kent Redford.[16] With hunters removing up to a million metric tons of bushmeat out of Congo Basin forests every year, biologists trying to census bird, mammal, or other species often found little but wind in the trees.

* * *

Because conservation history in Africa was so mired in colonialism, conservationists struggled repeatedly to move beyond the past, trying to develop programs that would break free of a century of injustice, illegality, and bad faith. In the 1980s, organizations pinned their hopes on "utilization"—using revenue from controlled hunting to benefit communities. At first, the notion seemed sound. Distribute revenue from trophy hunts—permits to shoot elephant or lion can generate tens of thousands of dollars—to local villages to compensate for agricultural damage caused by wildlife. In places where elephants can wipe out a family's livelihood or the sustenance for a village in a single overnight raid, these efforts were welcomed. Local people began to express greater tolerance for wildlife that came with a significant dollar value attached. Among the most famous such efforts was CAMPFIRE, or Communal Areas Management Programme for Indigenous Resources, in Zimbabwe, which conferred ownership of elephants and wildlife to villagers, who could then make money from the sale of hunting permits. In his 1993 book, *At the Hand of Man: Peril and Hope for Africa's Wildlife*, Raymond Bonner lauded CAMPFIRE as "grass-roots democracy."[17]

While the program got off to a promising start, yielding revenues that were distributed to communities, it was vulnerable to exploitation, in this case by the country's president, Robert Mugabe. Mugabe's cronies acquired hunting concessions for their own profit, and by the mid-1990s, CAMPFIRE was plagued by reports of kickbacks, misappropriation of funds, and poaching.[18]

A program in Zambia similar to CAMPFIRE also fell victim to nepotism, although this time to a more localized form. ADMADE (Administrative Management Design for Game Management Areas) empowered village chiefs to select men to be trained as wildlife scouts to patrol areas outside national parks; chiefs were also given the power to distribute revenue generated by trophy hunting. Like CAMPFIRE, the program aimed to prevent local people from shooting large game like elephant or buffalo, so that high-priced hunting permits could be sold, limiting the number of animals killed. Local hunters, however, simply switched their focus to smaller game, using hidden snares to evade detection. Chiefs hired relatives for scouting jobs and distributed revenues in ways that enriched themselves and their families. According to independent studies, neither conservation nor communities benefited.[19]

Biologists expressed alarm at the pairing of conservation and development, arguing that the programs were based on a sentimental stereotype of indigenous people living in harmony with nature. In "The Ecologically Noble Savage," an article about the behavior of Amazonian peoples before the arrival of Europeans, Kent Redford, the same biologist who had identified empty forests as a looming peril, demolished one of the sacred tenets of poverty alleviation, that native peoples act as responsible stewards of their land, taking just enough and never too much. Redford wrote that Amazonian peoples "behaved as humans do now: they did whatever they had to to feed themselves and their families."[20]

Indeed, the desire to yoke conservation with development had in it a strange amalgam of moral superiority and guilt. In *The Myth of Wild Africa*, conservationists Jonathan Adams and Thomas McShane argued that Africans were the only legitimate guardians of conservation on their continent. In response to the history of patronizing foreigners coming to the continent to save wildlife and stem poaching, they wrote that "Africans do care about wildlife. They live with it every day. They have been labeled as the problem; they are in fact the solution."[21] That the authors themselves were white Western conservationists and that the social programs they were promoting constituted yet another solution imposed from abroad seems not to have occurred to them. Despite their evident capacity to take over conservation jobs that were once the province of whites, Africans were as much a part of the problem of conservation in their countries as North Americans and Europeans in theirs. There were no benign human societies that took only as much as they needed.

In 1992, in response to increasing pressure to link conservation and development, the term *integrated conservation and development projects* (ICDPs) was coined at the World Bank, launching a new era in conservation.[22] Planners eager to avail themselves of grants promised by the bank pledged to set aside the punitive model of "guns and fences" that had early on defined African subsistence hunters as poachers, in favor of programs that would reduce poverty and raise income while breaking the destructive reliance on wilderness resources. Henceforth, conservation and human aid programs would seek shared goals, advancing human development and addressing conflict between people and wildlife. They would abandon the outdated park model—based on setting wildlife aside in small, strictly protected reserves—and promote conservation while simultaneously engaging local people, offering them the economic benefits of ecotourism, hunting revenues, and other forms of sustainable development.

However unrealistic or poorly designed the early programs were, the problems they were meant to address were only too real. Poverty perennially undercut conservation, sending people into the bush to cut wood, hunt, or till arable land. Despite strong parks systems in a few countries, conservation in Africa continued to be underfunded, shorthanded, and overwhelmed, crippled by weak governments, poverty, and civil war. It was in this discouraging climate, after this history of bitter debates and conservation debacles, that peace parks and transboundary projects came to Africa. No continent needed them more, but no continent presented as many obstacles, as many wars, as many refugees, as many disputed borders. By 1995, Africa's population was over seven hundred million and growing rapidly; 58 percent of its people were living in poverty, on around seventy cents a day. Transboundary projects offered a new solution, a new way to combine conservation and poverty alleviation. They quickly became the latest trend in large-scale conservation.

Southern Africa: Southern Africa has become the test case for transfrontier peace parks designed to counter over a century of fragmentation and fencing by restoring migratory corridors for elephants and other wildlife.

National Parks

Transfrontier Areas

ANGOLA

Kavango-Zambezi
Transfrontier
Conservation Area

ZAMBIA

ZIMBABWE

Great Limpopo
Transfrontier
Conservation Area

MOZAMBIQUE

NAMIBIA

BOTSWANA

Great Limpopo
Transfrontier Park

Kgalagadi
Transfrontier
Park

Johannesburg

Maputo

SWAZILAND

Lubombo

LESOTHO

SOUTH
AFRICA

Durban

INDIAN
OCEAN

Cape Town

0   150   300 Miles

0   250   500 Kilometers

"An Africa without Fences"

Throughout the 1990s, transborder projects were launched across sub-Saharan Africa. In Central and West Africa, plans were jumping off the drawing boards, including the Tri-National de la Sangha, an ambitious plan to join three national parks in Cameroon, Congo Brazzaville, and the Central African Republic to create a landscape-scale rain forest region protecting forest elephants, lowland gorillas, and the ancestral territory of the Ba'Aka pygmies. Perhaps the most pressing of these efforts was the Virunga Volcanoes Transboundary Conservation Area, one of the last two places where the mountain gorilla survives. Mountain gorillas cannot breed in captivity, so the three national parks involved, which join at the international borders of their respective countries—the Mgahinga Gorilla National Park in Uganda, Volcanoes National Park in Rwanda, and Virunga National Park in the Democratic Republic of Congo—were the species' last hope. The only other population lay in Uganda's Bwindi Impenetrable Forest, some twenty-five miles north. Because the parks were contiguous and relatively small, and gorillas ranged freely across the mountains, what affected one park affected all. One of the chief goals of developing a single transboundary management plan for the three adjoining parks, together with Uganda's Bwindi, was to ensure that every country and organization involved was sharing data, communicating about evolving threats, coordinating policy on tourism, and making effective decisions.

Virunga was originally established in the Belgian Congo as an addition to Albert National Park, the first national park in Africa. The nineteenth-century explorer John Hanning Speke, while searching for the source of the Nile, had heard tales of humanoid monsters "who could not converse with men."[23] After Europeans located the creatures in 1902, Virunga was specifically created to protect them. The park attained prominence in the West after long-term behavioral studies revealed that the animals were not terrifying carnivorous giants but peaceable and engaging vegetarians. Dian Fossey furthered the study for eighteen years, and her 1983 memoir, *Gorillas in the Mist*, and the film made after her murder in 1985 conferred international fame on the animals, ironically inspiring the descent of thousands of tourists on the area, a traffic she had hoped to curtail: tourism presents threats to the apes, exposing them to potentially fatal respiratory and skin diseases.

In 1994, after the genocide in Rwanda, three-quarters of a million people fled that tiny country, the most densely populated on the continent, seeking safety in the Democratic Republic of Congo. Thousands of refugees traveled through the Virunga mountains, and many settled in and around the Congolese park for the next two and a half years, destroying the remaining forested buffer zone around it and creating the largest refugee camps in history. The refugees chopped down trees for fuel, built shelters, and hunted wildlife. Over seventy-five square kilometres of the park were completely stripped of vegetation. Gorillas were poached.

Hoping to forestall such severe threats to the parks in the future, the three neighboring governments agreed to establish the Virunga-Bwindi Transfrontier Park, to manage and protect all remaining reserves with mountain gorillas. While the transfrontier status of the park is still being discussed, collaborative management has taken hold. For example, the International Gorilla Conservation Programme—a coalition of conservation groups meant to facilitate communication between park authorities—convinced the Rwandan military to stop construction of a road through Volcanoes and Virunga parks that would have led to further poaching. They successfully urged the DRC to stop deforesting an area of the park for security reasons. A shared database has been established, and a ten-year transboundary strategic plan has been written. A unified system of regulations governing tourism has been put in place and joint patrols and ranger-training programs have been organized to ensure that all parks benefit from effective law enforcement.[24]

The results have been surprising. Even in the midst of continuing civil war—with Congolese rebels seizing parts of Virunga and roving

bands of former Rwandan militiamen clashing with the DRC army—
the mountain gorilla population has increased. In 1989, there were
620 individuals in the parks; by 2009, there were an estimated 720.
In Virunga, the most disturbed of the parks, an eight-week census
carried out in January 2009 revealed that the area's population had
increased from seventy-two to eighty-one.[25] In 2007, after Congolese
insurgents seized part of the park, looting park headquarters and
forcing rangers and their families to flee, seven adult gorillas were
killed, apparently by charcoal traders in league with corrupt park officials trying to conduct illegal logging. But within a year guards were
able to resume monitoring gorilla groups and crack down on charcoal
manufacturing. While the gains were gradual and the situation fragile, success in Virunga made a powerful argument for transboundary
management.[26]

But it was southern Africa that became synonymous with the approach. In the 1990s, South Africa, Mozambique, Botswana, Zimbabwe, Zambia, Namibia, and the tiny landlocked kingdoms of
Swaziland and Lesotho became the site of multiple transboundary
plans, including some of the most ambitious in the world. Political
fortunes were changing rapidly in the region: by 1994, South Africa
had held its first multiracial elections, abandoning apartheid and rejoining the world economy after thirty years of economic sanctions.
Mozambique, for its part, was emerging from twenty years of civil
war. The region was finally poised to take full economic advantage of
its exemplary natural resources and beauty, and governments were
urged to cash in on the so-called peace dividend by shifting military
personnel and equipment to safeguard parks.[27]

Aside from these immediate factors, there was also a long-
suppressed yearning to restore a kind of unity to a region that had
been radically carved up by competing powers and damaged by mining and grazing. As early as the 1920s, out of an urge to reunite wildlife, General Jan Smuts, a prime minister of South Africa famous for
his exploits during the Boer Wars, proposed a "great fauna and tourist
road through Africa," joining Kruger National Park and Rhodesia's

hunting grounds.[28] Like-minded officials hailed the creation of Go-narezhou, a wildlife reserve in Southern Rhodesia, as an extension of Kruger, and urged the Portuguese, who controlled the colony that became Mozambique, to establish a neighboring reserve.[29] In 1938, perhaps inspired by Waterton-Glacier, a Portuguese ecologist, Gomes de Sousa, proposed that South Africa's Kruger National Park and the land north of the Limpopo River in Mozambique should ideally be managed as one park, one ecosystem. His was an early vision of re-wilding, and although nothing came of the idea at the time, it lingered.

Despite these tantalizing images of a unified south, the fragmentation continued. In the ensuing decades, after recurring outbreaks of zoonotic disease, fencing became something of an obsession in the region. By the 1960s, veterinary control fences had been constructed throughout the south, stretching between and across entire countries, designed to keep wildlife from passing diseases to domestic animals. Restricting seasonal migratory movements and access to water, the fences caused an immediate, sharp decline in the numbers of antelope and buffalo and confined growing numbers of elephant to small areas.[30]

As South Africans faced this broken landscape, another man appeared on the scene, dreaming the same dream as de Sousa. One of the richest men in the country, he was looking for a legacy—and perhaps absolution for tacit acceptance of the tainted regime that ruled while he made his millions—and he embraced that ecologically radical idea.[31]

Born in 1916 of English and Afrikaner descent, Anton Rupert came of age during the Great Depression. Unable to afford medical school, his first choice, Rupert trained as a chemist and became a businessman, investing in dry cleaning and branching out into liquor sales and cigarettes. Studying the marketing techniques of American tobacco companies, Rupert eventually acquired a collection of luxury brands known as the Rembrandt Group—Cartier, Dunhill, Roth-mans, Montblanc—that made his family the second wealthiest in South Africa after the gold- and diamond-mining Oppenheimers.

In private interviews, Rupert argued passionately against the "homeland" system of segregation, which isolated blacks in econom-ically marginal rural areas.[32] But he never used his influence to op-pose it actively. In later life, he expended his energies on conservation, promoting a grand agenda that would reach beyond the borders of South Africa: restoring migratory corridors throughout the region, the idea that had been kicking around since de Sousa. Rupert had heard of the earlier proposals to connect Zimbabwe's Gonarezhou National Park to South Africa's Kruger and to Mozambique, extend-ing them east to the Indian Ocean. He had the financial and political wherewithal to make it happen.

Rupert first tried to convince the Southern African Nature Foun-dation to implement the plan, but the organization refused, fearing competition with its own projects, a series of new reserves that were being established around the country and stocked with wildlife from neighboring Botswana and Namibia. For years, he visited with heads of state, meeting with the president of Mozambique, Joachim Chis-sano, who begged off because of the ongoing war, and eventually proposing the transborder idea to South African president Nelson Mandela, who embraced it. After the conclusion of Mozambique's war in 1992, Rupert again broached the proposal with Chissano, and he agreed. Finally, in 1997, Rupert established the Peace Parks Foun-dation (PPF) with a personal donation of $260,000.[33]

The idea behind the Peace Parks Foundation was stunningly am-bitious, even romantic, holding out hopes for a reunification of Africa through conservation. An early vision put forth on the PPF Web site invited supporters to:

> Dream of an Africa without fences. Dream of ancient migration trails trodden deep by an instinct that time has never contained. Dream of a wilderness where the ele-phant roams and the roar of the lion shatters the night. Dream, like us, of experiencing Africa wild and free, where people can reap the benefits of nature and in turn support her. This is the dream of Peace Parks Foundation.

A dream that will only be realised through the establishment of peace parks.[34]

But the foundation's first director, elephant specialist John Hanks, outlined a less dreamy agenda, one of purchasing land, negotiating loans and legal issues with governments, and promoting the development of transfrontier conservation areas "on a commercial basis."[35] Rupert himself saw peace parks driving the economy, saying: "Sustainable conservation can attract millions of tourists to peace parks in Africa and by involving rural populations in these initiatives, the continent's greatest problem, unemployment, can be beaten."[36]

For years, conservationists in Africa had been working to strengthen the link between economic development and wildlife protection, believing that people would be more inclined to protect wildlife if there was something in it for them. It was a commonsense philosophy, devised with the best of intentions. But in 1997, at an international conference on the wildlife trade, Robert Mugabe put a more sinister spin on the formula, which has since become almost universally accepted within Africa. Countries like his, he said, were preserving wildlife "at great expense and sacrifice," pumping water for animals "at great cost."[37] He delivered a bizarre ultimatum: "Elephants, especially because of their huge bodies, consume large amounts of this underground water and, we believe, *every species must pay its way to survival.*"[38] A dubious point, made by a despot, but it expressed the rigidly utilitarian attitude many African leaders held about conservation, seeing it as a luxury no developing country could afford unless it could be manipulated to yield a return. The statement was embraced by government leaders, including Thabo Mbeki of South Africa, who were longtime friends of Mugabe. It is now commonly asserted by African governments, land-use planners, and conservationists as received wisdom: wildlife must pay its own way or "pay to stay."[39]

The formula influenced the development of the peace parks. Rupert outlined four "pillars" necessary for their success.[40] They were the criteria of the businessman, the hardheaded tobacco magnate:

secure space for parks through signed international agreements and protocols establishing large-scale protected areas; train wildlife managers to ensure efficient ecotourism management and produce income for all countries involved; train guesthouse managers to provide appealing tourism services; and improve access by relaxing visa requirements and customs regulations at border crossings.

These were all things Rupert knew how to do. He arranged for the PPF to provide substantial support to the Southern African Wildlife College, which trained students from all over the continent to be conservation managers, park rangers, and field guides. He expanded a wildlife veterinary training program to work on the difficult issue of disease transmission across borders. He bought a hotel near a nature reserve and turned it into the South African College for Tourism, a facility devoted to training young rural women in hotel and lodge management, cooking, food and beverage operations, and hospitality skills.

While not as absolutist as Mugabe, Rupert was insistent that peace parks could and should pay for themselves through ecotourism. In many ways, peace parks represented another grand run at poverty alleviation: they were ICDPs writ large, ambitious attempts to marry conservation and development.

Curiously, although Rupert imagined the peace parks restoring migratory routes for elephant, he never spoke about conservation in the parks in detail. Establishing an Africa without fences would be enough to resolve conservation issues: the wildlife would take care of the rest.

Within a year of its founding, the Peace Parks Foundation proposed six transfrontier projects in which South Africa was a partner. Notwithstanding the name of the organization, not all were peace parks. The largest of the six was the Great Limpopo Transfrontier Park (13,513 square miles), which sought to deal with elephant overpopulation by expanding the crown jewel of South Africa's park system, Kruger, joining it to parks in Mozambique and Zimbabwe; plans for a surrounding Transfrontier Conservation Area would expand its size to 36,679 square miles, the size of Portugal. The Kgalagadi Transfrontier Park joined two existing parks in a desert region shared with Botswana, producing a park of 14,384 square miles. The Lubombo Transfrontier Conservation and Resource Area, which would stretch from South Africa's eastern coast inland, crossing into Mozambique and Swaziland, was also a hugely ambitious proposal, intended to open up fragmented elephant habitat and transform a three-nation region into a major tourism destination.[41]

The three remaining projects were more modest. The Limpopo-Shashe Transfrontier Conservation Area (1,881 square miles) pulled together state land, national parks, and private land owned by De Beers Consolidated Mines in a complex centered around an important archaeological site at the confluence of the Limpopo and Shashe rivers where South Africa met Botswana and Zimbabwe. The |Ai-|Ais / Richtersveld Transfrontier Park (2,333 square miles) joined national parks in Namibia and South Africa in an arid desert region often compared to the Grand Canyon. And the Maloti-Drakensberg Transfrontier Conservation and Development Project (3,127 square miles) protected a spectacular alpine region shared by South Africa and the tiny kingdom of Lesotho, which lies entirely within the borders of the larger country.

Early on, a seventh project was added to the list, the only one of the original plans that did not involve South Africa. It was potentially the largest of all, three times the size of the Great Limpopo Transfrontier Conservation Area, at around 116,000 square miles, involving five countries: Botswana, Zimbabwe, Angola, Zambia, and Namibia. It would place nearly the entire Okavango Delta, a huge seasonally flooded wetland in the Kalahari Desert, a major source of water for several countries, under some form of protection. While the Kgalagadi, Limpopo-Shashe, Richtersveld, and Maloti-Drakensberg projects occupied deserts, canyons, or mountainous regions with relatively sparse populations, the three largest proposals—the Great Limpopo, the Lubombo, and the Okavango Delta—were planned for areas where there was a significant human presence.

Much of the future of rewilding in the developing world hung on the Peace Parks Foundation's success, particularly concerning its three largest projects. If the Great Limpopo truly joined conservation and development in a way that both solved the elephant problem and improved the economies of the countries involved, if the Lubombo project freed elephants from border fences and created jobs, if the Okavango Delta scheme became a model for conservation of one of the continent's major wetlands, then the transboundary idea seemed sure to become a solid basis for conservation across Africa. How the organization handled the conflicts that would arise from people living in planned conservation areas would be critical. But perhaps the most important goal, in a region where people and wildlife were still suffering from poverty, stagnation, and constraint, was to make measurable progress.

CHAPTER 6

THE GREAT LIMPOPO

The Elephant Problem

SOUTH AFRICA HAD THE CAPACITY TO LEAD THE TRANS-
boundary movement, but it was also obliged to do so: It desperately
needed jobs. The country was struggling to lift millions out of poverty, failing to provide even basic services. While racist policies were
banished, economic apartheid remained all too evident. With a population of forty-eight million, wealth remained in the hands of the white
minority, less than 10 percent of the population. The country had one of
the highest rates of income inequality in the world. Life expectancy for
men was forty-five, for women fifty. Average annual unemployment
was as high as 40 percent in some parts of the country, fueling the crime
rate. Former president Thabo Mbeki's botched response to the AIDS
crisis—he banned antiretrovirals for years, promoting quack theories—
cost 300,000 lives. Seven and a half million people contracted HIV,
one person out of every six. Nearly a thousand were dying every day,
with six million expected to follow in the next decade. Throughout the
region, the illness took a significant toll in conservation circles, killing
trained staff in government agencies, conservation organizations, and
parks, many of which lost rangers and other workers to the virus,

allowing poaching to escalate. Forests were falling to the coffin trade, with tons of firewood consumed during funeral preparations.

The human toll was compounded by an environmental one. Rural areas near parks were hard hit by the crisis: people living in extreme poverty—their situations made perilous by lack of access to health care—surrounded most parks and protected areas in southern Africa. In the extremity of their need, they depended on the meager government welfare system, supplemented by whatever resources—water, fish, bushmeat, wood, wild plants—they could scavenge for themselves. As they stripped the land of native plants and wildlife, they were pushing themselves to the brink of environmental destruction that threatened to become permanent.

As the Peace Parks Foundation (PPF) forged ahead with its projects, the complexity of the tasks it faced became clear. Its first project was deceptively simple: The Kgalagadi Transfrontier Park was established with the sweep of a pen in 2000. Its transformation into a single transboundary entity joining two national parks, one in South Africa, the other in Botswana, could not have been easier, in part because the two countries had maintained a pragmatic relationship even during apartheid. The two long-established parks, preserving sections of the vast Kalahari Desert, had never been fenced and were separated by only a sandy riverbed, dry most of the year. There had been an informal agreement between the two conservation authorities since 1948, and a joint management plan had been under development since the early 1990s. For the first time, enterprising tourists not put off by the scarcity of water or services could drive, if they wished, from South Africa into Botswana through the park.

The PPF's next project was as convoluted and complex as the Kgalagadi was straightforward. The Great Limpopo, a vast "superpark," as the South African media dubbed it, sought to expand Kruger, one of the largest national parks in the world, by joining it with Zimbabwe's Gonarezhou and a former hunting concession, known as Coutada 16, in Mozambique. But the new park, the "Great Limpopo Transfrontier Park," launched in 2001, faced long-standing political tensions. Two of the countries—South Africa and Mozambique—had engaged in hostile machinations in each other's politics for decades. Apartheid South Africa had armed and supported rebel groups opposed to Mozambique's Marxist regime beginning in 1975; in return, that government trained and outfitted members of the African National Congress, the party of Mandela, in its fight against apartheid. Zimbabwe was sliding into economic collapse and social catastrophe. Even after overt hostilities ended, the countries eyed one another with suspicion, and the poorer members of the triumvirate—Mozambique and Zimbabwe—resented their rich neighbor.

From the beginning, virtually everything about the Great Limpopo Transfrontier Park was in dispute. It involved unequal partners, with South Africa holding all the cards: comparative wealth, power, infrastructure, political stability, and an established reputation for high-quality tourism. By comparison, Mozambique had been decimated by civil war and natural disasters, including flooding in 2000 that killed eight hundred people and twenty thousand cattle and left thousands homeless. Zimbabwe, once the breadbasket of Africa, with excellent wildlife parks and facilities, was on the brink of ruination.

The parks were also unequal. Kruger, one of the most successful and profitable national parks in the world, was the antithesis of a paper park, offering visitors close encounters with the world's most spectacular wildlife, from the traditional hunter's "big five"—elephant, buffalo, leopard, lion, and rhino—to impala, wildebeest, zebra, giraffe, hippo, hyena, and crocodile. Coutada 16, on the other hand, was an alternate universe, an empty one. During Mozambique's civil war, from 1975 to 1992, the hunting concession had been a killing ground not only for human refugees trying to escape but for virtually every other living creature. Not long before it was transformed into a park, the reporter Peter Godwin described it from the air in *National Geographic*: "You can fly over it for hours . . . and see not a single animal, not a solitary game trail."[1] An elephant specialist told of witnessing the carnage during the war, visible across the twelve-foot-high, 220-mile-long electrified border fence erected by the apartheid government during the 1970s to protect the park and cut off a wave of refugees: "Vehicles drove up and down the fence, shooting anything that moved, irrespective of size or sex or species. AK-47s were common, so any impala, kudu, or duiker was fair game. Even smaller animals such as genets and porcupines got shot."[2] During apartheid, the fence itself was lethal, killing over five hundred people. After apartheid, it protected wildlife from subsistence hunting and organized poaching.

Soon after the Great Limpopo was launched, Mozambique and Zimbabwe began grumbling bitterly about whether Kruger, with its 700,000 tourists a year, would share its enormous revenue with the poorer parks. With no mechanism for revenue sharing, South Africa's neighbors would see little of the benefit that Anton Rupert had promised. There were also logistical issues. With Kruger in easy reach of tourists, there was no good reason for visitors to drive to Mozambique, where roads were primitive, game viewing was sparse, and facilities were virtually nonexistent. Likewise, Zimbabwe's Gonarezhou received only a few thousand visitors a year and had little prospect for more as that country's situation worsened. Zimbabwe also feared dropping its fences and allowing buffalo from Kruger, carriers of bovine tuberculosis, to cross the border, potentially infecting its cattle herds.

But peace parks were supposed to heal such divides, while improving conservation. The Great Limpopo, for example, was designed to solve South Africa's elephant problem. Even though Kruger was enormous—around the size of the Netherlands, 236 miles long from north to south and 33 miles wide—it seemed to have an overpopulation problem.[3] South Africa had long confined elephants to parks, where they were beginning to burst their bounds, and Kruger was no exception. But in addressing the elephants' needs, moving them into areas where people were living, the Great Limpopo created another problem, a people problem. The superpark's fortunes would depend on how well it solved those vexing issues.

The African elephant is the largest land animal, the only terrestrial mammal with a life span comparable to that of humans, up to seventy years. Males measure thirteen feet high and up to twenty-four feet long, weighing nearly eleven tons. An adult drinks thirty to fifty gallons of water and eats anywhere from a hundred to a thousand pounds of vegetation daily, including grass, shrubs, leaves, and the pulpy wood and bark of trees.

Elephants are the fertilizers of Africa. Conservation biologists consider them keystone herbivores, essential to the maintenance of forested areas and open savannah grasslands. Their huge appetites and inefficient digestion—they digest only around 40 percent of what they eat—account for a tremendous impact on the environment. In the course of traveling their large home ranges, of up to 500 square miles, and migrating long distances in search of food and water, they are responsible for maintaining and restoring myriad ecosystems. The elephants that frequent the rain forests act as "forest engineers," creating openings in the understory called "light gaps," encouraging decomposition by trampling rotten logs, aerating soil as they dig for minerals, providing the mechanism for the dispersal and germination of many plants: up to a third of tree seeds in West Africa must pass through an elephant's gut to germinate. Their capacious deposits of dung spread the seeds of hundreds of species; their feeding habits, rooting out tender saplings and consuming young trees and shrubs, keep the savannah open. In the absence of elephants, woody plants would have shaded out grasslands and forests would have claimed the Serengeti long ago.

The social dynamics of elephant herds are only beginning to be understood. It has long been known that elephants form tight bonds in breeding herds, led by an older matriarch who monitors and teaches younger females, subadults, and infants. Adult males may wander on their own or travel with other bachelors, coming together with females when they enter the breeding phase known as musth. Recently, evidence has emerged confirming that elephants possess a previously unimagined capacity to communicate over long distances through low-frequency infrasonic rumbling, sounds pitched below human hearing but apparently detectable through vibration, which elephants feel through their feet. Adults are believed to pass on knowledge of migratory routes and the location of water sources to the young. Our growing understanding of elephant intelligence and complexity has only deepened feelings, shared by biologists and animal rights advocates, that the species deserves to be treated with sensitivity and respect. Even scientists who support culling, the selective reduction of a population through shooting individuals or entire herds, often express reluctance about the process.

Westerners find elephants wonderfully charismatic and photogenic, but to the people who actually have to coexist with them, they can be frightening and destructive. It can be difficult for people in the urbanized, developed world to imagine living alongside some of the largest, heaviest, most dangerous mammals in the world: elephant, hippo, rhino, and buffalo. Elephants cause most of the conflicts on land, and in day-to-day life in rural Africa, there are constant points of friction. Elephants find crops tasty and easily accessible. They destroy water pumps and knock down fragile housing made of straw, cardboard, or mud. They can move about silently, startling people in the bush and triggering dangerous encounters. They compete with people for plants and fruits. While there are no continent-wide statistics on the number of people killed by elephants, the frequency of attacks seems to be growing and their severity worsening.

Biologists and behavioral experts have begun to believe that elephant aggression, expressed in rampages that have destroyed villages and schools, has become epidemic in recent years, in response to systematic poaching and culling. In a 2005 Nature article, Gay Bradshaw, a psychologist in environmental science, and other experts, including Cynthia Moss and Joyce Poole, theorize that elephant populations in Africa and Asia may be experiencing "psychobiological trauma" caused by habitat loss and disruption of the healthy dynamics of stable herds.[4] As humans poach and kill more elephants, the beleaguered giants are lashing out. Bradshaw has called the deterioration of the relationship between people and elephants "extraordinary. . . . Where for centuries humans and elephants lived in relatively peaceful coexistence, there is now hostility and violence."[5] Recently, rangers in Pilanesberg National Park in South

Africa reported that young male elephants had attacked, raped, and killed sixty-three rhinos, hitherto unheard-of and aberrant behavior.[6] The three elephants responsible, subsequently shot by park officials, were found to have been adolescents who had witnessed the shooting of their families in other parks, where it was a common practice to tie the young to the bodies of the dead before they could be collected and moved. Bradshaw and her colleagues argue that neurological development of young elephants subjected to trauma and maternal deprivation can be damaged in ways similar to that seen in human victims. They call for "new conservation strategies that promote normal social patterns."[7] One of those strategies is to restore large-scale elephant habitat.

Southern Africa currently has a population of 250,000 elephants; by 2020, that population is projected to reach 400,000. The Great Limpopo's attempt to deal with this jumbo problem made it the first "major test case" for South Africa's peace parks.[8]

Historically, South Africa attempted to solve its elephant problem by the most draconian means possible, wiping out the elephant in the wild. This did not sit well even with those who had participated in the slaughter. In 1898, Paul Kruger, an enthusiastic hunter, a Boer resistance leader against the British, and a president of the Republic of South Africa, convinced the legislature to proclaim an area of lowveld between the Sabie and Crocodile rivers as a game reserve. After the Boers lost and the British annexed the area, James Stevenson-Hamilton, a Scotsman, was appointed as the reserve's first warden. He oversaw the expulsion of several thousand people from the area and the reserve's expansion into a national park in 1926.*

---

* As the Sabie Game Reserve was enlarged, other villages were allowed to stay on, but inhabitants were forced to pay rent to the government. Jane Carruthers, author of *The Kruger National Park: A Social and Political History*, reports that Africans residing in game reserves became "a source of considerable revenue" for the South African government. The Makuleke people were forcibly removed from the northern part of Kruger in 1969; in 1998, they successfully applied for compensation and restitution of their land. Their agreement with park authorities stipulates that the land be used solely for wildlife conservation but grants them the right to tourism and lodging concessions.

Soon after the reserve was established, a herd of elephants migrated from Mozambique into Kruger, which was not yet fenced, perhaps seeking the plentiful water in the rivers that formed the original core of the park.

Ever since that small herd wandered from Mozambique into Kruger early in the last century, the reestablished elephant population has boomed, in part due to the park's own policies. In the early days, the park was run by hunter-naturalists in an era predating a sophisticated biological understanding of ecosystems. Kruger's first warden, Stevenson-Hamilton, was famous for asserting that nature could take care of itself. "Science," he once opined, "with its classical approaches and verbose jargon . . . can be very dangerous."[9]

More dangerous, however, was management uninformed by science: Kruger built its reputation for unparalleled game viewing on an unsound foundation. Drilling hundreds of "boreholes" (artificial water holes) and damming rivers, Kruger's planners established a pattern of attracting wildlife to the accessible southern reaches of the park, where extensive tourist lodging and services were built in the 1950s. In doing so, they set themselves up for ecological disaster. Those dams and boreholes helped fuel an elephant baby boom, and the fenced-in pachyderms began decimating vegetation around water holes. Tourists, too, were concentrated in these areas, and park managers vigorously resisted the idea of changing their successful formula.

But confining elephants to parks did not solve the elephant problem; it only concentrated it. Inside their fenced enclaves, elephants were stretching Kruger's resources and also getting into trouble with villagers: sometimes elephants walked out of the park through unfenced areas and private game reserves to the west. Kruger's answer was culling—shooting entire herds from the air, including mothers and infants. Beginning in 1967, Kruger killed more than fourteen thousand elephants over the next twenty-seven years, keeping the population at around seven thousand. It developed an on-site meat processing facility and sold canned elephant as a souvenir. But culling proved counterproductive: scientists compiled data suggesting that culling raised rather than lowered the birth rate among surviving herds, potentially nullifying the desired outcome. The slaughter grew so repulsive to visitors and animal rights proponents that it was finally halted in 1994.

The peace parks seemed to offer a solution, promising to restore the free movement of wildlife throughout southern Africa. Scientifically, enlarging the available range for elephants made more sense than trying to manage isolated and confined populations artificially, as Kruger had been doing. Biologists cited metapopulation dynamics, which derived from the equilibrium theory of island biogeography: with sufficient connectivity in an ecosystem, local populations may expand in some places or decline in others, with numbers remaining relatively stable through immigration or recolonization. Give elephants more space and restore their migratory corridors, scientists believed, and their numbers would stabilize naturally. Together, the three major transfrontier conservation areas would open up enough room so that elephants across the subcontinent could begin to function again as a single population.

In 2000, the Peace Parks Foundation and other donors supported a major project undertaken by the Conservation Ecology Research Unit at the University of Pretoria, part of its "Megapark Research Programme" collecting data on the home ranges of southern Africa's remaining populations of elephant, which roughly conformed to the proposed peace parks, including the Great Limpopo, Lubombo, and the Okavango Delta, where veterinary and border fences continued to limit movement. Fitting a sample of elephants in each area with satellite collars, seventy-three animals in all, the study produced maps of the daily positions and home ranges of elephants during the dry and wet seasons (in the dry season, elephants travel longer distances in search of food and water).[10] The study has since been used in designing the parameters of the core areas of the peace parks and linkages between them; it was also seen as a tool to predict and address potential conflicts between elephants and human settlements.

While focusing on elephants as a "flagship" species—representative of the ecosystem as a whole—the study was also applicable to other "co-occurring" species; it provided a snapshot of ungulate movement in the most important conservation areas left on the subcontinent.

Conservationists saw that the peace parks might offer opportunities to protect the most vulnerable villages, those that lay in elephant migratory paths. Simultaneously, the parks could promote ecotourism as a sustainable way for communities to benefit from wildlife. These projects could do for southern Africa what the Serengeti and Masai Mara had done for East Africa, branding a region with a highly marketable wildlife phenomenon, as elephant and other species resumed their ancient migrations. After decades of mismanagement, persecution, and culling, the peace parks—or so it was hoped—could finally solve the elephant problem.

On 4 October 2001, at a ceremony in Kruger National Park, former president Nelson Mandela opened the border gates between his country and Mozambique to allow seven elephants to walk from Kruger into the former hunting concession, Coutada 16. This simple act represented the first step in reversing decades of deep divisions, political and ecological. Mandela was jubilant at the prospect of a unified future for the neighboring countries' wildlife, joking that the gift of the elephants represented a dowry for his third wife, Graça Machel, the widow of a former Mozambican president.[11]

As it was designed to do, the event, scheduled on Rupert's eighty-fifth birthday, attracted a good deal of media attention. The PPF apparently felt a high profile for its iconic peace park would translate into fund-raising gold. The elephant translocation took place against the advice of ecologists, who wanted to reintroduce smaller grazing species first, and planners who wanted to consult with the people living in the park over how they could protect themselves and their crops.[12] The PPF held its party regardless.

Rupert spoke hopefully about the potential for the new peace park to alleviate the elephant problem, rejoicing that the thousand ele-

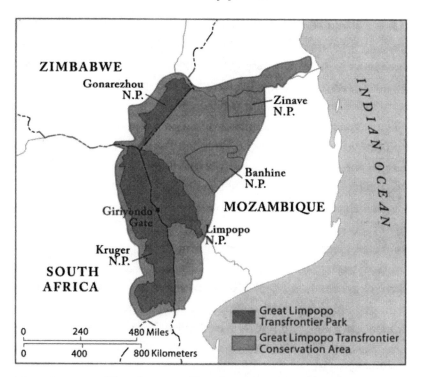

Great Limpopo Transfrontier: The Great Limpopo Transfrontier Park greatly expanded Kruger, the jewel of South Africa's park system, but has yet to come to grips with disgruntled villagers living alongside dangerous wildlife.

phants that were to be moved to Mozambique would be saved from culling. (The government clearly agreed: in 2002, the South African Department of Environmental Affairs and Tourism described the Great Limpopo as "particularly exciting because it will re-establish to some extent, ancient migratory routes for various animals, and relieve wildlife population pressure, notably for elephants.")[13] Rupert predicted that tourism would be the force that saved wildlife for future generations: "Tourism, which provides employment for one out of every ten people in the world, is the key to the successful protection of wildlife in the peace parks and to its sustainable survival."[14]

But the elephants—two adults, two calves, and three infants (eighteen others were driven on a flatbed truck to a fenced enclosure inside Mozambique)—had their own ideas. With their prodigious memory for danger and a marked indifference to the pomp and circumstance with which they had graduated to Coutada 16, the elephants decided that it was the last place they wanted to be. After the festivities, they wasted no time heading back to Kruger. One bull elephant trudged over 150 miles, walking north along the fence until he came to an opening at the Limpopo River. Another walked into Kruger through an unfenced section of riverbed and then returned to Mozambique to lead a group of females back to South Africa.

The elephants' reluctance proved a minor setback, and the wildlife of Mozambique has slowly been repopulated. Since 2001, 4,148 animals—elephant, buffalo, giraffe, white rhino, antelope—have been translocated into a sanctuary in the Limpopo National Park. Perhaps drawn by them, more elephants began migrating across the border on their own initiative, finding gaps in the fence. Giraffe, rhino, buffalo, wildebeest, kudu, and impala have followed suit. Markus Hoftmeyr, head of Kruger's wildlife veterinary services, hopefully suggested that the elephants were "signalling each other . . . that it [was] safe to return to their old stomping grounds."[15]

But several years later, the elephant problem had still not been solved. Kruger park managers were alarmed at the growing populations. Since culling had ceased, the population of seven thousand had nearly doubled, to twelve thousand by 2004. In their view, the peace park was not working fast enough: neither translocations nor individual elephants sifting out of the park through gaps in the fence were providing relief from what park managers believed was a severe overpopulation. They pointed to the park's unique flora, claiming that elephants were killing large numbers of trees and threatening to cause local extinctions of some plant species.

In October 2004, Kruger hosted a weekend meeting, the "Great Elephant Indaba"—*indaba* is a Zulu word meaning conference or gathering—to find "an African solution to an African problem."[16] Participants were representatives from Kruger, as well as relevant government agencies, NGOs, biologists, and representatives from neighboring countries. Four methods of elephant management and control were considered: culling, fertility control, translocation, and the development of transfrontier parks. The fifth and final option was to do nothing, to wait and see what happened. Kruger managers favored culling within their park, but elephant specialists argued against it, asserting that it could affect population dynamics, accelerate the growth rate, and stress the animals in ways deleterious to tourism. Biologists questioned the argument that Kruger had exceeded its carrying capacity for elephant, criticizing park data. There was no clear consensus on fertility control, while translocation, moving animals individually to other parks, was deemed too expensive.

At the end of the debate, the only argument against transfrontier parks was the possibility of increased elephant conflict. Biologists argued that increasing the range for elephants would allow habitat to regenerate and elephants to resume a natural process of self-limitation, based on available food and water. But the solution that was actually adopted was the least popular: doing nothing.

I visited Kruger a year after the *indaba*, and it was clear that the conference had failed to convince park officials of the advantage of transfrontier parks. "They are not a solution to the elephant management problem," Raymond Travers, media spokesman for the park, said flatly. "Opening the borders and increasing the size of your conservation area will only bring temporary relief to a species. It won't solve the problem."[17]

Travers was laying down the law, something for which Kruger has quite a reputation. Defiantly contradicting a chief argument for transboundary parks, he defended Kruger's long-cherished belief that elephants must be culled. By this point, the park was pulling in 1.3 million visitors a year and revenues of $40 million. Theoretically, if Kruger had been willing to eliminate artificial water sources in the southern reaches of the park, if it had been willing to cope with the dissatisfaction of tourists who might find it more challenging to see wildlife without boreholes, if it had been willing to give the peace park time to succeed, then it might well have begun to solve the elephant problem. Of course, taking that course of action might have affected the park's revenue; Kruger might have lost, at least temporarily, a little of its popularity.

Instead, Kruger threw its considerable weight against the biologists and transfrontier proponents. Like the ranchers in Catron County, Kruger would not be swayed by science, despite the fact that it was a world-renowned locus of research. The peace park got a taste of something known in South Africa as "Krugerisation," the bullying and insensitive management policies for which the park is famous.[18] In 2007, Kruger announced that culling would begin again.

No one knows what Anton Rupert might have made of Kruger's intransigence: He died on 18 January 2006, at the age of eighty-nine. Obituaries reported that he began his business career with an investment of ten pounds in 1941, leaving an empire worth ten billion.[19] Only one of his three children, Anthonij Rupert, had been an enthusiastic backer of the peace parks vision, but he died in a car accident in 2001.[20] With Anton Rupert's death, the PPF lost its most powerful voice in South Africa.

So it seemed that the Great Limpopo was not going to solve the elephant problem. Kruger's attitude toward the peace park, and its decision to return to culling, undercut the chief biological rationale for the project. From a conservation point of view, if the Great Limpopo did not solve the elephant problem, what was the point of having a transboundary protected area at all?

While Kruger was issuing its elephant fiat, another question swirled around the Great Limpopo, arising from social forces as ponderous and balky as elephants themselves. Could the peace park drive the kind of economic benefits that could reconcile people to being displaced?

The People Problem

"Transboundary conservation is a very slow and deliberate process."[21] So cautions Charles Besançon, a peace park specialist who worked on plans for the Virunga Volcanoes transboundary area. But the Great Limpopo was launched in haste and without deliberation. Its precipitate start exposed a discomfiting fact: no provisions had been made to safeguard people living within the Great Limpopo Transfrontier Park, much less inform them of the radical changes taking place. Recalling the dark history of conservation in this part of the world, the oversight suggested that the cavalier attitude toward human rights once common in South Africa still prevailed. Not only did it inflame fears and suspicions among those inside the park, it shook the confidence of the World Bank and donors supporting the project.

To make matters worse, as negotiations evolved in 1998, the original plan, focused on developing a more flexible transfrontier conservation area, was altered. The new plan, reportedly pushed through by the South African minister of the environment, would be for a transfrontier conservation area *and* a park. According to IUCN definitions, a conservation area could remain open to utilization and habitation; a park precluded human use. The revised plans recast Kruger and Coutada 16 as the core protected park, with a larger conservation area surrounding it as a buffer zone.

The Great Limpopo's people problem was immediately apparent. Mozambique had formally renamed Coutada 16, declaring the Limpopo National Park the month after elephants were released in October 2001. Critics suggested that the rush to recategorize was an attempt to justify the release of wildlife into the area before local people had been informed; government sources later complained that they had not been given sufficient time to contact them. The Peace Parks Foundation, nonetheless, radiated confidence. "This is not going to be a people problem, this park," Willem van Riet, then its executive director, confidently told *National Geographic* that same year.[22] The people who lived there felt differently.

In Mozambique alone, twenty-seven thousand people (and five thousand cattle) were living in the area assigned to the new park. Their settlements stretched along the Limpopo, Shingwedzi, and Olifants rivers within Limpopo National Park, where they grew maize, pumpkins, and beans. Many were refugees, uprooted or relocated before, during the civil war and construction of a nearby dam; many suffered agricultural and other losses from the torrential flooding in 2000. After Mozambique's declaration of the Limpopo, the park's buffer zone alongside the Limpopo River, on the southern edge of the new park, was realigned so that twenty thousand of the villagers living there would not need to be relocated. That left sixty-five hundred people inside the park, along the Shingwedzi River and near the Massingir Dam, who would be endangered as elephant, leopard, lion, and buffalo began making their way into the area. These people, as a park planning document noted dryly, constituted "a challenge for park management."[23]

By all accounts, both the government of Mozambique and the Peace Parks Foundation failed to communicate with villagers about their plans. Legally, the government bore responsibility, but as everyone involved knew, it was weak and ineffectual. In 2002—the same year that the three heads of state met, amid much pomp, to sign the international treaty establishing the Great Limpopo—the Refugee Research Programme at the University of the Witwatersrand discovered that news of the park had seeped out through rumors, misleading radio reports, or bullying comments from park rangers. The university conducted interviews with eighty-four heads of households representing over a thousand people inside the Limpopo National Park. They found that 40 percent had never heard of it, a year *after* the name and status of the area had been changed.[24]

Human rights advocates and academics familiar with the history of apartheid in South Africa, where conservation was often used as an excuse to push people off ancestral lands, were alarmed. "Community development issues have become secondary to conservation," an aid worker in one village told the Inter-Press Service of Africa News. "This is colonialism by conservationists."[25] Donors developed cold feet. At the end of 2002, the World Bank reversed its decision to fund the Great Limpopo, and the German Development Bank, sensitive to any suggestion that it might be involved in forced relocations, withheld its first payment.[26]

Academics examined the promises made regarding community participation and benefits and found that villagers had been sidelined and marginalized. The Great Limpopo became the chief focus of their attention in papers, PowerPoint presentations, and books. Few had a good word to say about it. There were cries of "economic imperialism," "land grab," and the "globalization of conservation." In the *Journal of Modern African Studies*, two scholars declared that the Great Limpopo Transfrontier Park had contributed to the downfall of a potential "African Renaissance."[27]

While funding was eventually restored to the park, donors remained sensitive to inflammatory charges that the project was inflicting misery on the poor and stipulated that resettlement must be voluntary, in line with World Bank and IUCN guidelines. Villagers living off subsistence farming, earning between $15 and $110 per year, were presented with several choices. While elephants were being translocated in, residents could be translocated out, to villages outside the park, where alternative housing would be constructed. Or various fencing schemes could be employed around villages within the park to provide some safety. According to the Refugee Research interviews, most villagers found all options unappealing: 84 percent said they would refuse to move, citing their attachment to the land, ancestral sites, and "sacred trees."[28] Fencing would restrict access to agricultural land and the river, their livelihoods. Some of those interviewed proposed "a curious third option," that of staying and taking their chances with the lions and leopards, saying they "would adapt by killing the wildlife."[29]

There were protests in 2005, with villagers placing rocks on the roads to stop park rangers long enough to ask them questions. With the dropping of seventy miles of fencing between Kruger and Limpopo, wildlife had been filtering into their settlements, and people were becoming fearful, with hyenas killing their goats and elephants trampling their fields. There were threats of sabotage against the park. The inhabitants of one village, Makandezulu, were terrorized for two weeks by a lion before the villagers snared and killed it. Some elephant and buffalo met the same fate. Many felt that park authorities were simply trying to drive them out by releasing wildlife into their midst. They were told that they could no longer practice crop rotation, since it would utilize land outside what had been allocated. As one villager told a reporter, "They say that the resettlement is not forced, but that is not true. We are forced because we are no longer allowed to live our lives as before, we can no longer cultivate where we want, we can no longer take our cattle out to graze. Yes, we agreed to move, but we did not do so freely."[30]

In the years since, committees have been formed and development plans drawn up. By 2008, 70 percent of villagers in the park had apparently agreed to move. Makandezulu and two other villages have chosen to stay where they are, inside the park, waiting to see what happens to the people who move. Various relocation schemes have been proposed, and have fallen through. As of 2009, no one had moved. When I spoke to Arrie van Wyk, the South African project manager, he wearily acknowledged that villagers are skeptical that they will receive the promised benefits: "People don't believe things will happen," he said, and they may be right.[31]

Much of the private tourist investment promised in early planning for the Limpopo has not materialized. By 2006, $8.7 million had been spent on infrastructure development in Mozambique's park: mines had been cleared from the area, and park headquarters and ranger housing built. A rustic tented camp in the Lebombo Mountains and a fishing camp on the shores of the Massingir Dam were constructed, for a handful of tourists. Over the past five years, 150 jobs have been created in Mozambique.[32] But the development has fallen short of accommodating the thousands of overnight visitors envisaged by the Peace Parks Foundation, based on projections of 486,180 visitors per annum.[33]

Meanwhile, organized poaching has increased. Seventy rhino have been killed in Kruger in the past six years, and there are reports of rhino poachers driving in from Limpopo National Park.[34] In 2003, a white rhino was poached within Limpopo, reportedly by a park ranger. In 2005, Kruger rangers discovered seventy-nine snares set near a spring at the northernmost end of the park, where the fence had been taken down; a buffalo calf, hyena, and impala were killed. A few months later, a poacher was captured with 129 skins of genets, small, catlike predators, which one park official admitted was "virtually the entire population of genets along the Limpopo."[35] The 150 jobs included seventy rangers, some hired from among those living in the park. Working closely with Kruger, they have apprehended thirty-six poachers equipped with AK–47 assault rifles, leftovers from the war.

Had the Peace Parks Foundation not raised such expectations and made such promises, these halting steps might have been seen as a welcome sign of things to come. As it was, when I spoke recently with the new CEO of the Peace Parks Foundation, a young man named Werner Myburgh, he sounded plaintive when he spoke of the prospects for the Great Limpopo. "These are early days," he said, pleadingly. "Early days."[36]

The Giriyondo Gate

Hyped as changing "the conservation map of a continent," the transfrontier park seemed as tantalizing as a mirage. It first appeared during the splashy trilateral signing ceremony. It wavered into sight again when Mandela opened the gate for those seven puzzled elephants. Then it disappeared.

No one in South Africa was eager to admit that the Great Limpopo was not fulfilling its promise. When I requested permission to travel to the Giriyondo Gate, the customs facility and gateway to the Limpopo National Park being built on the Mozambique border, I was met with a silence that lasted some weeks. Finally Kruger's headquarters grudgingly acquiesced. Supposed to have been completed by 2004, the gate still wasn't open in late 2005: rumor had it that Mozambique's resentment of South Africa—over revenue sharing and news that the wealthier country was planning to build a new airport near Kruger to fly more tourists to its side—was the hitch.

When I arrived at the Giriyondo Gate, the facility was deserted on the South African side. It seemed outscale and out of place in the desolate border area, with the high border fence marching into the hills on either side. Costing the fantastic sum of 6 million rand (around half a million dollars), the substantial, heavily thatched, and fully furnished facilities sat silently, ringed by rock walls. A wide multilane driveway on the South African side awaited large crowds. Satellite dishes sprouted from the roof; the rooms inside were stocked with computers and telephones. The landscaping included large baobab trees, propped up with poles, and an assortment of palms trucked in and transplanted at great expense. On the other side of the fence, workers were huddling over a fire and drinking tea in Mozambique.

At the moment, it was a gate to nowhere. Of the South Africans I met, only Oupa Modirwa, a park ranger accompanying me, seemed excited about the opening. He said people who saw him in his ranger uniform were always asking him about it: Mozambicans who could afford a car or taxi were eager to bring goods to sell in South Africa, and the gate promised a quicker trip. The use of the gate as a commercial portal was not exactly the kind of conservation boon the PPF had imagined, but in the event, that was what happened. Since the gate opened, most of the trips through it have been not "tourism related," according to Kruger's statistics, but for the transport of people and goods in and out of Mozambique.

Mike Stephens, the white South African guide who was with us that day, was scandalized at the expense. The gate seemed like a boondoggle, the border post equivalent of the grand mansions beloved by Africa's strongmen. The money could have gone to house and relocate any number of people within the park; it could have sent people for training or higher education. But then, money wasn't really the problem, judging by the foundation's fund-raising success.

The real problem was that no one who lived in or around the Great Limpopo had any reason to support it. Kruger had never shown concern about the villagers, and the feeling was mutual; that bitterness extended to the new peace park. Unless they were among the fortunate few to get a job as a ranger, they had seen no revenue from tourism, and few jobs had been created. The lack of support in Mozambique was mirrored in South Africa. Six million South Africans were living along the southern and western borders of Kruger and had seen few benefits from the original park. Why should they support a larger one?

I visited one town, Welverdiend, a village of around ten thousand near one of Kruger's western gates. Welverdiend was only minutes from Kruger, but the town seemed impossibly remote. Women and girls walked wearily along the road, carrying plastic water jugs, on their way to or from a communal tap. Goats and cattle were penned inside crazily leaning corrals, bleached branches lashed together. At a local nursery school, the only toys were a jagged metal slide and two truck tires embedded in compacted soil. Inside, the children played with old detergent boxes.

The Conservation Corporation of Africa, affiliated with a nearby private game reserve, was contributing money for scholarships and infrastructure at the town schools, building a computer center in the high school and badly needed restrooms. The elementary school's principal, Police Jeroboam Manzini, showed me his crowded classrooms and dangerously unsanitary facilities, stretched thin by an influx of refugees and orphans from Mozambique. An overwhelming majority had seen no benefit from tourism. He said of the town, "Only CC–Africa is giving back to the community. If Kruger was doing as much as CC–Africa, they would be feeling happy."[37]

While the charitable organization had made arrangements for children to visit the private reserve, few residents had ever been to Kruger. Imagine how much support Yellowstone or Yosemite would inspire if the people who lived nearby never got a job or a salary from tourism, if they had never had an opportunity to go to the park. Kruger could easily contribute to the transboundary effort by investing in jobs and education in surrounding communities.[38] Instead, the parks agency that runs Kruger has refused to allow community members from Welwerdiend to participate in decisions regarding the management of the private reserve.[39] As with their stubbornness over the elephant culling issue, the park seemed determined to maintain control over its fiefdom, frustrating attempts to bring communities into the dialogue over protected areas and their management.

Meanwhile, in imploding Zimbabwe, 750 settlers moved into Gonarezhou, the park at the northern end of GLTP, clearing, burning, and planting crops with the support of the government. While Kruger was restocking Mozambique with wild animals, Zimbabweans were killing theirs and South Africa was strengthening fences against millions of refugees.[40]

The View from Cape Town

Cape Town is a city of emotional whiplash. Gifted with one of the most striking natural sites in the world, Cape Town occupies the vertiginous slopes of Table Mountain, a flat tableland of dramatic views and dense biodiversity. Over a thousand species of heaths, proteas, lilies, and succulents grow on the mountain, the richest botanical area of its size on the planet, looming thousands of feet above the city, patrolled by baboons and rock hyrax that sometimes disappear in "ghost fog," a cloud of mist. Its historic Victoria and Albert waterfront glamorously restored with luxury hotels and shopping malls, its neighborhoods flush with boutiques and chic restaurants, Cape Town is also home to the Cape Flats, southeast of the city center: mile after mile of miserable slums, metal and cardboard shacks lashed together by the hundreds, flanked by rows of outhouses, lines of people queuing at a single water tap. The dramatic contrast between wealth and want is always visible, always part of the view. Beggars walk the streets, and squatters live in cardboard boxes in parking medians. Cape Town is the top tourist destination in the country and the hub of conservation planners who argue that a rising tide of transboundary ecotourism can float all boats, lifting people out of poverty while restoring southern Africa's fractured habitats.

Tourism is the world's largest industry, with annual revenues of around $3 trillion, yielding almost 11 percent of global gross national product. At the turn of the twenty-first century, Africa lagged behind the rest of the world in tourism, with only 4.5 percent of the market share. There was nowhere to go but up, and ecotourism was the leader, growing by 30 percent annually as of 2007 and generating the equivalent of twelve dollars per acre in South Africa, compared with agriculture, at only three dollars.[41] Nearly 80 percent of tourists in South Africa participate in nature-based tourism; altogether, the industry in South Africa, Zimbabwe, and Mozambique brings in $2.4 billion per year.[42] Defined as "responsible travel to natural areas that conserves the environment and improves the welfare of local people," ecotourism has been embedded in the peace parks formula, a consistent feature in calculations of how wildlife can be made to pay for itself and one of the only sustainable forms of development compatible with conservation.

Recently, the South African government ratcheted up its focus on ecotourism, taking up the promotion of the peace parks. Anticipating its hosting of the soccer event of the decade, the 2010 World Cup, South Africa decided to "brand" the transboundary parks. In 2006, environmental affairs and tourism minister Marthinus van Schalkwyk released "The TFCA [Transfrontier Conservation Area] Route," a plan to turn everything transboundary into an "international product," through brochures and online publicity.[43] The TFCA Route—a full-color map of southern Africa with big circles around the seven peace parks—is a marketing, not a conservation plan, breathlessly promoting "1 Trail / 2 Oceans / 9 Countries / 7 Transfrontier Parks," suggesting that tourists can somehow follow a "coast-to-coast" itinerary.[44]

Although the oceans and countries are real enough, there is no trail. Arguably, there are not yet seven transfrontier parks either. Despite the map's promise of "established amenities, sound infrastructure, extraordinary adventures," most parks lack international airports or air connections, roads, tourism accommodations, border posts, and, in general, access. Even if such facilities were to appear, the route's ambitious itinerary would be physically and financially punishing. A version of the route now appears on a Web site and in brochures distributed by "Boundless Southern Africa," a slick marketing initiative sponsored by South Africa's tourism and environment department, touting "investment opportunities" associated with the parks. The Boundless campaign has even sponsored a trek along this fanciful route—from the Atlantic to the Indian Ocean—undertaken by the eccentric South African explorer Kingsley Holgate, known as the "grey beard of African Adventure" for his hirsute facial stylings and televised Cape-to-Cairo journeys. Accompanied by a Land Rover pulling a grader, with which he would clear seven new soccer fields for communities adjacent to the parks, Holgate announced, "This one is all about conservation."[45] Inspired by this bold foray, copycat "routes" have proliferated. De Beers announced a "Diamond Route," a proposal for ecotourists to visit eight "conservation properties" surrounding former diamond mines. In conjunction with a 2008 entomology conference held in Durban, De Beers also promoted the world's first "Insect Route," although its parameters were not made clear.[46]

But for all the excitement attendant upon these Voortrekker fantasies, the economic argument on which the marketing plans were based was fraught with problems. Tourism in Africa has been plagued by "leakages," or profits that end up back in the developed countries where ownership of hotels, lodges, or resorts often resides, rather than remaining in the region and buoying local economies. The UN Conference on Trade and Development estimates that up to 85 percent of tourism profits in African countries leak back to wealthier countries, leaving only a fraction behind.[47] Assumptions about ecotourism's economic potential may be overly optimistic.

Then, too, as experts point out, a reliance on ecotourism carries risks. Anna Spenceley, a research fellow at the Transboundary Protected Areas Research Initiative at the University of the Witwatersrand, characterizes the industry in South Africa as "highly volatile," requiring considerable capital investment.[48] While interest in ecotourism has been growing, many ecobusinesses incur significant start-up costs and may never turn a profit. They are also subject to market fluctuations, vulnerable to severe economic downturns like those caused by 9/11 and the banking collapse of 2008. For all these reasons, it seems unlikely that ecotourism can sustain conservation fully and indefinitely. Even in Costa Rica, where nature-based tourism is the second-largest industry in the country, ecotourism revenue does not pay for conservation.

Leo Braack, a veteran biologist with long experience working in Kruger, acknowledged as much, admitting that peace parks could take decades to bring in significant revenue and that ecotourism was unlikely to foot the bill for conservation. Currently the director of Conservation International's South Africa Wilderness Program, Braack cautioned against expecting too much. He had hopes for small-scale improvements, better roads, longer hours at border posts. He laughed at the 2010 TFCA Route. "Very optimistic," he said. "Maybe 2020."[49]

The view from Cape Town, where the inequities of South Africa lie exposed like the ancient layers of sandstone that make up Table Mountain, was sobering. Ecotourism promised to generate some income for remote rural communities. But there was no proof that it could meet the great expectations that had been attached to it. There was no proof that wildlife could pay for itself.

It appeared that Anton Rupert's original criteria for the peace parks, the criteria of the businessman, were not enough. The expectation that wildlife could "pay to stay," the overreliance on ecotourism, the crutch of exaggerated marketing hype—these created an insecure foundation for what should have been a scientific enterprise. Judging the Great Limpopo Transfrontier Park against Anton Rupert's own expectations, it has so far achieved little. It has not made peace. It has not made money, building accommodation for only a few dozen people on the Mozambique side. And it has not made friends, leaving people alienated. The elephant problem has not been solved. At the Giriyondo Gate, easy access—one of Rupert's pillars—has been secured, but few tourists seem to be taking advantage of it.

The mistakes made in launching the Great Limpopo—undue haste, lack of conservation planning, cavalier treatment of people inside the park—could have been avoided. Once committed, those mistakes served as hard lessons, providing a course correction for future plans. The Great Limpopo called into question basic assumptions—that launching parks and raising money were enough, that conservation and economic development could take care of themselves. They weren't, and they couldn't.

Nonetheless, the Peace Parks Foundation accomplished something unprecedented. It put large-scale transboundary planning on the African map, investing more money and resources in conservation than had ever been seen in that part of the world. It funded and expanded colleges that were turning out trained rangers, wildlife guides, veterinary specialists, and lodge managers. It inaugurated the first organized effort to manage elephant populations regionally, throughout the subcontinent. Its projects won worldwide attention and inspired more countries to begin transboundary planning.

Within South Africa, park authorities and private organizations launched a project in the Cape region to expand and link protected areas through a series of "mega-reserves" designed to consolidate the rich biodiversity of the Cape floristic region while putting people to work through the government's Civilian Conservation Corps–style programs, "Work for Water" and "Work for Woodlands." One of the flagship projects, the Baviaanskloof Mega-reserve Project, avoided the early missteps of the Great Limpopo by developing an elaborate "stakeholder engagement program" that encouraged voluntary participation and collaboration by neighboring urban and rural communities, farmers, and private landowners.[50] It has created hundreds of jobs in varied occupations, including ecotourism, removal of invasive plants, and carbon sequestration, an innovative program in restoring native vegetation.

As the Peace Parks Foundation began moving ahead with other ambitious transboundary plans, there were signs that the organization itself was evolving in the direction taken by Baviaanskloof, learning from mistakes in the Great Limpopo and rethinking the balance of conservation and development in its programs. There were signs that at least one of its major transboundary complexes might be succeeding on both fronts.

CHAPTER 7

THE LUBOMBO TRANSFRONTIER

Tembe

THE MOST PROMISING OF THE PROJECTS SO FAR—CREATING the most new conservation areas, the most jobs, and the first corridor for elephants in southern Africa—is the Lubombo Transfrontier Conservation Area, a wide stretch of country running up the coast between the Indian Ocean and the mountains of Swaziland, from South Africa into Mozambique.

The Lubombo project differed in basic design from the Great Limpopo. While the peace park joined large contiguous protected areas, the Lubombo, a more flexible transfrontier area, was designed around five discontiguous sites, each with its own challenges. The projects also differed in priorities. The Great Limpopo, for example, arguably began backward: it started by moving elephants and dealt with people only belatedly. The Lubombo, on the other hand, put people in jobs at the beginning, creating the very protected areas where they might ultimately find permanent employment. Crucially, the two projects have so far produced different outcomes.

Since Lubombo's individual transfrontier areas cropped up in densely populated rural countryside, there was no question of asking people to move or relocating villages. For Lubombo to succeed,

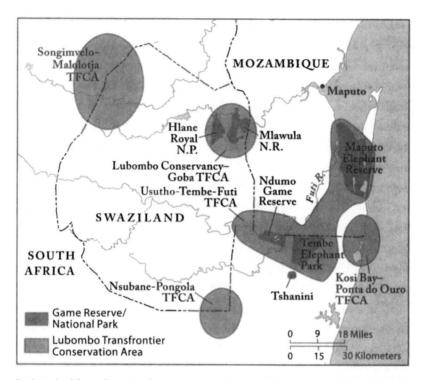

Lubombo Transfrontier Conservation Area: In five projects linking Swaziland, Mozambique, and South Africa, the Lubombo Transfrontier Conservation Area has created jobs in areas crippled by unemployment while reuniting an elephant population cut in half by war and border fences.

people would have to accept it, cooperate with it, and be trained to work for it. They would have to resolve long-standing disputes over land tenure going back to the apartheid era; they would have to rely on promises of tourism revenue and trust government agencies that had disappointed them in the past. Lubombo would be an experiment in creating a conservation beachhead in one of the most densely populated and impoverished places on the continent.

Economically, the region was struggling. Mozambique, of course, had been shattered, its infrastructure leveled and parks stripped. The neighboring Kingdom of Swaziland was ruled by a monarch known to be among the most corrupt in Africa, notable for his flock of wives, his fleet of limousines, and the near-total misery he had inflicted on over a million subjects, nearly half of whom were infected with HIV/ AIDS. Life expectancy fell from sixty in 1997 to thirty-one less than a decade later. KwaZulu Natal, the northeastern South African province containing many Lubombo reserves, was among the poorest in South Africa, with high unemployment, a high rate of HIV/AIDS infection, and a long history of illegal border activity, prostitution, bushmeat hunting, and gun and alcohol running (alcohol was illegal in townships during apartheid). The province was famous as the home of the Zulu, whose nineteenth-century leader Shaka conquered and organized warring clans to resist the British, but its people had been confined into remote "homelands" during apartheid, employed in mines that had recently closed, and neglected by the government. The Lubombo Transfrontier promised to bring welcome change, jobs, tourism, commerce, and attention.

The existing protected areas of the Lubombo Transfrontier were so far apart that planners grouped them into five separate "mini-TFCAs."* Scattered across the borderlands of the three countries, they lacked a large-scale core area like Kruger to expand from, only a handful of game reserves and beaches. Yet, like the Great Limpopo, the Lubombo was potentially massive, encompassing five Ramsar sites, linking to the largest estuary in Africa, Lake St. Lucia, and one of the most endangered sand forest ecosystems on the continent. The five mini-TFCAs had not yet been mapped out in 2000, when a general agreement establishing the Lubombo was signed by the three governments. Taken together, the individual transfrontier areas eventually covered some 1,621 square miles. The full area to be affected added up to something much larger, thousands of square miles.[1]

---

* The five mini-TFCAs of the Lubombo Transfrontier Conservation Area are: Usuthu-Tembe-Futi TFCA (Swaziland, South Africa, and Mozambique), Nsubane-Pongola TFCA (southeastern Swaziland and South Africa), Kosi Bay–Ponta do Ouro TFCA (a transboundary coastal area on the Indian Ocean, in South Africa and Mozambique), Lubombo Conservancy–Goba TFCA (northeastern Swaziland and Mozambique), and Songimvelo-Malolotja TFCA (northwestern Swaziland and South Africa).

The largest of the mini-TFCAs and the first project to be actively developed was the Usuthu-Tembe-Futi Transfrontier Conservation Area. It was a boot-shaped region: the toe was the Usuthu Gorge, a spectacular river canyon in eastern Swaziland, which would be linked with corridors to South Africa's Ndumo Game Reserve (the instep), the Tembe Elephant Park (the heel), and the Maputo Elephant Reserve in Mozambique (at the top). Both Ndumo and Tembe were small, isolated fragments. From a conservation standpoint, the solution was obvious. Join them to create a larger, connected reserve that, in turn, could be reconnected with the Maputo reserve in the north, the traditional elephant breeding grounds. The project also offered immediate ecotourism benefits. Each reserve had its own unique draw: indigenous elephants at Tembe and rare birds at the park nearby.

The scheme had a strong emotional appeal, based on its potential to break down barriers between long-fractured populations of people and elephants. It planned to reunite scattered remnants of the last native southern African elephants, whose migratory corridor had been cut in half by civil war and border fences. Some of these elephants were stranded in Mozambique, in the Maputo Elephant Reserve. Others found themselves fenced into Tembe and were so traumatized by land mines and poachers—24 percent bore visible wounds—that they reacted aggressively to the slightest disturbance. "Elephants with grudges," a park official called them, and jeep drivers entering the park after it was reopened had to be prepared to throw their gears into reverse at the first sign of jumbo tantrums.[2] After the war, the animals sometimes gathered on either side of the border fence, trumpeting and waving their trunks at one another. The study of elephant home ranges, completed in 2005, revealed that the Tembe elephants' was the smallest recorded for any in the subcontinent. If the fence could be taken down and a safe corridor reestablished for elephant traffic along the Futi River, a seasonal river flowing from the mountains of Swaziland through Tembe and emptying into the coastal floodplain at the Maputo reserve, the old annual migratory path that bulls had followed for centuries could be restored.

People would be reunited as well. In 1875, when the border was drawn between the Portuguese colony that became Mozambique and the British one that became South Africa, the line bisected an existing kingdom, that of the Tembe-Thonga people. Subsequently, the people living in the border area between South Africa and Mozambique were, like the elephants, cut in two, first by the colonial border and then by the recent war. They, too, were displaced and victimized; they, too, survived an era characterized by furtive border crossings and persecution by various militias. Many Tembe who live in South Africa cross the border regularly, if illegally (due to the hassle of reaching border posts) to visit farms and families to the north.

Some things had to be worked out in the real world before the project could come together. These reserves, like Kruger, had their own dark past. Tembe was relatively new, its sandy, agriculturally poor land voluntarily donated by the Tembe tribe in the 1980s, in the hopes of developing tourism. But the Ndumo Game Reserve, established in 1924, was built on apartheid. West of Tembe and separated from it by a strip of land only three miles wide, Ndumo was prime farming land, with year-round water in the Pongola River, fish, and rich wildlife. People living there were forcibly removed in the 1940s, and those defying the order were subjected to tsetse fly spraying by the government to enforce submission. In 1947, the boundaries of the reserve were extended to include the Pongola River, cutting off access to water, fish, and farmland. Only women were allowed access to the river (men were assumed to be poachers), and some were killed by crocodiles at the access points. Productive agricultural land within the Pongola floodplain became off-limits, and farmers were forced to use poor areas of sandy soil. Fishing was not permitted in the nearby Ndumo pans.

So the three-mile-wide strip between the two reserves, home to a community known as Mbangweni, remained a barrier: Thousands were living in it, seething with resentment at the land grab that kicked them out of Ndumo in the first place, for which conservation had been the excuse. In the 1990s, the community filed land claims. In 2000, with the claims still not resolved, people from Mbangweni burned several kilometers of the Ndumo fence line in frustration, calling the action a "fence telegram" to authorities, asking them to resolve the issue.[3] If the protestors did not receive a response, they said, they would invade the park. A poacher was shot and killed, an off-duty ranger attacked. Park authorities promised to donate hippo meat from culled animals to convince villagers to stay out of the reserve, handing over a couple of carcasses before reneging altogether.[4] Proposals to restore agricultural areas of the floodplain to the communities were made, then withdrawn when environmental concerns were raised. While the land claim was resolved later that year, giving the community "nonoccupational title" to the area and providing financial compensation, people continued to view conservation as the enemy. These were the obstacles proponents had to overcome.

"CAUTION: Dung Beetles Have Right of Way" is the sign greeting visitors to Tembe Elephant Park. Only a hundred and twenty miles south of Kruger as the crow flies, on the border with Mozambique, not far from the Indian Ocean, Tembe is a different world, a small, intimate park without any of Kruger's teeming crowds. Until recently, Tembe was closed to the public, and even now it allows only a limited number of visitors per day. The park's most famous residents are the group of war-refugee elephants victimized by the long conflict in Mozambique, who remain distinctly skeptical about human beings.

A five- or six-hour drive from Durban on recently paved roads, Tembe contains a single tented camp, owned jointly by the Tembe tribe and a Durban businessman, Ernest Robbertse, a former private detective who fell in love with the area while investigating alcohol running on the border. All of the twenty or so employees of the lodge come from the tribe—cooks, staff, guides—and I saw some raised eyebrows and sidelong glances among them when I asked to talk to Roelie Kloppers, a young Afrikaner in charge of the Peace Parks Foundation's work in the area, organizing a jobs-creation program.

After a long history of running things in South Africa, Afrikaners have a lot to reconcile.

Without the throngs of tourists and photographers in the larger parks, Tembe offers a more direct experience of wildlife, from cunningly built hides near water holes and the sand tracks themselves, winding through dense thickets of vegetation that barely conceal quivering suni (an endangered rabbit-sized antelope), nyala (rare striped antelope), black and white rhino (Tembe is one of the few places on the continent where both species are commonly seen), and elephant, often standing in the road, staring at the tiny jeep confronting them. Wherever visitors go in Tembe, they are followed by flying dung beetles, astonishingly pretty with iridescent colors, whirring like tiny mechanical windup toys. Jeeps veer to avoid piles of elephant dung, crawling with the creatures, who are fashioning and pushing dung balls that reach startling dimensions.

Roelie Kloppers showed me the human side of the community, just across the road from the park. Kloppers had an innocent, wide face topped by a straw-colored mop of hair. A native Afrikaans speaker, he hadn't learned English until he was in high school but spoke it fluently. He also knew some isiZulu and interviewed over four hundred refugees, smugglers, fishermen, and tradespeople for his dissertation, a fascinating and detailed ethnographic portrait of the culture around Tembe, specifically the byzantine border commerce at the Puza Market, on the Mozambique side of the border. South Africans, he reported, sold mainly processed goods and clothing; Mozambicans, homegrown foods and crafts. Vendors clipped bags of potato chips to the wire border fence and offered apples, mangoes, peanuts, sweet potatoes, bottled beer, sodas, palm wine, home-brewed maize liquor, sugar, cell phones, cosmetics, jewelry, candy, fried dough or "fats," and "Drostyhof wine by the glass."[5] Women dredged and deep-fried a local speciality known as "Kentucky Fried Fish."[6] Visitors could consult a Mozambican doctor in a reed consultation office, or make phone calls at a phone attached to a car battery, or get their hair cut by a traveling barber whose most popular cut was known as "the Afghanistan."[7]

Behind a stand of palm trees near the market, conveniently out of sight of conservation officers who sometimes patrolled the market, vendors also butchered and sold illegal bushmeat, largely smaller antelope, such as impala, but also warthog, cane rat, and monkey, much of it reportedly caught near the Maputo Elephant Reserve. Kloppers on occasion saw an entire suni carcass on sale for fifty rand.[8]

The Tembe area, he said, had remained undeveloped—and unspoiled—because of the war. "Political problems do have their benefits," he said wryly.[9] Kloppers was determined to use that situation to aid the community in establishing new and potentially lucrative ecotourism attractions.

And they were taking full advantage of the opportunity. We stopped by a recently completed project, the stout new fence built around the Tshanini Community Conservation Area, just south of Tembe, set up by the community to lure serious birders to see some of its hundreds of rare species. An aerial birding boardwalk was planned. Farther into the bush, we came to a schoolyard where a crew of workers, men and women, were painting a low cement building; another crew was working with plumbing equipment. Abashed, Kloppers explained that this was a makeshift school in the construction arts, painting, electrical work, bricklaying, and plumbing installation. The workers weren't really building anything; they were being trained. "It's a forty-five-day course," he said. I realized that this community, like so many others across the country, had been systematically starved of education and training for decades. They had to start somewhere.

Virtually all of Kloppers's efforts were concentrated on poverty relief. "It's a welfare state," he said of his homeland. "Forty-five million people. If you're over sixty-five, you get 750 rand a month ($100), if you have AIDS, 750 rand a month. With kids, 150 rand per kid." Poverty was directly linked to subsistence hunting. While there were enough fish in the area to guarantee protein in most diets, people hunted for the money it could bring in, for the traditional medicines it provided, and for variety. "In Ndumo, it's very bad," he said. "There are guys who steal crocs, hippo." Local *sangoma*, witch doctors, used the organs of creatures believed to be evil, such as crocodiles, to hex victims. "They take croc brains, put a piece in a matchbox, throw it in front of a taxi, throw it in the fire. It's voodoo."

The *sangoma*, who also provide natural remedies made from native plants and animals, encourage poaching and the unsustainable gathering of plants: a Wildlife Crime Unit in KwaZulu Natal recently found that hundreds of tons of wildlife products—mainly tubers, bark, and leaves, but also tortoise shells and the skins, bones, teeth, and claws of lion, leopard, and other species—were being sold every month in a single market in the Lubombo transfrontier area. The persistence of such rituals provides another compelling reason for transfrontier projects to support basic education and conservation programs that clarify the importance of wildlife in local ecology.

After I left Tembe, I followed the project's progress in a newsletter. By 2007, 1,800 jobs had been created in construction as people were hired to put up fences, drill boreholes, and prepare campsites at Tembe, Ndumo, Tshanini, and the Usuthu Gorge, earning over 10 million rand (around $1.35 million).[10] A local man who had worked monitoring a pride of lions in Tembe was sent on scholarship from the Peace Parks Foundation to the Southern African Wildlife College, where he excelled, earning a certificate in natural resource management and a conservation award along with a full scholarship to return for a diploma. He would later become the conservation manager at Tshanini.[11] Twenty-two people from nearby communities were hired as firefighters. Four people were trained by Southern Africa Birdlife as accredited birding guides, and Tshanini celebrated its first Zululand Birding Day. In 2006, two rangers were taught radiotelemetry, to track three Natal hinged tortoises released into the reserve.[12] If the Lubombo newsletter's headlines sounded fervent—"More Than 400 Participate in Training Programme to Date!"—there was reason for pride.[13]

Creating jobs and enlisting community support were key. At first puzzled by the necessity for a new fence around Tshanini—weren't peace parks supposed to be taking fences down?—I realized soon enough that it was the only measure that could protect the birding area and its tourism potential. With thousands of people living nearby, the temptation to hunt or gather firewood was too high. Tembe was fenced on all four sides for the same reason, but instead of removing fences, the solution was to create a gap on the northern border for elephant migration. While Kloppers's emphasis on poverty relief was not, strictly speaking, conservation, I could see no alternative to it: the support of the people was essential. While not purely dedicated to conservation, to reconnecting Tembe and Ndumo, the jobs were building a foundation for future conservation. Of course, the question remained whether the wildlife, particularly the Tembe elephants, could wait that long, or whether the villagers could.

Skeptics were questioning whether transfrontier conservation areas could bring enough jobs to meet the demand and the expectations that had been repeatedly raised. Jennifer Jones, an American academic, quoted an exasperated tribal leader who expressed the prevalent view: "'I am tired of people coming here and talking about development, making promises they don't keep.'"[14] He was waiting, he said, "'to see that thing they call tourism because I don't know what it is.'" But according to the Tembe *induna*, or head man, Mabhudu Israel Tembe, a tourism economy nevertheless represented their only hope, "the only source of jobs for our people."[15]

I heard the same frustration from Thom Mahamba, the lodge manager, and Sipho Sibiya, a longtime wildlife guide at Tembe. It is impossible to overstate the depth of knowledge of Africa's home-grown naturalists and guides, who typically learn the signs and behavior of dangerous predators as young children entrusted with the family herd, the family's future. Failure to note the approach of lion or leopard is unthinkable, and those who mature into wildlife guides often function as a crucial bridge, ambassadors between the local culture and conservation outsiders, explaining them to each other.

Sibiya and Mahamba had plenty of criticism of the peace parks in general, illuminating a vast divide between themselves, poor rural blacks with little education, and these grandiose plans that originated somewhere far away. Their remarks showed how keenly they felt the unfairness and disappointment of having been shut out of the kind of jobs available to someone like Kloppers. If a few of their complaints seemed minor—they felt that plans to widen the roads in Tembe and install new directional signs were silly—the irritation behind them was not. Sibiya was dismayed at the lack of knowledge of wildlife behavior displayed by some transfrontier planners. He was particularly upset about the additional housing for park rangers and students being built behind the remote Ponweni Hide above Muzi Swamp. I had gone there with him one afternoon and watched a breeding herd of elephant emerge from the bush across the valley and make their way toward us, a stately single-file progression of females and infants. As we left, we could hear hammering in the bush, the sound of a radio playing. He grimaced at the sound, fearful that it would alarm the wildlife, and said, "That hide is very important to us. We use it for people to go see lions, and they built the camp right next to it."[16]

Over dinner one night, both men revealed the intensity of the struggles they had faced, trying to provide for large families. They had ventured to the big cities to look for jobs, without much luck. Ultimately, they had no choice but to return to Tembe. Having found well-paying jobs in tourism, they were the exceptions, the ones who could comprehend the value of conservation. "In my mind, I can see the advantage, can see how good the park is," Mahamba said. "But people outside, they don't understand. If I go to them to say this, they think it is because I am already working. But what about people who are not working? If they can get more opportunities, I think people will slowly, slowly realize the park is important."[17]

The sheer numbers of people living around the reserves made it difficult for the project to supply even a fraction of the jobs that were needed. Sibiya, an authoritative older man and representative for his ward of around five thousand people, said, "There are no factories here. Nothing." He pointed out that many in the local community had already made sacrifices for conservation. One afternoon, as we drove through Tembe, we passed through a clearing in the undergrowth. He called back over his shoulder, "This was my family village! In 1981, we moved out. My father had seven wives. Twenty-five kids." They had moved, voluntarily, when the clearing became part of the park. I asked what the name of the village had meant, and he said, "All of us resting in one place together." Twenty-eight families moved out of the reserve and were promised infrastructure improvements for their community in return, including ready access to potable water. Some promises, however, were not kept. When we stopped at Sibiya's home one afternoon, his wife was outside, washing dishes in a rusting barrel. Women in the community still had to carry water significant distances to their homes.

Clearly, the tensions rooted in South Africa's past, between black Africans and Afrikaners, remained. The country had moved away from apartheid only a decade earlier, and the promise of a renaissance was achingly slow to materialize. Neither the local community members that I spoke to nor Roelie Kloppers, as the representative of the PPF, seemed entirely fair to each other or cognizant of the pressures and strains felt by the other. Pressing ahead with building projects inside Tembe, Kloppers had not addressed the disappointment felt by people like Mahamba and Sibiya. They, in turn, seemed unimpressed by the time and effort that a young Afrikaner had put in, studying their culture and way of life.

In spite of these mutual suspicions and insensitivities, something good was happening. Jobs were being created, hundreds more than at the Great Limpopo, and progress was being made in carving out conservation areas and a new elephant corridor. Sibiya took me to a new community center down the road from Tembe where a group of women were making baskets, amassing a collection of crafts to sell. I was their only customer, and they welcomed me, helping me pick out something to buy. As everywhere in the transfrontier area, I could feel the expectations. Everyone was waiting, waiting, waiting to see this thing called tourism.

Breakthrough at Ndumo .

In 2007, there was finally a breakthrough in the impasse between the TFCA and the people living between Tembe and Ndumo. Perhaps inspired by the jobs created at nearby Tshanini, the community gave its blessing to a conservation corridor, the Bhekabantu Community Conservation Area, directly linking Ndumo and Tembe with a 2,000-hectare swath of land.[18] The Wildlands Conservation Trust donated 1.45 million rand to pay for fencing and ranger training for local people. Revenue was expected from commercial hunting of antelope, and an official with the trust estimated that the project would create 150 temporary jobs and 50 permanent ones. There was talk of building a new, community-owned lodge on the Pongola River, and the government conservation agency reviled for expelling people from Ndumo donated a training center to the Mbangweni community, to use as its first local primary school. Before, young children had walked four miles to go to school. The gesture was greeted as a step in reconciling the community to a long-distrusted foe.

Visiting Ndumo with Sipho Sibiya, I could see immediately why this land had touched off such bitter battles. A paradise of pans—seasonally flooded wetlands full of hippos, crocodiles, and waterfowl—it presented stunning vistas. Ndumo is often called "the little Okavango," referring to the enormous wetlands that lie northwest. Joseph Gumede, a local guide, met us at the park headquarters and accompanied us as we drove along the edges of the pans. He softly murmured the names of the "specials," rare or coveted species on birders' lists, as we passed through groves of yellow-green fever trees. "This is the black-crested lourie. . . . This is the red-faced mousebird . . . the goliath heron . . . the African green pigeon." Across the Nyamithi Pan, a line of hippo backs stretched out like islands. Crossing a stream via a small bridge, we looked down into the mouths of crocodiles below, frozen open, waiting for fish to flow in. The water played over their throats, creating a strange, predatory fountain. If the TFCA found a way to make this exquisite landscape pay the people back for what had been taken from them, it would be a triumph.

But while there was progress on reuniting Tembe and Ndumo, the Tembe elephants remained behind bars. The fence was still up, and the clock was ticking. It was becoming painfully obvious to Wayne Matthews, Tembe's resident ecologist, that the elephants enclosed within Tembe's chain-link walls were beginning to put significant pressure on the sand forest itself, a fragile ecosystem and the last haven for its endemic plants. "The elephants do not utilise the sand forest for food but for shelter when it is hot and cold," he told a Durban newspaper.[19] He said that over the thirteen years he had been studying the sand forest, it was obvious that parts of it were being lost. "Patches of sand forest cover only 25% of Tembe," he said, "and there have been recent alarming declines of important species like suni." Something had to be done, he said, and quickly. In 2007, he launched a program of darting seventy-five of Tembe's elephants in an experimental contraception scheme, to slow the growth rate and try to keep numbers down until the corridor could be opened.

Still, plans to reunite the Tembe and the Maputo elephants limped along. KwaZulu Natal officials were afraid that if they took the fence down poachers would strike. Mozambique was afraid that elephants would wreak havoc in communities near the Futi River. The PPF proposed funding a "fenced-off corridor" along the river, thirty miles long by twelve-and-a-half miles wide, to link the two reserves, protect wildlife, and theoretically keep people and homes from harm. As in the Great Limpopo, the plan was not popular with the people living between the Futi and Maputo rivers, given the density of the growing elephant population.* A cross-border game count in 2006 revealed 320 elephants (including 42 yearlings) in the Maputo reserve. With Tembe's 200, there would be at least 500 in the new corridor.

---

* There are an estimated six hundred to eight hundred people living in and around the Maputo Elephant Reserve; an additional eighteen households, around eighty people, live along the Futi corridor.

In February 2008, cows in Tembe received birth control "boosters" and were watched closely to see how the treatment might affect the elephants' social relationships and behavior. It was, as Matthews acknowledged, "a holding pattern."[20] Administering vaccine and booster shots was expensive, requiring pricey helicopter time and stressing the animals. "The dream," he said, "is to have more space for elephants. We are all trying to work on that. . . . We very much hope we do not have to go the culling route before elephant habitat is expanded."[21]

Slowly, slowly, progress was being made. In 2009, the Maputo Elephant Reserve completed a new headquarters and upgraded its basic camping facilities. Its boundary fence was strengthened against poachers, and the elephant "restraining line" of two electrified wires was strung along both sides of the Futi. A Lubombo progress report hailed this attempt at elephant dissuasion: running six-and-a-half feet above ground, the strands were high enough so that people could safely walk under them to get to the Futi River but ideally placed to prevent elephants from slipping out into the communities to raid crops.[22]

Now it remained only for the gap in the Tembe fence to be opened. In 2009, when I spoke to Werner Myburgh at the Peace Parks Foundation, he explained that there would be a further delay while workers completed a fenced "sanctuary" for the elephants around the opening. "It's a phased approach," he said apologetically, arguing that the sanctuary would provide an additional layer of separation between the population along the Futi River and the elephants.[23] "The last thing we want," Myburgh said, "is for a child to be trampled by an elephant." So the elephants of Tembe would have to wait another year or two before rejoining their long-lost relatives.

Ernest Robbertse, for his part, was tired of waiting. In early 2009, he sent me a photograph of Thom Mahamba and the explorer Kingsley Holgate—on his "Boundless Southern Africa" tour—pacing alongside Tembe's northern boundary fence. Robbertse captioned it, "TAKE DOWN THIS WALL."[24]

* * *

Meanwhile, another popular tourist "route" was launched. Unlike the more dubious TFCA Route, the Lubombo Tourism Route presented an actual, driveable tour of Lubombo beaches and game reserves, particularly after improvements to a northern section of the road near Tembe were completed.[25] Beginning in Durban, the loop swung up near Tembe, then over to the South African and Mozambican coastline before heading across country to Swaziland and then turning south, past the other Lubombo protected complexes. It passed through some of the most important archaeological and historical sites in South Africa, offering the possibility to combine freshwater fishing at the Greater St. Lucia Wetland Park, a big game safari at Tembe, and a stop at coastal areas for swimming and whale watching.

The Lubombo Route was itself a kind of breakthrough, quickly becoming one of the most popular driving routes in the region. Other mini-TFCAs in the Lubombo project were also gathering speed, generating jobs and income. South of the Tembe project, at the Nsubane-Pongola Transfrontier Conservation Area on the border between South Africa and Swaziland, nearly 3 million rand were spent upgrading camping facilities for 120 visitors, employing 330 people. Roelie Kloppers announced a transfrontier turtle project to be launched along the entire Maputaland coastline.[26] Local people would tag and monitor several species, including loggerhead, green, and leatherback turtles, the largest reptiles in the world, weighing up to one and a half tons. The effort was expected to produce the first report on turtle populations on the coastline, providing the basis for a management plan and tourist regulations for the beaches where the females lay their eggs.

Not all of the transfrontier areas seemed unambiguously devoted to conservation. In southeastern Swaziland, next door to the Nsubane-Pongola TFCA—where modest camping facilities had just been installed—construction began in 2007 on a dense development of over six hundred residential sites, including ninety waterfront, two hundred mountain, and hundreds of additional bush camp vacation homes. Some of the homesites exceeded forty thousand square feet in size and were perched alongside and above Lake Jozini, created by a dam built in the 1960s. Funded with an astonishing 500 million euros by a private South African real estate venture called the eLan Group, the project—dubbed the "Royal Jozini Big Six"—also included a waterfront hotel, marina, conference center, and casino.[27] Two hundred private homes were to be built around an eighteen-hole golf course, along the eastern side of a new Big Five game reserve. Descriptions of the reserve seemed more akin to a private zoo: wildlife would be "stocked" in the fenced reserve, and its movement would be accommodated by "allowing for free passage of animals between [home]sites."[28] On the developer's Web site, investors were promised the opportunity of witnessing "lion capture and release programmes" or the translocation of black rhino.

The developer launched two cruisers on the lake, one seventy-eight feet, the other a hundred feet long, with air-conditioned cabins and lounges furnished with satellite television, surround-sound music systems, a wet bar, and Jacuzzi. A new international airport was slated for completion in 2010, built to accommodate the enormous new Airbus A380. According to admiring coverage in the South African press, the Royal Jozini would "form part of" the Nsubane-Pongola TFCA and was supported by the Peace Parks Foundation.[29]

By definition, rewilding is never compatible with dense commercial development. Siting development adjacent to a conservation area violates important tenets, particularly those concerning buffer zones. In addition, golf courses, which require obscene amounts of water, fertilizers, and pesticides to preserve their greens, top the list of unsustainable facilities, particularly in arid environments like that of Swaziland, which had been suffering from a thirteen-year drought. While the project promised to create two thousand jobs, it seemed more likely to enrich wealthy South African investors and to further widen the gap between rich and poor. Far from a sustainable form of ecotourism, the resort's density of development and its plan to use water from the lake to supply its hotels, residences, and swimming pools meant it was likely to be a permanent drain on resources that were already scarce. While hardly surprising that the ruler of Swaziland should have approved such plans, it was baffling that an organization like the PPF could.

Werner Myburgh allowed that the Royal Jozini was "not the typical conservation setup." He argued, disingenuously, that the development was technically next door to the formally designated protected area agreed upon by the two governments; it was not part of it. He declined to condemn casinos or golf courses. "In our opinion, it's in the best interests of the people and the area to remain apolitical," he said.

Ultimately, despite the aberrant Swaziland development, the Lubombo TFCA appears to be succeeding where the Great Limpopo, so far, has failed. Bringing education and training, creating employment, linking protected areas, overcoming the divisiveness of past conservation decisions, attracting an influx of tourists, setting up regional elephant management for the last indigenous elephants in southern Africa, the Lubombo project was working. It neither sought nor received the corrosive attention that paralyzed the Great Limpopo. Its projects were smaller and more manageable in scale, so that their completion helped inspire local support and international funding. It had managed to pace itself, balancing caution with results. Transboundary conservation was indeed a very slow and deliberate process. Now, the Peace Parks Foundation pushed forward in the Okavango Delta, trying to bring those small-scale successes to the biggest transboundary project ever imagined.

CHAPTER 8

LOOKING FOR KAZA

"It Looks Great on Paper"

THE TRANSFRONTIER CONSERVATION AREA PLANNED FOR THE Okavango Delta was the Peace Parks Foundation's most ambitious venture yet. The largest proposed protected area in the world, encompassing nearly 116,000 square miles, the project was the size of Italy. Covering parts of five countries—Angola, Botswana, Namibia, Zambia, and Zimbabwe—the boundaries encircled thirty-six protected areas, including fourteen national parks and an impressive array of game reserves, forest reserves, community wildlife areas, and the spectacular Victoria Falls, a UNESCO World Heritage Site and one of the "natural wonders of the world." The region held the largest aggregation of elephants on earth.

This project, too, was dedicated to solving a number of problems, through both conservation and development. But it brought to bear a more broad-scale method than that of the Great Limpopo: it called for realigning priorities across countries, enlisting governments, park rangers, conservationists, and communities. With water so precious in southern Africa that it may eventually be worth more than diamonds, the plan would put under conservation protection the chief water sources of several states involved, including arid Botswana and

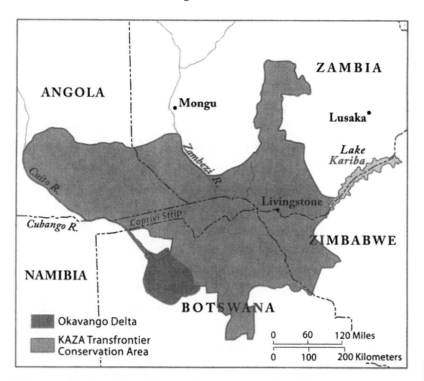

Kavango-Zambezi Transfrontier Conservation Area: The largest proposed protected area in the world, the Kavango-Zambezi Transfrontier Conservation Area calls on five countries—Angola, Botswana, Namibia, Zambia, and Zimbabwe—to collaborate on wildlife and wetland management.

Namibia. Because of the critical importance of this resource, and barely contained national urges to dam it, the plan called for cooperation and collaboration on every level of society. For the first time, conservation sought to reorganize human society around transborder wildlife management, in order to accommodate the migratory needs of elephant, the keystone species that relies on the delta as a haven in the dry season. It also sought to reconcile a dubious public to the top predator in the wetland, the crocodile. If it succeeded, it promised to restore and revivify ecosystem services—fisheries, forests, water—crucial to every living thing.

This, too, was a hastily conceived project, envisioned in the early days of the PPF and launched only a few years later. In 2005, a glossy coffee-table book published by Conservation International acknowledged that "the precise boundaries of the TFCA and the design and shape of the corridors joining protected areas together are yet to be defined."[1] A year later, in 2006, boundaries had been drawn by somebody, somewhere: The five countries involved signed a memorandum of understanding.[2] The U.S. Agency for International Development and the Swiss Agency for Development and Cooperation signed on to support the first planning phase of development, which promised to be more inclusive of local communities. Elaborate and detailed "integrated development plans," featuring input by all affected communities, were to be produced by each of the countries before work was undertaken. The plan was so new, so fresh from the drawing board, that even its name was in flux. By 2007, the name had jelled: "KAZA," the Kavango-Zambezi Transfrontier Conservation Area, another major Peace Parks Foundation initiative.[3]

The KAZA area was populated with an estimated 2.5 million people, spread out across an enormous territory. Its centerpiece was the Okavango Delta, the vast fragile wetland that annually attracts waves of wildlife from the desert while drawing more human use each year as farmers clamor for water for crops. The challenges it faced were legion. People in the area were reliant on the delta's water and fish. In addition to the breakdown of Zimbabwe, the project had to grapple with intensive poverty in all five countries, minefields left by civil war in Angola, and the eagerness of Namibia to dam the Okavango River. KAZA had also to come to grips with a teeming overpopulation of around 150,000 elephant—the largest and densest remaining concentration in all Africa—bottled up in a narrow strip of Namibia by Angola's minefields and Botswana's veterinary fences. Releasing those confined populations of elephant and other wildlife to range over a continental-scale wilderness could take pressure off the entire delta ecosystem. Making good on the initial promise of the Peace Parks Foundation—re-creating an Africa without fences—was critical to this plan. More than any other, this region was crisscrossed with barriers, protecting 47 million cattle in southern Africa, half of which were considered at risk of contracting diseases from wild animals. It was not at all clear, at the outset, what the countries involved were willing to do about that.

Botswana was the economic power anchoring the project, one of the few stable democracies in sub-Saharan Africa. With a population of 1.8 million, this former British protectorate—now the world's largest exporter of gemstone diamonds and a major beef supplier to the EU—achieved independence in 1966. Soon thereafter, its leadership negotiated an advantageous fifty–fifty ownership share of its mines with De Beers, the South African diamond mining consortium (a previous deal gave De Beers an 85 percent share and Botswana only 15 percent). Since then, the country has consistently reinvested diamond profits in education, health care, social services, and infrastructure. Poverty in Botswana has declined, and access to potable water, electricity, and education has improved. Although AIDS infection was widespread, a high percentage of rural citizens had access to health care. In 2008, gross domestic product was $13,300; in Kenya, it was $1,600. One of the least corrupt nations in Africa, with one of the world's highest rates of per capita growth, the country had an A credit rating from Standard and Poor's, the best on the continent. In recent years, as diamond prices fluctuated, the country diversified its economy. It negotiated preferential contracts with the EU for its beef, maintaining its historic veterinary fences—including one along the border with Namibia and others running from east to west through the northern delta—to separate cattle from wild buffalo and other species, potential carriers of foot-and-mouth and other diseases. But the country's most promising economic sector by far has been tourism: Botswana focused on high-end, low-volume ecotourism with stunning results. With a million visitors and over $300 million in revenue a year, tourism accounted for 12 percent of GDP.* A full 95 percent of tourists headed for the Okavango Delta.

---

* By comparison, South Africa draws nearly ten million tourists a year.

Seeping into the center of the Kalahari Desert, which occupies two-thirds of Botswana, the delta is a freak of nature, a river that never reaches the sea. The largest inland delta in the world, it forms a vast wetland oasis filled by the Okavango River, rising in the highlands of Angola and emptying out into the sands of the Kalahari in a spectacular, ever-changing fan of reed-covered islands, shallow channels, and lagoons. It swells during the winter, flooding to two or three times its size, offering world-renowned fishing and birding. The most critical wetland refuge for wildlife in southern Africa, the delta supports rare species, including two endangered antelope, sitatunga and red lechwe, and genetically unique subspecies of black rhino and elephant. It is the center for conservation research on Africa's most fascinating pack hunter, the wild dog. It also contains one of Africa's major concentrations of the Nile crocodile.

The Peace Parks Foundation and the governments involved deliberately kept a lower profile with this initiative to avoid the attention, speculation, and criticism that greeted the Great Limpopo. They may have gone too far in keeping it under wraps: everywhere I went in the delta, I found that few communities, tourist guides, or scientists had even heard of KAZA. Alison Leslie, a South African crocodile specialist at the University of Stellenbosch, had been working in the Okavango Delta for years and knew people in government agencies there. She was startled to learn about KAZA from a South African newspaper in 2006, and her concerns reflected those of many scientific and tourism specialists. "It was kept very quiet," she said.[4] "I remember reading one of the first articles and thinking, how on earth are they going to do this? There are thirty-six parks involved, never mind all the game management areas. It looks great on paper, but a lot of those parks aren't managed properly now, so how are we going to manage one this big?" She wavered between hope and skepticism. "They say it's going to start happening in 2010," she said. "Well, *what's* going to start happening? Nobody knows. Suddenly it's just going to be there." But she acknowledged that "it would be fantastic if it works."

* * *

The countries around the delta have been struggling to improve their management of this resource for some time, and the Ramsar Convention on Wetlands recently called for an intensive push to gather baseline data on its flora and fauna: it was not even known how many endangered species existed there. What was known was that the delta was shrinking as increasing amounts of water were drawn off to support irrigation, mining, and urban development, critically endangering the wetland. Neighboring countries would have to develop cross-border management to address extraction and other threats before the delta was lost forever.

No creature illustrates the need for transborder management more than the crocodile, which together with the hippopotamus plays a complex role in regulating a healthy, functional ecosystem. In the absence of human predation, *Crocodylus niloticus* is a long-lived reptile, with a long-lived effect on its environment. Its lifespan, sixty to seventy years, approaches that of the elephant. Adult females fashion incubatory nests of sand and soil on the riverbank in which they bury their eggs during the wet season; the temperature in the nest, influenced by atmospheric conditions, determines the sex of the infants that emerge. During the incubation period, around eighty days, nests are vulnerable to flooding, fire, predation, and attack by humans or other animals.

The crocodile is the apex predator in African rivers, affecting every trophic, or nutritional, level. Baby crocs feed many species, from herons, storks, and eagles to big fish and bigger crocs. Crocodiles also remove waste and recycle nutrients. With their massive bodies and constant activity, hippos and crocodiles together are credited with maintaining the system of open-water canals and channels in the Okavango Delta, which might otherwise become clotted and overgrown with reeds. Ecologically, the delta may not be able to function without them.

The crocodile is the delta's canary in the mine, considered an "indicator" species, the first to suffer from environmental poisons, contamination, pollution, or other stressors. As in all other reptiles and amphibians, the nervous system of the crocodile is exquisitely sensitive to endocrine disruptors, compounds found in plastics and pesticides that mimic hormones and can affect reproduction. Many people working in the delta began noticing a decline in big crocodiles over the past few decades; the population may have lost up to 60 percent of breeding females. That loss is particularly worrying since the crocodile acts as a control on the delta's chief nuisance, catfish. One of the most important segments of tourism in the delta has been sports fishermen, lured by the chance to land a coveted freshwater trophy, the tigerfish, or *Hydrocynus vittatus*, which means "striped waterdog" in Latin, a fish with dark stripes across a white body, bright orange tail and fins, and impressive carnivorous teeth.[5] But the delta's tigerfish have fallen victim to an exploding population of catfish that has been gobbling the fry of tigerfish, bream, and tilapia.

Catfish are a problem that crocodiles can solve. Unfortunately, people in the region are continuing to enthusiastically and indiscriminately kill crocodiles, the predator of their worst pest. Fishermen are still using gill nets, snagging crocodiles by accident, killing the animals and destroying the nets; heavy commercial gill netting further diminishes valuable fish populations. As the delta shrinks, as the crocodile declines, the entire ecosystem—from its main predator on down the food chain—is becoming imbalanced, reinforcing the breakdown in fish stocks and, potentially, water quality.

This has an important bearing on the region's economy. I traveled around the delta with the Okavango Crocodile Research Group, a team of Alison Leslie's graduate students, who were collecting data on every aspect of the crocodile population—distribution, abundance, home ranges, dispersal patterns, diet, gene flow, health status—while also investigating human-crocodile conflict and ways to alleviate it. These data would provide some of the biological groundwork for KAZA, much as Leandro Silveira's work in Brazil or the Prombergers' in Romania provided the underpinning to projects there. Information gathered by the research group would ultimately form part of the Okavango Delta Management Plan being developed by Botswana's Department of Environmental Affairs, contributing to an overall strategy to manage Nile crocodile biology and the health of the delta population. In turn, that plan would be a central feature of revised fishing and tourism regulations that would take effect during the new KAZA era. The research was thus part of the regional effort to manage the entire delta wetland system, on which millions are dependent for food and agriculture.

Alison Leslie started from zero. Virtually nothing was known about the delta crocodile population. The lack of studies is just one of many gaps in our ecological knowledge. A major scientific publication in 2005, *Large Carnivores and the Conservation of Biodiversity*, has a lot to say about grizzly bear, wolverine, tiger, lion, and wolf but mentions the crocodile only in passing, to confirm that the "depletion" of crocodilians has "gone mostly unreported."[6] The crocodile population in the delta was not considered endangered, but so little information was available that it was impossible to know scientifically how crocodiles were faring.* Leslie and her team set out to remedy that state of affairs.

Night Shift to Namibia

How do you catch a crocodile? With your bare hands.

Two students were in charge of the croc project: Kevin Wallace, a gangly graduate student from the British town of Ipswich, and Audrey Detoeuf-Boulade, his professional and romantic partner, a beautiful blond Frenchwoman who grew up in New York City. They were a biological odd couple. Kevin had a degree in zoology from the University of Liverpool and had worked for a year on a meerkat project in the Kalahari, spending nine months in the wilderness habituating meerkats to people for *Meerkat Manor*, a television series, and filming dramatic sequences, including the blinding of a young meerkat by a

---

* A fire destroyed Botswana's records on crocodile numbers as they existed prior to an intensive hunting period, between 1957 and 1969, when fifty thousand animals were shot.

spitting cobra. He was completing a diet study for his master's, while Audrey was collecting sperm samples from adult male crocs for hers, trying to determine when males become mature, an important fact in understanding the viability of breeding populations.[7]

"We don't drug any of our crocs," Kevin said. A few miles south of Shakawe, one of the northernmost towns in the narrow panhandle of the Okavango Delta, a handful of American and British Earthwatch volunteers were standing in the sand a few yards from the water, staring with concern at Kevin's crocodile noose, a flimsy-looking contraption of wire loosely lashed to a pole with duct tape.

The noose was necessary for big crocs, anything up to two and a half meters, or about eight feet long. Slipped around their neck, Kevin explained, the noose held the animal next to the boat while it spun around in a death roll, tiring itself. Like all other reptiles, crocodiles have a limited energy budget. The spent croc—"tired enough but not too tired," Kevin said—could then be lifted aboard the project's sixteen-foot-long aluminum skiff with a couple of straps of thick canvas webbing passed around its body. Crocodiles longer than eight feet could not be accommodated; the boat wasn't big enough.

Smaller crocs could be plucked from the water with a hand strategically placed behind the triangular jaw. "We call them 'hand jobs,'" said Kevin. "That's a scientific term." He got a few worried laughs. Steve Irwin—the manically cheerful Australian crocodile wrangler—had recently been killed by a stingray. Crocodiles played no role in his death, but as the volunteers contemplated an essential service they would perform—it could take the combined heft of everyone on board to lift, subdue, and release a big croc—the specter of freak accidents was in the air.

Completing the last month of the Okavango crocodile project—the research would then shift to the Zambezi River in Zambia—we would be heading out on night shifts on a "recapture" mission, to see how many of the crocs we caught bore the markings of previously captured animals. So-called mark-recapture (or capture-recapture) projects are a staple of biological research: individual animals in a defined study area are caught, marked, and released after basic information has been collected. Although age in crocodiles is difficult to assess, various body measurements place the individual in a "size class," which allows an estimate of age. Eventually, once the study area has been surveyed repeatedly and a substantial proportion of the population has been marked, the recapture phase begins. When marked animals have been recaptured one or more times, that data can be used in constructing statistical distribution patterns that show the average number of yearlings, juveniles, subadults, and adults that populate the area. Ultimately the data allow biologists to estimate overall population size and growth rates, as well as patterns of distribution and dispersal. Those patterns tell scientists how healthy and viable the population is.

Between 2000 and 2004, Leslie's team had captured, processed, tagged, and released 1,159 crocodiles of various sizes, including 993 different individuals; 112 were recaptured once, 21 twice, 1 four times. One particularly unlucky individual was snatched five times. Already, the data suggested that the crocodile population was in trouble: there was a relatively low number of hatchlings, hardly a surprise given the common local practice of stealing eggs and destroying nests. There was a higher number of yearlings, suggesting that individuals were making their way into the system from upstream, in Namibia. And there were lower numbers of juveniles, subadults, and adults: few were surviving to propagate the species. This first study of survivorship and growth rates of wild Nile crocodiles had already compiled groundbreaking information on movement, reproduction, and the genetic structure of the population, which may have been weakened by inbreeding due to the number of individuals removed during Botswana's intensive hunting period. As we set out on our mission, we were collecting information that underscored the crocodile's vulnerability to changes in the ecosystem and the need for transboundary management and cooperation between Botswana and Namibia. Of course, if we found large sperm-emitting males for Audrey, all the better.

Crocodiles bask during the day and hunt at night, poised motion-lessly near riverbanks. In the dark, they can be found and hypnotized with a powerful spotlight. Heading upriver before reaching the GPS point where the night's work began, we were in an alien world. The panhandle, a permanent swamp, is the chief breeding ground of the delta's crocodile population. We had seen the channel during the day, framed in towering swaths of papyrus reeds, waving in the breeze, forming a backdrop to herons and egrets plunging for catfish. But in the dark, the water was black, and papyrus loomed overhead, casting weird shadows. There were seven of us in the boat, several veteran Earthwatch volunteers, myself, and Kevin and Audrey, muffled in all the clothing we possessed as Audrey steered out into the main chan-nel. The delta is in a desert, blazing hot during the day, cold at night.

The plan that night, unless we caught a lot of crocodiles, would be to head north to the Namibian border, to finish the recapture survey for this section of the river. The volunteer manning the spotlight panned the light in front, scanning the banks ahead of the boat. The trick was to hold the light in such a way that it illuminated the bank or reeds where they met the water. That's where crocodiles hang out while hunting. We were looking for a single red gleam—the reflective layer behind the crocodile's retina—shining back at us.

The peril on these trips, surprisingly, was not crocodiles. Croco-diles are predictable, and Kevin and Audrey were confident in han-dling them. The most dangerous animal on the delta at night is that ill-tempered creature the hippo. Although hippos are not predators—they are nocturnal grazers—they reportedly kill more people than any other species in Africa, including lions, leopards, or elephants. Like an angry fat man, the hippopotamus swaggers around the con-tinent, wreaking havoc. Its vast bulk hidden underwater, plagued by a powerful irritability born of the territorial imperative, it tramples the unwary and is capable of snapping in two a person or a boat—say, a sixteen-foot aluminum skiff. Even crocodiles give them a wide berth. Audrey was the most audacious woman I had ever met, but she was afraid of hippos.

We sped upriver for nearly an hour, passing the town of Shakawe, a few lights visible on the shore. It was not long before we sighted our first croc; Kevin signaled Audrey, who sped up to the young animal—crocs are seemingly paralyzed by the light and the young ones, particularly, remain fixed in place—and Kevin snatched it out of the water. It was assigned an identification number, and Audrey snipped off a piece of black electrical tape to secure his snout, probed his vent for urine, and soon he was flat on his back on a blanket on the deck, being measured and weighed by us with tape measure, calipers, and scales: total length, from snout tip to tail tip; snout to vent length; neck circumference; base of tail circumference; head length, breadth, and depth. He was a miniature dragon, his tail studded with "scutes," horny triangular appendages, but also a surprisingly vulnerable little creature, with soft creamy white scales across his stomach, a contrast to the hard bony plates, or osteoderms, on his back. Visible in each scale was a sensory pit, the exact function of which has yet to be determined, although it is believed that the pits around the jaw detect changes in water pressure caused by movement. A snap of the jaw, an instant meal.

Putting a croc on its back has a narcotizing effect caused by compressing the animal's lungs. But when we lifted him to weigh him, he began squeaking loudly in protest—a nasal sound like air squeezed out of a half-inflated beach ball—lifting his legs spasmodically like a wooden jumping jack, his legs pumping up and down as if we had pulled a string. The most ticklish process was when Kevin or Audrey took blood; the boat had to be kept completely still, everyone frozen in place. Then the animal was marked with a readily identifiable code: with a razor blade, tiny cuts were made in the tail scutes in a sequence corresponding to a code devised from the identification number. If he was recaptured, the marking would reveal his number along with all the other previously collected data. After a flurry of picture taking, the mouth tape was peeled off, and the creature slid into the black water. He hung there for a moment, as if in shock, then shot off into the depths.

This quickly became routine. All the crocodiles we caught that night were hand jobs. Whenever we spotted something larger, it gave us the slip. The big crocs hadn't gotten big for no reason. They were masters of the evasive move. For crocs, Kevin said, this was the equivalent of alien abduction. "They go off and tell their mates and no one believes them. Till it happens to them."

Hippos broke the tedium by providing moments of stark terror. We were pulling into the reeds, preparing to process a croc, when suddenly Audrey heard an angry bellow nearby and we sped off to another spot, the croc held inside Kevin's jacket for warmth. Halfway through blood drawing, another bellow, another dash to safety. Crocs must be returned to the spot they came from—they are cannibalistic and will prey on others in their territories—which caused a lot of backtracking. During one high-speed maneuver into the reeds, with Kevin clutching an infant croc, I saw that we were bearing down on a jewel, glowing in the light from the boat: a malachite kingfisher, one of the tiniest, most beautiful birds in the world, iridescent turquoise on his back, orange cheeks, and a red bill. He was all tucked up, roosting for the night. In the chaos, I was afraid we would crush him, but he shot instantly into the reeds.

The tea break was perhaps the most surreal refreshment I have ever taken, a spot of domesticity on an aluminum deck adrift among the papyrus somewhere in a vast African marsh. Big flying bugs with grotesque bodies that curled spasmodically—Audrey called them "sausage bugs"—flew into the light. Audrey produced special cookies for these occasions—lemon creams, shortbread. Crouched on the deck, we drank coffee and tea as the sausage bugs writhed companionably around us. Audrey leaned back next to Kevin, smoking a cigarette, looking at the stars. Kevin whiled away the time inventing new constellations, finding a Mickey Mouse riding a bicycle near Cassiopeia. Others took the occasion for a bathroom break off the back of the boat. Someone recalled a horrifying true story: an American man had recently been snatched from a canoe by a crocodile and was never seen again. Crocodiles are capable of an astonishing range of motion in addition to their métier, swimming. On land, there's the belly crawl, the high walk, the gallop. In water, they can lunge, propelled by their tails, exploding upward with great force to snatch animals off a bank. Or out of a boat.

I looked the incident up later. It happened in eastern Botswana, on the Limpopo River. Richard K. Root, sixty-eight, a professor emeritus of medicine at the University of Washington and an expert on infectious diseases, had arrived in the spring of 2006, to train doctors and nurses in treating HIV/AIDS–related illnesses.[8] On a wildlife tour, he was traveling in a canoe with a guide, followed in another canoe by his wife of eighteen months, who witnessed the attack.

Crocodiles are not what you'd call friendly. Pliny called them "a curse on four legs . . . equally pernicious on land and in the water." They are ambush hunters, with one of the most powerful bites of any animal, exerting five thousand pounds of force per square inch, compared to four hundred for a great white shark. Alistair Graham's *Eyelids of Morning: The Mingled Destinies of Crocodiles and Men*, illustrated by the Africaphile Peter Beard, features an unforgettable photograph of the remains of a Peace Corps volunteer—twenty-five-year-old William Olsen—who, along with a group of friends and fellow volunteers, refused to heed warnings not to swim in the croc-infested Baro River in Ethiopia on a hot day in April 1966. He was last seen standing on a rock midstream. The next day, a local hunter shot a huge croc basking nearby, with a strip of pale flesh hanging from its mouth. Inside the croc, the hunter found "his legs, intact from the knees down, still joined together at the pelvis. We found his head, crushed into small chunks, a barely recognizable mass of hair and flesh; and we found other chunks of unidentifiable tissue."[9] Originally printed in *Time* magazine, the photo showed a bloodstained cardboard box, two human legs flopping out at crazy angles.

That night we caught and processed five small crocs—one a recapture—and went all the way to Namibia, even into it, accidentally, until Audrey recognized the border marker. After a long, cold trip back to camp, it was near dawn. We still needed to label vials, spin blood in a manual centrifuge, separate plasma from red blood cells, and perform a hematocrit blood measurement test to determine the ratio of red blood cells in the plasma, or whole blood, a measurement of croc health. A normal plasma reading for crocodiles is 32 percent red blood cells; lower readings suggest anemia. Over time, blood samples would provide a picture of the overall health of the population, its susceptibility to disease and parasites.

While we huddled onshore, spinning their blood in the predawn, the crocodiles remained hanging, suspended in the warm dark water of the delta. Sensing movement, perhaps they lashed out for a moment, catching something in their jaws. Fish were swallowed, bigger prey carried to the bottom, until the thrashing stopped. Waiting a few hours, a few days, until the dead thing was softer, deliquescent, they descended again to eat. Human beings are primitive in garbage disposal, throwing things in pits, scratching dirt over them. Crocodiles are organic perfection, polished by two hundred million years of evolution. Like dung beetles and vultures, they are a critical cog in nature's recycling system, turning rotten hippos or excessive catfish back into nutrients, returning their elements to the food chain. They can be a curse, but also a blessing, millions of years in the making. The species has survived great extinction events of the past, but it remains to be seen if they can survive the likes of us.

Early one morning, Kevin took us to a small woodland reserve near the camp, a rich display of Africa's strangely threatening yet endearing flora: strangler figs; camelthorn, with big, furry ear-shaped pods; knobthorn, bristling with massive hooked spines; sausage trees, covered in dangling rock-hard sausage-shaped fruits that have killed unwary campers. Because it does not crack, the wood of the sausage tree is used for making local canoes, called *makoros*. We often saw men in *makoros* engaged in illegal night fishing, a hazard to the fishery and to the men in the tippy wooden crafts, a hippo or crocodile snap away from a watery death.

The forest was noisy, humming with bees, flies, mosquitoes. "Ah," Kevin said, "the soothing sound of stinging insects." The birding was astonishing, every minute another species. Along the water, egrets, squacco heron, black-crowned night heron, swamp boubou, African jacana, African pygmy geese, black-shouldered kite. Flying overhead, collared sunbird, carmine bee-eater, bleating warbler, and the heart-stopping blue waxbill, with a breast the color of a powder blue lounge suit. In Africa, even starlings are gorgeous: the greater blue-eared starling, with a yellow eye and glossy iridescent blue green plumage, stared us down impudently.

Spectacular though it was, it was nothing to what we saw the following dusk. Leaving late in the afternoon, heading far south downriver, Audrey turned into a lagoon and edged the boat up against a clay cliff, home to a wall of carmine bee-eaters, gorgeous rose-colored, long-tailed birds with a turquoise crown. Hundreds of them were swooping around the cliff. In flight, their bodies glowed rosy orange in the setting sun. The only sounds were the lapping of water against the cliff and the cries of the bee-eaters, perching in trees, taking to the air by the dozens, diving in their holes. Nearby, a fish eagle looked down impassively from a limb over the water.

Later that night, I caught my first hand job, Capture No. 1897, a young male of 77.6 centimeters, about two and a half feet long. I poised myself over the floating creature, which at that moment was looking rather large, and grabbed it behind the jaw, clasping its lashing tail to my stomach. He sported two valuable leeches, which were removed for Audrey's collection. She kept them alive—"in a tube with water, and every day I open and give them a little air"—for a colleague who was working on fish parasites: leeches may be a vector for parasites in the blood.

Later still, at 23:20, we caught the big one. Kevin had signaled Audrey forward, reaching for the noose. As he slipped it over the neck, all hell broke loose. The crocodile whipped around in a death roll, thrashing as Kevin struggled to hold him and Audrey yelled, trying to figure out what was going on. Abruptly, she jumped from the controls to the side of the boat and spun a towel around the croc's mouth, tying it shut and covering the creature's eyes. She ordered

James, an able-bodied youngster, to lift the heavy tail and get a strap around the croc's body when, suddenly, a crisis. "Bollocks!" she cried. Then she tipped halfway out of the boat herself, and someone grabbed the back of her jacket to keep her from falling in. Muffled but triumphant, she yelled, "I've got it!"

Somehow, with its mouth taped shut, the croc had managed to break out of the noose. Not about to let the animal sink with its mouth taped shut—a death sentence—Audrey leapt and caught it before it slipped under. By now, it had expended much of its energy, and we were able to lift it in, whereupon we all sat on it. Before it gave up and played dead, it emitted a bloodcurdling sound, a deep, prolonged, menacing growl, not unrelated to the anguished squeaking of the youngsters but octaves lower, decibels louder, and many times angrier. The hair stood up on my arms.

Capture No. 1901 was male, and his total length was 247 centimeters, around eight feet long. His neck circumference was 63.2 centimeters, or two feet.

Over six night shifts, we covered sixty miles, sighted seventy-one crocs, and captured twenty-seven, including one recapture. Without any particular love of reptiles, I found myself astounded by their power, their cunning, and their evolutionary polish. Asked why she had chosen to study perhaps the most intimidating of animals, Alison Leslie said, "What draws me to them is that they are, nine times out of ten, the underdogs."

Despite the small triumphs of the research project, signs that the crocodile population was in trouble were indisputable. Night after night, adult crocs were few and far between. We saw more adults in captivity than in the wild. Leslie was worried about their future. "They're not *critically* threatened," she said. "But they're definitely threatened. The changes we've noticed in the last four, five years are pretty scary. The amount of water extracted from the delta, less rainfall, the numbers of humans, the human population burning vast tracts of land, converting reeds to grassland. I've seen projections that the delta will possibly be dry in twenty-five years' time. Hopefully, this KAZA park might help things out a bit."

The Demon Croc

Kevin and Audrey often joked about her proclivities, her avidity for reptilian sperm, but, in fact, they were both devoted to crocodiles. Audrey sadly acknowledged that the animals were not as charismatic as "furry and cute" creatures. The couple had been frustrated in every effort to persuade locals of the importance of crocodiles in the ecosystem. "They think the devil put crocs here," Kevin said gloomily.

Indeed, the cultural aversion to crocodiles in the region, fueled by a fervent belief in witchcraft, had taken on aspects of religious fundamentalism, reinforcing the natural fear and loathing of a species that may have killed a family member or neighbor. Crocodiles are seen as supernaturally powerful and fundamentally evil. People who live around the delta zealously destroy crocodiles at every opportunity, stealing eggs, burning nests, hunting problem animals that go after livestock. The decline in the fishery has hit people hard, affecting the local diet and the tourism industry, and crocodiles, however wrongly, were bearing the brunt of people's wrath.

Throughout the years of the project, the team visited some twenty schools in the delta region to educate students about the importance of crocodiles. "At the end," Kevin said, "the teacher would say, 'We should still shoot all the crocs' and 'Could you ask the government to cull more crocodiles?'"

They talked to communities about ways to protect against crocodile attacks, suggesting fenced-in areas for swimming or watering cattle. There was little interest. "They said, 'Our kids, we don't let them go swimming.' But of course they do."

They hosted a meeting with a chief and his elders in one village, explaining the crocodiles' role in controlling catfish. After serving a hearty dinner, they took the chief, dressed in his finest bowler hat, white shirt, and pressed corduroy trousers, for a spin in the project's boat, which he thoroughly enjoyed. "He said, 'Faster! Faster!'" Kevin recalled. They thought they were getting somewhere, but by the end of the evening there was no breakthrough in convincing the elders that local reptiles were an indispensable part of the ecosystem. The prejudice was simply too ingrained. "They didn't understand it at all," Kevin said. The chief told them, witheringly, "Thank you very much, but you're wasting your time."

The best prospect for forging a mutually beneficial relationship between people and crocodiles in the delta was a local business that the croc project helped establish, the Krokovango Crocodile Farm. Croc farming can be lucrative, but as a business it presents some unusual start-up challenges. Female Nile crocodiles breed readily in captivity, but wild females used to start a breeding colony must have reached 2.7 meters in size to produce viable eggs. Crocodiles that large are around thirty years old and can be taken legally only with a permit. The Okavango Crocodile Research Group had helped the farm acquire its breeding stock by contributing "problem crocs," animals that had threatened livestock or people, satisfying the security needs of the local populace.

Krokovango's pools and hatcheries now held around five thousand crocodiles, bred for meat and the skin trade in Asia. Leslie and her students were helping to launch a business that could create permanent employment, benefit the community, and take pressure off the wild population. The Crocodile Research Group also planned to use the site for education and tourism. Pending permission from Botswana's Department of Wildlife, it hoped eventually to use some of the farm's stock to release 5 percent of the young hatchlings into the wild, to help restore the population.

On the day we stopped by, we saw a local man, Amose, often employed by the team for his skill in finding crocodile nests, feeding dozens of midsize crocs with juicy red chunks of fresh elephant, from a licensed hunt. Before the feed, Amose, in green overalls, walked around the pool, tapping sluggish crocs on the noggin with a pole, conditioning the animals to be intimidated by humans. "He's trying to make them think that anything in a green overall is not nice," Kevin said. When workers tossed the meat, the big dominant crocs piled in, grabbing chunks and tossing them down their gullets. The smaller reptiles hung back, darting in only when the others had their fill.

Krokovango's star attraction was Sam, a crocodile so big he looked fake. Over sixteen feet long, between fifty and sixty years old, Sam had terrorized this part of the delta for years, killing cows and lunging at people. Too big and lethal to live with the other crocs, he was suffering solitary confinement in a separate pool too small for him. We watched as Amose calmly walked into Sam's enclosure, standing perilously close as he held a chunk of elephant on the end of a pole, rubbing it gently against the croc's partially open jaw. No response. Amose maneuvered it into Sam's mouth over and over again, only to have Sam drop the meat. Amose patiently picked it up, dipped it in the water to wash off the sand, and tried again, crooning softly, "Eat, Sam, eat, eat." He finally coaxed the animal to swallow a piece, but for an animal that size, it was a drop in the bucket. One of the luckless men who lost cattle to Sam told a Botswana newspaper, "This crocodile is big enough to eat a whole person and still not be satisfied."[10] Sam, who once commanded a mighty stretch of the Okavango with his pick of the females, had lost his will to live.

Annelise Langman, manager of the Krokovango, sitting in her office nearby, talked about the difficulty of persuading locals to switch from watering livestock at the river—which inevitably triggered attacks—to using stock tanks or boreholes in the bush. Otherwise, she said, "the cattle all go to the same place, and the crocs just choose a hiding place and sit and wait." The presence of the farm had not yet inspired the necessary changes in livestock handling. "It's become so easy for crocodiles to go for easy prey," she said. "Sam had been living here for I don't know how long. It got to where he was taking five cows a week. The villagers started trying to kill him, shooting at any croc they saw."

Meanwhile, however, Leslie's group began to reap the rewards of the Krokovango Farm's success. Having waited four years—much of that time spent pouring resources into communities and schools to win their acceptance of the release of crocodiles into the Okavango—the team was overcoming local resistance. The message, that crocodiles could help restore indigenous fish populations, was getting through. In 2008, with the permission of the Botswana Department of Veterinary Services, the group released eighty-eight farmed crocodiles of 1.2 meters (around three years old) into the delta. Each had been weighed, marked, and its DNA sampled; the progress of these individuals would be monitored by members of the team who remained in Botswana, even while the project moved on to Zambia. The local newspaper, the *Ngami Times*, heralded this "remarkable and commendable example of how agriculture, wildlife and academia could work together." Its headline read, "Crocodiles Finally Released!"[11]

"The Elephants Are Going Home"

While Leslie and her team were moving their gear to Zambia and while plans for KAZA continued to mystify the scientific community and local people of five countries, the elephants made their move.

In 2008, a female with matriarchal aspirations named Letsatsi, which means "sun" in Setswana, gathered her herd of seventeen, including four youngsters, and led them on the "long walk to freedom," as it was termed in the local press, echoing the title of Nelson Mandela's autobiography.[12] Starting around the first rains of October, the elephants left Botswana and the northern stretches of the Okavango Delta, heading north. They filed across the "Caprivi Strip," the narrow slice of Namibia between Botswana and Angola. Much of the Caprivi is taken up by parks and game reserves, their beautiful riverine and forested habitat laid waste in recent years by thousands of elephants congregating and snacking on the local foliage, frustrated in their desire to follow seasonal migratory paths to the north by land mines, human settlements, and veterinary fences.*

But this time, Letsatsi kept going. She and her herd crossed the border into Angola, into the Luiana Partial Reserve—part of KAZA—in the Cuando Cubango province. The elephants walked over 110 miles, and the biologist Mike Chase, who was tracking them by satellite, watched them move safely through the minefields, as astounded by their journey as Paul Paquet once was, watching a wolf.

Chase is a fifth-generation Botswanan and founder of Elephants Without Borders (EWB), a conservation group dedicated to tracking elephants and identifying possible conservation corridors. Studying the region's elephants since 2000, Chase and his EWB team embarked in 2007 on the first effort to record and map elephant movement across the international borders of southern Africa. Matriarchs make decisions about where to lead the herd, and Chase had pegged Letsatsi, a large, spunky female, as a potential matriarch, darting and fitting her with a satellite collar.

Chase had also begun to suspect that elephants had learned or were sensing—perhaps through an ability to smell explosives, proven in rats and dogs—where it was safe to walk. He had seen them carefully navigate areas where elephants had previously been maimed and killed. Now they were surviving. He said that he had never seen such "dramatic movements" by a family group in seven years of constant monitoring. "The elephants are going home," he said, "telling us which areas to conserve, which areas to look after to safeguard their future."[13]

Chase's work was providing astonishing insights into elephant migratory movements, behavior, and home ranges, critical information for the success of KAZA. He had recorded journeys that exceeded

---

* Fifteen major veterinary fences have been constructed throughout the KAZA area; many of them affect wildlife access to and movement within northern Botswana and Namibia's Caprivi Strip. While adult elephants can often knock down or step over fences, infants cannot, and mothers will not leave their young. Thus breeding herds are especially stressed by the long fences; such herds have been observed repeatedly pacing their length, searching for routes around them.

anything seen before, including males moving 285 miles over seven days across four countries. "Elephants are highly intelligent," he said. "They know and identify certain barriers and won't cross others where they know they'll be persecuted."[14] Using the data EWB collected, Chase focused on documenting the parameters of four major wildlife corridors that would allow elephants to resume migration north into Zambia and Angola. In the future, he hoped to develop a program to mitigate bottlenecks created by fences and to persuade villagers to shift fields or settlements in exchange for developing community eco-tourism projects, photographic safaris, and cultural tourism. Having seen Chase's maps showing the movement of elephants across the landscape, Werner Myburgh, then PPF's project manager for KAZA, said he could foresee a time when the project could advertise the restored migration of thousands of elephants as a tourist attraction: "Come to see Africa's greatest elephant herd."[15]

Mike Chase was markedly less optimistic. About to take off for the Caprivi Strip to resume monitoring the collared elephants when I spoke to him, he was frustrated with KAZA's progress thus far. Recruited years earlier to study migratory behavior in preparation for KAZA, Chase was both an advocate and a skeptic, waiting to see if the project would pay off. "I don't see any difference to when I started," he said. "It's a signed document, but that's it. Fences are still up, there are segregated wildlife policies, there's no regional attempt to manage elephants or to work together to undertake contiguous aerial surveys to determine spaces for elephants, there's still considerable red tape and bureaucracy." He was concerned that governments were not sufficiently committed to KAZA. "I suspect nothing will happen for the next five, ten years," he said. "Maybe that's too harsh, but in all honesty, there's no practical management. And there are very few projects benefiting people on the ground, especially in southeast Angola. Angola's still on the back burner."

When I asked whether Botswana's veterinary or border fences would be taken down to facilitate the elephant corridors, a central vision of KAZA, he said, "I don't think so." He had made extensive recommendations for the realignment or removal of the Caprivi border fence, as part of a study showing the fence's effect on elephant movement. "I suspect that fence will stay up, and it will take considerable lobbying and pressure to realign or decommission it. I don't think the TFCA is going to alter the Botswana government's stance, especially considering its contracts with the European Union for the export of meat. You never know. But there have been no incidents of the Botswana government decommissioning or realigning any government fence they have constructed. So I would be very surprised if they did do that." He added, "Once these fences are up, they tend to remain up." The most he was hoping for was the chance to create strategic gaps in some of these barriers.

Myburgh, however, remained enthusiastic about the potential for transboundary wildlife management, pointing to an unprecedented meeting in October 2008 of veterinary specialists from all five countries, who were beginning talks on how to deal with cross-border threats of foot-and-mouth disease and bovine pleuropneumonia; in future, vaccines might obviate the need for fences. "At least they're engaging in dialogue," he pointed out. That same year, Zambia finalized its integrated development plan after a yearlong participatory process in which every community affected by KAZA was invited to submit a list of priorities.[16] Participation was reportedly high, with some community groups walking for four days to Livingstone, Zambia, to submit their proposals.

The development plan, like the Baviaanskloof Mega-reserve's stakeholder program, showed a sincere effort to solicit community collaboration. In it, villages were offering detailed information on wildlife movement and how it affected them, a valuable source of data in designing wildlife corridors. Both conservationists and communities outlined what they wanted: better monitoring of human-wildlife conflicts and improved methods of avoiding them; better communication between road planners and park authorities; improved law enforcement around protected areas; the development of alternative sources of income and farming methods that would ensure food security, including bee-keeping, fish farming, and permaculture. The idea was to elicit such à plan from every KAZA country; next up was Angola.

As elsewhere, of course, there were disturbing signs that people might not be willing to wait for the peace parks and their long-promised ecotourism revenue. Zambia announced in 2007 that it had approved the building of a multimillion-dollar development inside the national park protecting the three-hundred-foot Victoria Falls, a part of KAZA. Plans called for two hotels to be built only a few miles upriver from the falls, a clubhouse, 450 villas, and an eighteen-hole golf course. There was an immediate outcry by environmental organizations in Zambia against the desecration of the park, and conservationists pointed out that the development would destroy a corridor used by hundreds of elephants crossing from Zimbabwe into Zambia each year. They also argued that the development could potentially destroy the falls and the tourism economy: hotels already built further upriver, in violation of regulations prohibiting building on riverbanks, had led to pollution and decreased flow to the falls.[17]

The Zambian government expressed indifference to these concerns, citing unemployment at around 70 percent and per capita income that had fallen markedly since 1960. A local chief in the Livingstone area said that his community did not want elephants, because they destroyed crops. But here, too, there was progress: tourism operators threatened to boycott the developer, and the U.S. Environmental Protection Agency conveyed its alarm to the Zambian government. Ultimately, a compromise was reached, through negotiations with the Environmental Council of Zambia. The developer agreed to drop the golf course and the villas from its plan and to build the hotels on a modest scale, not "higher than the treetops."[18] No fencing would be allowed. The decision was hailed as a victory for conservation, the result of the most energized environmental campaign ever seen in Zambia.

Thus KAZA was moving forward, incrementally, inch by inch. In Botswana, Sibangani Mosojane, a district wildlife coordinator in the Department of Wildlife and National Parks, had begun the process of notifying, and meeting with, every community inside the KAZA boundary, doing a "house-to-house campaign to put them in the picture."[19] Having studied in New Mexico, where chili peppers are an important commercial crop, he was experimenting with hot peppers as both an agricultural product—chilies are also beloved in southern Africa—and an elephant deterrent, to keep them away from crops and settlements. "We mix it with elephant dung and burn it, and it irritates them," he said. "They try to avoid it." He was realistic about the challenges. "One big thing is that conflict will *increase* in trying to solve this problem," he said. "We will have to educate communities. If they see tangible benefits, they will support it."

Embarking on her hazardous trip to a home registered in elephantine memory, Letsatsi may prove the spark that makes those "tangible benefits" possible. She and elephants like her are the stars of a 2008 BBC/Animal Planet documentary, featuring Mike Chase, that will doubtless inspire further tourism and donor interest.[20] She took the first step. Now communities, governments, the Peace Parks Foundation, and the cluster of groups and donors supporting these projects will have to do the rest.

Unsurprisingly, the vast social engineering experiment represented by transboundary and peace parks has hit some rough patches. Mistakes were made, beginning with misplaced priorities in the Great Limpopo and exaggerated promises about ecotourism, with the Peace Parks Foundation selling "tourism products" instead of focusing on conservation. For all his high-flown inspirational rhetoric, Anton Rupert put too much store in marketability and not enough in involving communities in plans for their future.

There were also troubling implications to the insistence that wildlife must pay for itself. At its most doctrinaire, the "pay to stay" principle trivializes the central importance of wildife and its role in ecosystems. Simply put, conservation is a necessity. If clean air, clean water, and access to adequate food are necessities, then conservation, dedicated to securing them in perpetuity—for people and wildlife—is a basic requirement. It is not a luxury, not a business proposition, not an excuse for golf courses or intensive development. It is not neocolonialism. Certainly, conservation should be integrated with economic benefits and jobs, but the practice of conservation must be the top priority. And it should be managed not by marketers but by biologists, park rangers, wildlife specialists, and, in the end, communities themselves.

Peace parks have become the dominant model of rewilding in Africa. In 2002, a PPF feasibility study identified twenty-two potential peace parks or transboundary complexes in southern Africa, many of which have been added to its roster of projects under development. Agreements have now been signed between Angola and Namibia. Discussions are ongoing regarding new protected areas and marine parks between Malawi and Zambia, and Mozambique and Tanzania.

No one should downplay the challenges to rewilding in Africa, a bristling array of impediments including fencing, poaching, poverty, minefields, corruption, and political deadlocks. But no one should bet against its potential to succeed. The resources are staggering. South Africa alone has five thousand private game ranches, four thousand mixed livestock and game properties, eighty-odd fenced areas with elephants. Gifted with expertise in wildlife conservation, from biologists and academics to unparalleled wildlife trackers, southern Africa can accomplish, in principle, what it set out to do.

The writer Paul Theroux, a former Peace Corps volunteer in Malawi, once described the frozen defeatism that hung about the continent: "No one expected very much, because this was Africa. . . . We could never succeed, nor could we fail."[21] Southern Africa has taken extraordinary steps to break out of that stasis. Like the elephants going home to Angola, peace parks are too big, too important, too precious to fail.

PART III

COMMUNITY CONSERVATION: "VERY TRICKY"

CHAPTER 9

THE CONSERVANCY MOVEMENT

Namibia's Experiment

PARKS ARE NOT ENOUGH. NOT BIG ENOUGH, NOT CONNECTED enough, not always in the right places. The central premise of rewilding holds that while parks are necessary as core reserves—and should be as vigilantly protected from human encroachment as possible, according to biologists—they are only a part of the solution. The vast majority of the world's biodiversity occurs outside of parks. To save it, conservation has to protect entire ecosystems, reducing fragmentation and isolation, which inevitably means rewilding across landscapes dotted with human populations and private property. So rewilding, by definition, is dependent on cooperation and collaboration.

Peace parks attempted to ensure such cooperation by imposing conservation from above. Negotiating treaties and advising government agencies, the Peace Parks Foundation was betting that the approach would work by putting political muscle behind the initiatives. Only belatedly did the South African group recognize that top-down efforts, particularly those that sought to remove people from inside parks, engendered resistance. While the PPF struggled to reformulate its programs, finding a balance between imposition and motivation, between conservation and economic development, a new movement evolved. Community-based conservation—another method linking conservation and development—allowed communities to take conservation into their own hands and put it to work outside parks.

Earlier attempts to link conservation and development were simplistic and poorly designed, based on a crude quid pro quo—save an elephant, get a water well—conferring little power or authority on communities. Nor did they translate into support for conservation: subsistence hunting, for example, was such an ingrained activity in much of Africa that it was not likely to be abandoned in exchange for minimal goods and services. But the early, failed experiments served a purpose, suggesting solutions, refinements, and alternatives.

While no single accepted definition of community-based conservation has emerged and the term can seem nebulous—applied to very different undertakings, involving distinct types of governance, levels of participation, or benefit sharing—the most successful community projects do share certain characteristics. Typically, the most significant programs conferred legal rights to wildlife and wild products on communities—the right to gather plants or cut trees sustainably, the right to develop businesses centered around ecotourism and sustainable agriculture, the right to a percentage of neighboring park revenues, even the limited right to hunt—triggering an important transformation in local attitudes. Suddenly, wildlife was worth something, and it was worth protecting.

The common form that community conservation has taken in countries where rural people have traditionally shared communal grazing or farming land without ownership or title is "conservancies," communally managed grasslands or forests. Conservancies are legal entities granting ownership and responsibility and allowing limited farming or grazing on lands predominantly managed for wildlife. Income from ecotourism, cultural tourism, trophy hunting, the sale of local crafts, and sustainable use of resources is pooled, and the community collectively determines how it should be spent.

Some community projects include a law enforcement component, recruiting local people to act as wildlife guardians or to join antipoaching patrols. This participatory approach also invests authority in communities, providing a corrective to the sense that an agenda is being imposed from outside or above. Oddly, however, organizations supporting community programs have often played down the enforcement component, preferring to keep such tactics under wraps. Perhaps any mention of security issues serves to recall the exclusionary history of parks or the colonial-era "guns and fences" model of conservation. For whatever reason, many projects have shied away from the issue of law enforcement, despite the fact that illegal hunting and organized poaching pose a constant threat to endangered wildlife, and to local people themselves. A recent 139-page document from the International Union for Conservation of Nature concerning "community conserved areas" never mentions these issues.[1]

But in a quiet corner of Africa, one country has doggedly pursued its own blend of community-based conservation and law enforcement. Namibia's program aims to serve all of its highly varied people: a majority of black pastoralists and a minority of white farmers, as well as foreign tourists. It is designed to restore Namibia's wildlife, decimated by political instability, warfare, and poaching. Without fanfare, without flashy international signing ceremonies, Namibia's conservancies have become the gold standard of community-based conservation.

The second-youngest country in Africa, the Republic of Namibia, on the southwest Atlantic coast between South Africa and Angola, won independence in 1990, after a twenty-year struggle to free itself from rule by the apartheid government of South Africa. A largely desert region and one of the most sparsely inhabited nations in the world, with two million people and arable land of less than 1 percent, Namibia perennially struggled with poverty, with most people living in rural areas. In 1990, it became the first and so far the only country in the world to write conservation into its constitution, pledging to safeguard "ecosystems, essential ecological processes, and biological diversity . . . for the benefit of all Namibians, both present and future."[2]

In 1993, it adopted the USAID's "Living in a Finite Environment" (LIFE) project, which provided funding for conservancies. In 1996, Namibia passed the Nature Conservation Act, granting communities the legal right to create conservancies managed for wildlife and to contract with private companies to develop tourism businesses. Conservancies were thus a right, granted by legislation, and they were entirely voluntary. No community was required to create a conservancy. But the legislation proved to be enormously popular: fifty-two conservancies had been registered with the government by 2008.[3] Fifteen more are seeking registration, and twenty established conservancies have reached financial viability over the past decade, earning enough money to continue their programs without additional assistance. Currently, 250,000 people, or one in four rural Namibians, are part of the conservancy movement; several thousand jobs have been created; and over 46,000 square miles of land are involved—around 15 percent of the country—much of it acting as buffer zones or corridors around or between the country's existing parks and reserves.[4]

To create a conservancy, a community must define the boundaries of the area, register all of its members—many areas include more than one village—write a constitution, and elect a governing committee that ultimately makes decisions about the development of ecotourism and management of wildlife. Once registered and approved, the conservancies become a legal entity with conditional ownership of wildlife and other natural resources within their boundaries.

Namibia's conservancy system also conferred other crucial rights and responsibilities. Conservancy committees were put in charge of law enforcement and given the task of hiring community game guards and rangers who were empowered to deter and monitor illegal hunting; they were also trained to report on the number and condition of wildlife populations. In addition, communities were granted a limited right to hunt. The government translocated wildlife into the conservancies to replenish species that had been extirpated, and conservancies were granted an annual hunting quota of prey species—ostrich, oryx, springbok, kudu—which provided an immediate and satisfying incentive for villagers to control illegal poaching or profligate killing. With wildlife increasing on their land, communities could see that wild animals, as well as cattle, could become a kind of bank, promising an annual return in meat, hunting fees, and tourism. The advantages were obvious to people burdened by unemployment and want. "Before the conservancy, there were absolutely no jobs," Vitalis Florry, who became manager of tour guides in the Torra Conservancy, told a reporter in 2005.[5] "Now we see a small economy developing. Now we see some benefits."

One conservancy success was Salambala, the second to be registered, in 1998, with nineteen villages and a population of around eight thousand. When it began, the 359-square-mile area in the Caprivi region had only seven impala, twenty kudu, and twenty warthogs; elephants, zebras, and other large game had been wiped out by war in nearby Angola and uncontrolled poaching.[6] With initial assistance from the USAID and WWF, the government brought in important game species, including giraffe, wildebeest, impala, kudu, and other antelope; more wildlife moved in from Botswana's Chobe National Park, across the river. A few years later, the area boasted six hundred elephants, fifteen hundred zebra, and three prides of lion.[7] In 1998, the community as a whole was earning less than a thousand dollars a year from tourism; by 2006, that income had jumped to over $37,000 a year, generated from trophy hunts, craft sales, and game viewing, much of it paid out in salaries.[8] Elevated platforms for game viewing were constructed, a water hole for wildlife was dug, and a local crafts market was opened for the sale of woven baskets, masks, and carvings.

In other conservancies, women belonging to the Himba tribe developed profitable businesses by sustainably gathering the resins produced by *Commiphora* shrubs, native species once known as frankincense and myrrh. Himba women have long rubbed the substance on their own skins as a protective emollient, but the resins are also valued in producing perfumes and cosmetics. In 2007, several conservancies with access to the plants sold nearly 30,000 Namibian dollars' worth of *Commiphora* resin, over 3,000 U.S. dollars.[9]

In addition to economic benefits, the conservancies demonstrably improved conservation throughout the country. Caprivi floodplain surveys showed a 47 percent increase in wildlife sightings between 2004 and 2007.[10] Namibia's population of black rhino grew to over 1,200 by 2003 (from 450 in the 1980s), the largest in the world. The country continued to be a primary stronghold for cheetah, with 95 percent of Namibia's 2,500 cheetahs roaming on privately owned farms and ranches. The conservancies facilitated a friendlier relationship between cats and farmers, providing accurate information on cheetah predation on livestock and implementing a "Cheetah Country Beef" ecolabel certification program, bringing premium prices to farmers agreeing to practice nonlethal predator control through the use of calving "kraals," or corrals, as well as guard dogs and donkeys.

The conservancies also reopened and restored wildlife corridors. For the first time in decades, they reconnected the Skeleton Coast National Park, Namibia's wild desert coastline—where lions feed on dead whales, elephants "surf" on the sand dunes, and thousands of Cape fur seals haul out—to the inland Etosha National Park. The seasonal pan that forms a centerpiece of Etosha attracts two thousand elephant, three hundred black rhino, and twenty thousand springbok traveling in enormous herds, patrolled by hundreds of lions, leopards, and cheetahs. Conservancies may one day reach their Caprivi Strip counterparts and parks in the east. While a few critics questioned whether conservancies benefited all biodiversity or simply increased populations of popular game species, most conservationists agreed that the program buffered and strengthened Namibia's protected areas. The country now seems poised to achieve one of the best-designed and best-managed conservation programs in Africa.

As with any other complex conservation initiatives, there were setbacks. In granting power to communities, the Namibian government was gambling that people would make decisions that would benefit themselves and conservation, preserving wildlife as a resource. Sometimes, they didn't. For a recent documentary, *Milking the Rhino*, a 2006 meeting of the northwestern conservancies was filmed. At one point, a tense debate broke out over an incident with lions. Several young men, members of the Purros Conservancy in northwest Namibia, had been guarding a herd of cattle when they discovered four lions nearby. They shot and killed all four, despite the fact that the lions showed no interest in the livestock. Philip Stander, a carnivore specialist who had studied the behavior of lions in the region for years, argued that these "were the wrong lions to shoot."[11] He pointed out that tour operators in the area had been thrilled to have lions there for the first time in years. An official from a Namibian conservation organization added that, "if a lion is a problem, the government allows us to sell it to a trophy hunter. A lion is worth 40,000 Namibian dollars (over 5,000 U.S. dollars)." In other words, the young men who took it upon themselves to shoot the lions had potentially cost the community 20,000 dollars in trophy fees or—had the lions been allowed to survive—thousands of dollars in tourist revenue.

A man from the community got up to defend the shooting, saying he was "glad" the lions had been killed: "I don't want them around. The cattle are my bank. If lions come near my cattle, then I know my bank is robbed." These remarks inspired nervous laughter and murmurs of agreement, while Stander grimaced in frustration. But the affair of the four lions ultimately played a role in changing community attitudes. Stander told me later that the man who had spoken approvingly of the shooting was ousted from his seat on the conservancy committee, and the local people had strongly affirmed the value of the lions. He subsequently worked with them to develop "Desert Lion Safaris," an exclusive lion ecotourism project, sharing his research on lion movement—where the cats go and how best to observe them. Together, he and the community established a "Lions and Tourism" training course for guides, staff, and management. The project charges clients a conservation fee that goes into a "lion fund" for the community, compensating those who lost livestock. If cattle once represented the only bank in Purros, there was a new bank in town.

The incident influenced Stander as well. Like many other conservation biologists, he had long harbored reservations about community conservation, fearing such programs could not adequately protect wildlife and resenting the resources lavished on them. But by 2009, he felt that the conservancies had made significant progress and that scientists could continue to add "sound scientific elements" to community conservation. "The bottom line is simply that we have no choice," he wrote to me. "Local communities cannot be excluded if we are to conserve large and viable ecosystems."[12]

It was a conclusion that communities and conservationists were embracing all over Africa. If conservancies were not perfect, they were working. Namibia made it all look easy: the country was sparsely populated, it was at peace, and its government had made an unprecedented commitment to conservation. But the true test of the ability of conservancies to succeed would come in countries with none of those advantages. In Kenya—crowded, corrupt, full of people convinced that conservation was a colonialist tool—conservancies would undergo a severe trial, if not by fire, then by drought.

Kenya and "the Government's Cattle"

Kenya is divided by the Great Rift Valley, a wide plain formed thirty million years ago when tectonic plates collided and then heaved apart. To geologists, the Great Rift is a fault line stretching some four thousand miles from the Dead Sea to East Africa. But to Kenyans, it is the defining physical feature of their country, giving rise to its most stunning and dramatic spectacles, high escarpments and volcanic mountains—including Mt. Kilimanjaro and Mt. Kenya—overlooking enormous vistas across the valley below, once covered in acacia forests and dotted with depressions that became soda lakes, shallow-water pools with no outlet. Constant evaporation leaves the lakes with a high mineral content that feeds clotted masses of algae, insects, and crustaceans. In turn, those creatures draw brilliant flocks of pink flamingos and white storks. From the deserts surrounding Lake

Turkana in the north to the grasslands of the Masai Mara in the south, the Great Rift Valley sheltered early man: fossils belonging to extinct hominid species, such as *Australopithecus afarensis*, have been found here, suggesting that our early ancestors divided their time between walking the Great Rift and climbing its trees.

Kenya is a small country, only a little larger than France, with uninhabitable desert taking up much of its northeastern region bordering Somalia. It lacks the resources, infrastructure, or arable land to support its population, which has a 2 to 4 percent growth rate, one of the highest in the world. Arable land measures only 8 percent, compared with 18 percent in the United States. Over the past few decades, many citizens looking for work have flooded into the cities, principally Nairobi and Mombasa, where they sleep rough or crowd into enormous slums that lack clean water and have no sanitation or sewage facilities. Unemployment is at 40 percent. In the 1950s, Kenya had a population of six million; it now holds thirty-nine million. With a population nearly twice that, France boasts a life expectancy of around eighty. Kenyans can expect to live to a ripe middle age of fifty.

Most challenges to conservation in Kenya arise from these cramped conditions. Land distribution, for instance, has been endlessly contentious. After the British wrested the territory from Germany in the 1880s, British settlers seized the best communal farming and grazing land in the Rift Valley and subjugated the Kikuyu, who were forced to work the land in exchange for the use—not the ownership—of ever-shrinking plots for their huts and crops. The 1952–1960 Mau-Mau uprising of the Kikuyu against the British led to independence, which arrived in 1963 but did not resolve the social inequities. Since independence, the Kikuyu, 22 percent of the population, have held greater political power than any other group. Of the three presidents since 1963, two have been Kikuyu (Jomo Kenyatta and the current president, Mwai Kibaki); the third, Daniel arap Moi, a de facto dictator for twenty-four years, belonged to the Kalenjin tribe but was beholden to the Kikuyu. The ruling tribe has long received a disproportionate share of university seats and civil service jobs, and after Kenyatta's election, it consolidated its control over much of the best farmland. Many tribes, including the Luo people (to which Barack Obama's father belonged), have long resented their lack of representation in government, which has translated into an inability to hold on to their traditional lands. These tensions exploded after the contested election of 2007, with Kikuyu attacked by members of other tribes. A thousand people died in retaliatory violence and half a million were displaced, their homes burned and land seized. Since 2008, an uneasy truce has prevailed.

Land issues confronted the nomadic Maasai, as well. For centuries, they ranged widely over the southern half of Kenya and northern Tanzania, following good grass for their cattle.* But British settlers seized parts of "Maasailand," claiming that the area was underutilized; in fact, many pastoralists had fallen victim to smallpox introduced by the colonialists. Having roamed freely for centuries, the Maasai were restricted in 1902 to a "Southern Reserve" along the border with what became Tanzania. The reserve was expanded and formalized in a 1911 treaty that promised to respect the boundaries, without further government annexation, "for as long as the Maasai shall exist as a race."[13]

Conservation brought its own sense of betrayal. Kenya's national park system, created by ordinance in 1945, further infringed on Maasai land, with the Masai Mara and Amboseli reserves created within its boundaries. The enormous Tsavo National Park also contained lands traditionally used by the grazers. The Maasai expressed their contempt for the parks by designating them in Swahili as *"shamba la bibi,"* or "the woman's garden," referring to the queen, and protested the annexations.[14] But to no avail.

There were further provocations to come. Amboseli, north of Mt. Kilimanjaro, offered extraordinary views of elephants and giraffes

---

* The Maasai Association estimates the current Maasai population at around one million people in Kenya and Tanzania but notes that the Maasai often refuse to participate in the annual government census in Kenya; Tanzania does not conduct a census based on ethnicity.

parading in front of the dramatic mountain backdrop and quickly became one of the most popular tourist attractions in Kenya, with sixty
thousand visitors by the 1960s. In the early 1970s, the reserve was declared a national park, in part on the recommendation of the famous
Kenyan paleontologist Louis Leakey. Blamed for overgrazing, the
Maasai and their cattle were evicted amid drawn-out negotiations for
land access, financial compensation, and development assistance on
water projects. Little of what was promised by the government materialized, and in retaliation the Maasai took to herding and grazing
within the park illegally and killing elephants, lions, and leopard cubs.
They speared a number of rhinos, including "Gertie," known to be a
favorite of tourists.[15] Historically, the Maasai had little animus toward
wild animals, regarding them as "a second cattle," a potential resource
during droughts. But during this period, they began bitterly referring
to wildlife as "the government's cattle."[16]

Frustrated, the Maasai turned a blind eye toward poaching,
which was becoming a huge problem throughout Kenya's parks and
reserves, exacerbated by overpopulation and poverty. With the numbers of big cats, primates, elephants, and rhinos plummeting as their
habitats disappeared, the scarcity of the resource sent prices for exotic
pets or products made from bones and fur—skins, tusks, horns—
soaring. Demand was growing at the same time: countries that
valued medicinal and ceremonial products made from wild animals,
particularly in Asia and the Middle East, were accessing greater
wealth. Chinese people who craved remedies made from tiger bone or
bear pancreas were prepared to pay astonishingly high prices, as were
Saudi sheiks coveting ceremonial daggers made of horn.

What was going on in Kenya was going on around the world,
endangering species from virtually every continent. Western collectors fueled a multimillion-dollar illegal trade in orchids, and British
gardeners were possessed by a lust for enormous tree ferns, stripped
from the forests of Tasmania.[17] Sturgeon, the fabulously long-lived fish
of the Caspian Sea, grew perilously endangered, as females were caught
and killed for their eggs, the most prized of caviars. Shahtoosh shawls

made from the wool of the Tibetan antelope known as chiru became all the rage in high society; it took three to five dead antelope to make a single shawl. Illegal traffickers turned up in countries across Asia and Africa, creating virtual private armies with automatic weapons and paying astronomical wages and bribes. Civil wars provided a fertile hunting ground for organized poaching, and rebel groups across Africa traded ivory for arms. Corrupt governments colluded, allowing huge shipments through customs in exchange for bribes. In 1970, there were an estimated five million elephants in Africa. Twenty years later, there were 600,000. The population of black rhino fell from 65,000 in 1970 to 2,475 in 1992.

In Kenya, a bad situation was made worse by a series of severe droughts that pushed both people and wildlife to the brink. With farmers working every acre of arable land, there was simply no space left for wildlife. The young American photographer Peter Beard, an heir to railroad and tobacco fortunes and a Yale art student, arrived in Kenya in the early 1960s just as thousands of elephants in Tsavo were starving to death, having denuded the park of most of its foliage while confined by surrounding development, human populations, and lack of water outside the park. Taking part in a research project at Tsavo, a study of overcrowding in the park, Beard documented the plight of the penned-in elephants.

Beard took thousands of photographs of some of the 30,000 elephants that died in successive droughts, capturing carcasses in every stage of decomposition, from the initial moments after death to the gradual emptying out of fluids, the feasting of vultures, and the parching of bones. Page after mute page juxtaposing historical photographs of Kenya's big game hunters' trophies with contemporary carnage, Beard's volume, *The End of the Game*, was a relentless record of the consequences of ecosystem fragmentation.[18] It was a picture of Africa coming apart, body by body.

Kenya's government policies were not helping. High numbers of livestock were being concentrated in small areas. During the 1970s, as part of a land management program designed to increase livestock production and curb the perennial plague of rinderpest and other diseases, the government divided what was left of Kenya's communal grazing lands into "group ranches."[19] The Maasai were enrolled as members of these ranches, which were meant to allow them to manage their land communally through elected committees. In some areas, the group ranches provided the Maasai with a way to enter into negotiations with the government agency later known as the Kenya Wildlife Service (KWS), in charge of parks, revenues, and law enforcement. In ranches surrounding Amboseli, for example, the Maasai pressured the KWS to share park revenues and used the proceeds to build electric fences to protect their crops and set up a system of game rangers to protect wildlife. But in others, the Maasai continued to amass as many cattle as they could, albeit on smaller plots of land than they had historically roamed. Cattle confined in small areas quickly compact the soil, inhibiting the growth of new grass: it was a recipe for disaster on land stressed by cyclical drought. In addition, some group ranches allowed subdivision, with committees selling off parcels of land, further reducing the grazing allotment for remaining members.

In other conservation matters, too, Kenya's government was the problem, not the solution. "Hiving off," a corrupt practice by which the Kenyan government carved off pieces of protected land to feed to developers and settlers who threatened violence, deforested much of the country, ratcheting up climate change, erosion, and flooding. Through inadequate regulation, conservation was also waging a losing battle against the forces of industry. Water-hungry industrial flower farming, keeping Europe supplied with roses, became a notorious polluter, fouling fragile lakes with fungicides, pesticides, and fertilizers. Wangari Maathai, who launched a Green Belt movement to plant millions of trees and restore Kenya's forest cover, was persecuted by Daniel arap Moi's police for years before winning the Nobel Peace Prize in 2004.

Eventually, both the Kenyan government and the international community took steps to deal with poaching. Kenya banned all hunting in 1977, a move that helped bring the trade under control but caused resentment, warping local attitudes toward wildlife, while the antipoaching watchdog group TRAFFIC was launched to investigate and monitor trade in wild plants and animals. CITES, the treaty governing such trade, entered into force around the same time, and the sale of ivory was banned worldwide in 1989.

Just as the ban took effect, with Kenya's elephant population cut from 168,000 to 16,000 and rhinos nearly extirpated, the country turned to Richard Leakey, son of paleontologists Louis and Mary Leakey and a renowned fossil hunter in his own right.[20] Leakey had no training in conservation, but he did have an international reputation, friends in the donor community, and a relationship with Kenya's president Moi. During his five years as director of the Kenya Wildlife Service from 1989 to 1994, Leakey overhauled the department, fired fifteen hundred employees for corruption, and built a new facility in Nairobi. He raised salaries and morale, equipping park rangers with better vehicles and weapons. He convinced Moi to institute a shoot-to-kill policy against poachers and to assign paramilitary forces to assist the rangers. Over a hundred people, including poachers, park rangers, and security officers, were killed during Leakey's first several years in office, but the campaign succeeded, in conjunction with the CITES ban, in bringing poaching to a near halt.

Leakey had a flair for theatrics. The year he took over, he convinced Moi to set fire to a huge stockpile of illegal ivory worth several million dollars. Ivory does not burn readily, so Leakey called in a Hollywood special-effects expert to ensure that Moi would be able to ignite a dramatically explosive blaze for the news cameras.[21] Pictures of the tusk inferno became the most iconic image of the fight against poaching, and a few years later Leakey's success convinced the World Bank and other governments to commit $180 million in aid for a five-year conservation program.

But Leakey's heavy-handed approach was hated by the Maasai. "We should get all the money from the tourists," a Maasai elder named Joseph Lila told *Maclean's* magazine in 1995. "It is our livestock which suffer and die because of Leakey's animals."[22] That year, when

Leakey helped launch a new political party called Safina ("ark" in Swahili), the Maasai threw their support behind Moi and his party.

From a conservation standpoint, Leakey's law enforcement solved only part of the problem. For the moment, it countered the worst of the organized criminality threatening Kenya's parks. But it did nothing to address larger issues: habitat loss, human overpopulation, severe deforestation, or the fate of the estimated 75 percent of Kenya's wildlife that roamed outside the parks.

These threats were increasingly visible, affecting even the great East African migration, one of the most lucrative wildlife spectacles in the world. Every year, over a million blue wildebeest, 300,000 Thomson's gazelle, and 200,000 plains zebra follow the "long rains"—the rainy season between March and June—north from the Serengeti plains into the Great Rift Valley, eagerly trailed by lion, hyena, leopard, cheetah, and tourist. During the "short rains" of October to December, they filter south again. As they made these treks, millions of dollars in foreign revenue followed them, from Tanzania's Serengeti to Kenya's bordering Masai Mara. But even that was not enough to keep threats at bay. Animals moving out of the Mara ran a gauntlet through villages, tea plantations, industrial flower farms, and Maasai warriors, who killed lions that killed their cattle.

Kenya needed a better model, something that combined the conservation goals of rewilding—large-scale core areas, corridors, and buffer zones—with targeted law enforcement, something that spurred local economies, encouraging rural people to support wildlife rather than fight it. In a strange turn of events, a solution arrived through an alliance between a white man, the son and grandson of British colonialists, and the very people who had come to hate "the government's cattle" the most: the Maasai.

The Cattle Ranch That Became a Conservancy

In 1922, Alec Douglas, a World War I veteran of the King's African Rifles, bought a property on the outskirts of northern Kenya through the government's Soldier Settlement Scheme. Located just north of the equator on the Laikipia Plateau, Lewa Downs was a vast semiarid lava plain on the edge of the Rift Valley, at an altitude of 6,500 to 8,200 feet, with spectacular views of Mt. Kenya. In 1952, Douglas's daughter Delia and her husband, David Craig, took over the management of the property.[23] At that time, life on the ranch was a fairly harrowing prospect. The area was frequented by numerous black rhino, prides of lion, and battling bands of the Ndorobo Maasai and the Meru people.* Hostilities between the Craigs and their Maasai neighbors continued to plague the cattle operation, with stock theft and damaged fences taking a continual toll on profits.

As was the custom, the Craigs sent their children abroad for schooling. One of the boys, Ian, went to school in Ireland for a time and returned to Kenya in the 1960s as a teenager. His family hired a local Maasai youth, Kinanjui Lesenteria, to accompany him on forays into the bush, to teach him about wildlife and hunting. Unlike some of the other white settlers in the area, the Craigs valued wildlife; while they shot lions on the ranch, they kept many acres open to wildlife. In a photograph from that period, Ian Craig stands grinning widely at the camera, fully kitted out in camouflage, jaunty black beret, and ammunition belt, with a rifle slung over his shoulder. Next to him, Lesenteria stares gravely, a tall Maasai man in a wool coat. The two spent ten years patrolling the bush together. "Back when he was just a schoolboy," Lesenteria has said, "we grew up together to manhood. We've never grown apart. Never."[24] The bond between them—and the knowledge of Maasai culture the young Craig gleaned—ended up altering the ambitions of both his family and Lesenteria's tribe.

In 1977, Ian and his brother William became the third generation of the family to run the ranch, and they experimented with tourism, setting up a few tents and offering walking safaris in the bush.

---

* The Meru are a Bantu-speaking people living along the northeastern slopes of Mt. Kenya; the Ndorobo Maasai may originally have been hunter-gatherers who assimilated to the Maasai pastoralist life and to the Maa language.

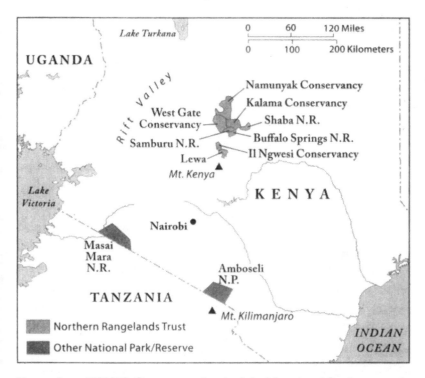

Kenya: Lewa Wildlife Conservancy inspired the Maasai and Samburu people to establish their own conservancies in the Northern Rangelands Trust, securing migratory corridors for wildlife while attracting tourist revenue.

Concerned by an escalation in poaching, the Craigs, who were struggling to adjust to the shifting realities of the cattle market as prices fell and costs rose sharply, developed a program to stop livestock theft, working with Lesenteria to create an intelligence network to track the sale of stolen cattle. Around this time, some miles north of Lewa, in the Matthews Range, Ian Craig witnessed the horrific spectacle of Somali poachers slaughtering a herd of elephants with automatic weapons. "The bandits," he exclaimed later. "They stole cattle, killed elephants for their ivory and anyone who got in the way. They had no hesitation in killing policemen."[25]

So in 1983, when the Craig family was approached about the possibility of converting part of their property to a rhino sanctuary, they already had the motivation and some of the skills to attempt it. Anna Merz, a retired British attorney from Ghana and honorary warden of the Ghanaian Game Department, persuaded the Craigs to fence five thousand acres—later enlarged to ten thousand—to create the Ngare Sergoi Rhino Sanctuary to establish a breeding herd of the critically endangered black rhino. Merz visited rhino specialists in South Africa and Nepal, taking a crash course in rhino care and breeding, and then organized the translocation of several orphans and vulnerable adult rhinos wandering outside parks, in danger of being poached.

In Ngare Sergoi, Merz pioneered the art and science of raising orphaned rhinos, adopting an abandoned two-day-old calf named Samia. To keep the animal warm, Merz slept in bed with her and finally tied a rug around the baby's body. Merz became a gifted rhino whisperer, although frustrated by an inability to fully decipher Samia's extensive vocabulary of alarm snorts, pleading grunts, and affectionate wheezing. She wrote in a diary, "Now I know that when she breathes heavily she is not out of breath but is telling me something, although stupidly I cannot understand."[26] Gradually introduced to the wilds of the sanctuary, Samia became a full-grown, independent rhino at the age of four but would still rush to greet Merz so enthusiastically that she often dented her truck. Sadly, Samia and her calf, Samuel, were killed in a fall from a cliff in 1995, a common cause of rhino mortality. Merz retired in 1995 but passed on her expertise to the Craigs; over the years, Ian Craig's wife, Jane, raised a number of orphaned rhino calves.

Ian Craig had kept an eye on the beginnings of the Namibian conservancy movement. As a businessman, he was convinced that tourism presented better prospects than cattle: the rhino sanctuary was attracting so many visitors and such interest from conservation agencies that the Craigs decided to convert the entire ranch, around forty-five thousand acres of savannah, wetlands, and forest, into a private wildlife conservancy. They retained some livestock but formed the nonprofit Lewa Wildlife Conservancy to manage the property in perpetuity for wildlife and ecotourism. With more endangered mammals and wild-life than any other reserve in Kenya except the Masai Mara, and add-ing adjacent properties as they became available, Lewa quickly became the largest private reserve in the country. It now covers over seventy thousand acres and features some of the densest populations of wild-life in Kenya, including the Big Five—elephant, buffalo, lion, leopard, rhino—in abundance. Its wetlands provided new habitat for a translo-cated population of sitatunga, a rare swamp-dwelling antelope. It also proved to be a safety zone for one of the world's most endangered Afri-can horses, *Equus grevyi*, the Grevy's zebra: Lewa contains over 20 percent of the world's remaining population of around two thousand animals.

To safeguard the conservancy, the Craig family perfected a state-of-the-art security system, staffed by 150 trained personnel, most from local Maasai communities. Surveillance teams checked on the where-abouts and health of every rhino on an almost daily basis. Lewa was also protected by dog teams, aerial surveillance, and a 140-kilometer-long, twelve-foot-high electric fence that ringed the entire property, except for an opening in the northern perimeter to let migratory elephants through.[27] The conservancy worked closely with the Kenya Wildlife Service and Kenya police on joint antipoaching patrols and routinely aided in tracking down and confiscating illegal firearms, snares, spears, and caches of poison used to kill lions and other predators. Since the rhino sanctuary's establishment, Lewa has never lost a rhino to poach-ing, an astonishing record given that the reserve borders Kenya's northern frontier, one of the most lawless areas in Africa, swept by warring tribes and *shifta*, armed bandits who steal livestock and prey on refugees from Sudan, Ethiopia, and Somalia.

For some time, Ian Craig had been talking to his old friend Kinanjui Lesenteria, who headed Lewa's security team, trying to con-vince him that his band of Maasai, on a group ranch just north of Lewa known as Il Ngwesi, could profitably develop its own ecotour-ism business and its own wildlife conservancy. "He used to tell me that conservation is the way to go," Lesenteria said. "Initially when he told me that, I didn't really believe him. After all, he's a white man. That's why I didn't really believe him."[28]

But after Craig took Maasai elders to the Mara to observe Maasai businesses there, they were convinced. The community of six thousand voted to commit itself to tourism, and the Il Ngwesi conservancy was born. The Maasai agreed to set aside 8,675 hectares—over half of their group ranch—as a core conservation area, where they would not graze livestock. Using local materials, they built a striking treehouse-style lodge on a lush hillside, with six individual cottages and an infinity pool that appears to run off a deck into the wilderness below, a design that won ecotourism awards and kept the lodge at a respectable 60 percent occupancy. The construction crew and lodge staff were all members of the group ranch. They habituated a troop of baboons to provide a focus for game drives and developed a cultural *boma*, or village, where visitors could witness Maasai dances and other activities. Several years later, an abandoned black rhino calf named Omni—raised at Lewa alongside his close friend Digby, a baby warthog who slept on the rhino's back at night—was translocated to Il Ngwesi. Omni often trots out of the bush to greet visitors driving up in the game drive vehicle, resting his massive head on the door and soliciting an ear scratch.

The switch from running cattle to running a business based on hospitality involved a major cultural adjustment for the Maasai. Strict gender roles in Maasai culture dictate that women cook, not men. But at the Il Ngwesi lodge, Maasai men were taught how to welcome guests and cook and serve food. For some, their new jobs represented the first time they had been in close proximity to white people, and in the documentary *Milking the Rhino*, James Ole Kinyaga, who eventually became the lodge manager, recalled how some Maasai were initially alarmed by what they saw. "We were worried about the pink skin," he said. "What is this pink person? It does not have a skin. . . . Maybe someone skinned it?"[29] Others worried that it was not right for men to cook for women. But as the Maasai saw the money coming into the community from tourism, they recognized that the lodge might allow them to continue to do what they loved best: care for their cattle.

As other group ranches neighboring Il Ngwesi saw its success, they gradually sought to establish their own conservancies, leading to a network known as the Northern Rangelands Trust. Lewa helped the communities organize a board of trustees, build airstrips, and amass equipment necessary for their antipoaching duties, including radio communications equipment, uniforms, weapons, camping gear, binoculars, and GPS units. So far eleven group ranches have joined the trust, and eight have established wildlife conservancies. Several have signed contracts with concessions to help build and run ecotourism lodges.

Like the conservancies in Namibia, the community conservancies in Kenya granted significant ownership rights along with control of the land and wildlife to the community. While the old group ranch system primarily conferred grazing rights, the new conservancies encouraged communities to officially register land title with the government and establish bank accounts. With a board of directors and several standard departments—security and ecological monitoring, community development, finance—each conservancy established a yearly budget, independently audited at year's end. The departments ensured that conservation goals were met; that security was provided for residents, wildlife, and tourists; that revenue from tourism was reinvested in education, scholarships, and other community priorities; and that rangeland was well managed in cooperation with neighboring conservancies. An annual general meeting was convened in each conservancy to vote on the distribution of revenues, which were never given out in the form of individual dividends but were instead targeted toward community development.

The Northern Rangelands Trust now encompasses a million and a half acres of northern Kenya, roamed by thousands of head of livestock and some of the most spectacular wildlife populations left on the continent. Unlike at Lewa, there are no fences in the rangelands, and trust communities have pledged to maintain wildlife migration routes while monitoring endangered species and the condition of the range itself.

Thus, Lewa, a private reserve, became a prototype for community conservation, providing a framework for the Maasai to create their own conservancies. In a country where political leaders had disappointed the Maasai again and again, the conservancies allowed them to administer their own communities and develop their own economies. The conservancies are now being copied around Kenya, and Lewa has also hosted official visitors from Ethiopia, Uganda, and Tanzania. In 2000, the Koija Group Ranch, on the outskirts of the Northern Rangelands Trust, with funding from the USAID and other partners, built a lodge similar to the one at Il Ngwesi, setting aside five hundred acres for its wildlife conservancy. The land had been severely degraded by overgrazing, with compacted soil and plant life reduced to the thorniest, most inedible plants. But within a few years, it was recovering, and a pack of rare African wild dogs has returned.

While fostering the conservancy movement, Lewa has become a major investor in community education, raising money through its annual Safaricom Marathon, thought to be one of the ten toughest in the world and the only race run in a wildlife reserve. (Spotter planes overhead ensure that lions and rhinos steer clear of the runners.) Since its inception in 2000, the marathon has raised over $1.7 million, money that helps pay for conservation but also for local hospital equipment, school supplies, library books, and water projects for nearby communities. In 2008, 10 percent of the $450,000 raised was donated to the Kenyan Red Cross, for those internally displaced by the postelection violence of 2007. Lewa sponsors eight schools, has built twenty-one classrooms in the last few years, and awarded scholarships to 206 children; it has also invested in a women's microcredit program, granting small loans to over 400 women to start craft businesses and other endeavors. Lewa's education director tells children in its sponsored schools, "When you sit at that desk or write in that book, it's the lion that bought that book, it's the elephant that bought that book."[30]

Lewa is the Africa imagined by every Westerner bewitched by *Out of Africa* and *Nature* documentaries. Its open vistas are so sweeping they make a herd of elephants look small. Everywhere, spectacular scenes unfold: cheetahs napping under a lone acacia on the breast of a hill, baboons surging up a cliff, giraffes galloping through thorn trees, lionesses crossing a road with their cubs in their mouths, sunsets of saturated hues.

But Lewa is more than a stunning landscape. Like Brazil's Emas National Park or Romania's Carpathian Mountains, Lewa's conservancy is an open-air laboratory. It provides an opportunity for researchers to gather long-term data examining the forces behind human-caused extinction and how to address them.

For the Grevy's zebra, one of the most endangered large mammals in Africa, Lewa has provided a secure refuge, and the wildlife conservancies in the Northern Rangelands Trust are expanding its range. The Grevy's cuts a more spectacular, horselike figure than its smaller cousin, the plains zebra, and is an important herbivore in the arid savannahs of the Horn of Africa. But the Grevy's has suffered severe habitat loss: since the 1970s, when there were populations of some fifteen thousand, largely in northern Kenya, its numbers have dropped to below three thousand.

Scientific studies have repeatedly demonstrated the powerful top-down regulatory role of predators, while the importance of herbivores, aside from elephants, has been relatively neglected. But scientists, including Daniel Rubenstein of Princeton University, one of the principal researchers studying the Grevy's zebra at Lewa, believe that herbivores such as zebras may play a key role in causing "rampant indirect effects" on their environment, controlling the density of trees and herbaceous ground cover.[31] In their absence, there may be sharp rises in populations of insects and lizards, which, in turn, could cause a cascading series of "perturbations," widespread changes to the plant and animal communities. In essence, biologists believe that maintaining existing populations of large herbivores may be as essential to the ecosystem as top predators.

Other studies have focused on how to integrate the human element. To ensure the Grevy's future prospects, scientists at Lewa and throughout the conservancy lands have been working to forge a beneficial relationship between local pastoralists—including northern tribes of Maasai known as the Samburu—and the zebras, which compete with livestock. As part of the Grevy's Zebra Scout Program, eighteen members of Samburu communities were recruited and trained to record the location, group size and structure (numbers of males, females, and young), and habitat of Grevy's zebras.[32] The information yielded maps showing where zebras grazed intensively and where they bred. Local communities then used the information to plan new settlements and water sources away from areas utilized by the zebras. As a result of the data, two communities were able to attract funding to establish their own wildlife conservancies.

Although it was a sanctuary, Lewa also posed some problems for the Grevy's. Like all other zebras, territorial Grevy's males acquire harems, but the females exhibit a unique behavior, stashing their foals in "kindergartens," groups of infants left in the charge of a minder, an auntie, while they graze farther afield. The behavior has advantages: foals save energy and grow rapidly while staying put. But at Lewa, where lions were no longer suppressed as they were in ranching days, the kindergartens proved to be the lions' favorite larder. Few foals were surviving past their first birthday. If the rate of decline continued for the species—already wiped out from its traditional range in Djibouti, Eritrea, and Somalia, and declining in Ethiopia—it might be doomed.

"That's what we're trying to prevent here," Joseph Kirathe told me.[33] Field coordinator for a long-term Earthwatch project monitoring the Grevy's zebra, Kirathe led teams in conducting field counts, vegetation transects, and behavioral studies and in photographing individuals, recognizable by their distinct stripe patterns. While the researchers were still gathering data to address the predation problem at Lewa, they had made progress on another front, experimenting with a method to improve the condition of the land, using "mobbing"—intensive, controlled grazing by cattle in a confined area—to smother tall grasses that have built up.[34] After the grasses are compacted and the soil churned up, new tender growth emerges, preferred by the zebra.

Driving around Lewa with Kirathe was like leafing through a wildlife encyclopedia: This was probably as close to what East Africa looked like before white settlers arrived as anyone will ever achieve. As we made our rounds, we passed through scores of ecological zones: open grasslands shimmering in the sun, sending waves of heat haze over grazing oryx and the startling Somali race of ostrich, with handsome black feathers and blue legs; swamps where elephant and buffalo backs were heaving above thick reeds; streamside woodland thick with giraffe and zebra; thickets of different species of acacia trees; scrubby areas full of sunken warthog holes, where the wily tusked hogs hunker down to avoid predators.

Everywhere, there were rhinos, napping in the road, trotting with their infants, standing patiently and flicking their ears as oxpeckers snatched ticks off their hides. We came across Lewa's famous fifteen-year-old blind black rhino, Mawingo, Swahili for "cloud." Mawingo has had many calves, two of which were almost immediately killed by hyena and leopard: Mawingo's blindness hampers her ability to defend her children, and she often misplaces them. Lewa stepped in. Omni was one of her infants. Her fourth, Tula, was found alone two days after birth and was subsequently raised by Jane Craig, on human baby formula (Nestlé Nan 1) mixed with porridge and vitamins. Several keepers tended her twenty-four hours a day. Like any self-respecting child of entitlement, she refused to go to bed without them. The Lewa Web site reports, "She is doing very well and is a big strong rhino girl."

Unquestionably, Lewa had established an effective core conservation area, and the Northern Rangelands Trust conservancies were following its lead, managing their land for conservation. But there were some significant differences at Lewa: it had fences and a sophisticated security system. The NRT conservancies were even more vast, but with no fences. Curious to see the migratory corridors that were being restored and how drought was affecting them, I went to the Northern Rangelands Trust.

In the Northern Rangelands

From the air, Lewa's swamps and riparian corridors were richly green, and even drier stretches of savannah were covered in grass, studded with different kinds of acacia. Close to Mt. Kenya, at high altitude, Lewa has the advantage of a cooler climate, and as a private wildlife reserve, it does not support a human population.

But as I could see during the short flight to the rangelands, the landscape to the north dropped down to an arid plain, bleaching out to desert, uniformly ocher-colored, almost burnt-looking, with sparse, stunted trees. Weirdly reminiscent of the ubiquitous drilling pads covering thousands of square miles of northwest Texas, rings of dead, bleached tree limbs could be seen from the air. These were *manyattas*, temporary settlements of the Samburu people, ringed by fences of thorny acacia limbs, heaped up to protect the inhabitants and their livestock against lion, leopard, and hyena. Some settlements, their visible round huts still in evidence, were occupied. Many were abandoned, just a ring of compacted soil left behind.

I visited three community conservancies in the Northern Rangelands Trust—Namunyak, Kalama, and West Gate—all adjacent to each other. All were supported by donor agencies and zoos, Namunyak by the Tusk Trust, Kalama by the St. Louis Zoo, and West Gate by the San Diego Zoo. Of the three, Namunyak was the oldest, established around the same time as Il Ngwesi, but Kalama and West Gate, dating to 2002 and 2004, respectively, were just getting off the ground. Both the West Gate and Kalama conservancies border a complex of three national reserves—Samburu, Shaba, and Buffalo Springs—that lies along the Ewaso Nyiro River, once the site of the research camp run by Joy Adamson, author of *Born Free*. These national reserves have long provided a core conservation area in the north, but the conservancies greatly expanded the buffer zone around them, increasing protected areas where Grevy's zebra and other species could recover their numbers. Namunyak, to the north of Kalama and West Gate, served as a critical migratory corridor for elephants moving seasonally between the Matthews Range and the area around Mt. Kenya and the Ngare Ndare Forest south of Lewa. After years of poaching, all three conservancies were crucial to the recovery of wildlife in northern Kenya.

Namunyak had already established a successful ecotourism business. In partnership with a private company, the conservancy's Sarara Tented Camp was perched above a water hole frequented by elephant, kudu, and leopard. This was where Ian Craig once watched a herd of elephants gunned down by poachers. The camp has proved to be a huge success, employing twenty people from the local area and contributing $1,000 a year to an endowment fund for educational scholarships, medical care, and compensation for livestock losses. Thanks to Sarara, Namunyak's annual operating budget, $120,000, was nearly twice that of Kalama's, at $65,000.[35] West Gate had signed a contract with a private tourism operator to develop the "Bedouin Camp," a luxury tented camp on the banks of the Ewaso Nyiro, and was beginning to see its revenues rise.[36] Kalama was in the process of completing an environmental impact assessment at the site of the ecolodge it hoped to develop.

The intense effort and pride invested in the conservancies was immediately apparent. Each had developed its own distinct emblem or crest. Kalama's crest pictured a Samburu woman dressed in blue, beads circling her neck and brow, seated on a stool, milking a startled-looking elephant. Sammy Latoona, warden of Kalama, pointed to the plaque, fixed to the new stone and brick headquarters in Kalama, and said it represented the greater prosperity the community hopes to derive from its wildlife. It also hinted at the social transformation ongoing throughout Kenya—the steady conversion of pastoralists to a more sedentary way of life.

The Samburu people live as the other Maasai do. Their language and customs are closely related, and they share the same tall, slim, straight physique. Samburu women wear heaps of wire and bead necklaces around their necks; young Samburu men, called *moran*, belong to a warrior class, assigned to protect livestock. They carry spears, sport ivory hoops set into their earlobes, and wear elaborate head-dresses fashioned out of beads, buttons, and incongruous plastic flow-ers. After the men marry, they wear long lengths of red cloth draped casually around their bodies. Children or *moran* herd cattle, goats, and camels. It is not uncommon, driving down the road, to see a Samburu man tapping blood from a goat's neck for a roadside snack. Samburu, like Maasai, rely on three foods: milk, meat, and blood.

The Samburu have walked with their cattle for centuries, but their way of life has run into terrible difficulties in recent years as drought after drought has diminished the ability of the land to sup-port their herds. Goats have replaced the cattle that have died, and goats are even tougher on the landscape: cattle eat grass, but goats notoriously eat anything, calmly consuming whole ecosystems. It was the aim of the Northern Rangelands Trust to convince local pastoral-ists to convert some livestock into cash deposits in banks and to diver-sify into ecotourism and crafts. Fewer domestic animals would give the land a chance to recover, would strengthen existing herds, and would allow owners to put something by for a (hopefully) rainier day.

Wamba, a town of 25,000 fifty miles north of Lewa, has been hard hit by drought. The main dirt road through town was slashed by an enormously deep, wide gully—too deep and wide to jump or drive over—that people and traffic negotiated with care in order to visit shops on either side. But that was only the beginning of the devasta-tion caused by severe erosion, the consequence of drought followed by heavy downpours. On the way to the Namunyak Conservancy head-quarters, a few modest wood and concrete buildings located just out-side Wamba, where security guards were trained, outfitted, and billeted, the van slowed to a crawl, pitching nose down into dry ar-royos and creeping back up their steep, eroded sides. A bridge that once spanned the local river had washed out. Jagged chunks of con-crete lay forlornly tipped sideways. The community itself was in perilous shape, with many dwellings patched together from scraps of tin, cardboard, and other flotsam, sure to be pulped in the next downpour.

From the conservancy headquarters, an Earthwatch team of biologists and volunteers drove out daily, interviewing locals and taking detailed censuses of wildlife and domestic animals and monitoring the condition of plant communities; I went with them. To support their new ecotourism economy, the conservancies had stopped culling lions and other predators, a shift that was inevitably increasing human-wildlife conflict in the area. The data being collected would serve to create an overall picture of where prey species were congregating, along with their home ranges and daily and seasonal movement patterns. Predators follow prey, so the data might help Samburu communities locate their own herds and livestock *bomas*, or corrals, out of harm's way.

The area seemed painfully dry, with many water holes parched, only greenish ooze left on top of baking clay, pitted with the hoof-prints of a thousand animals. Even so, we saw a startling amount of wildlife—a leopard panting in a patch of shade; huge flocks of vul-turine guinea fowl, iridescent blue green feathers winking as they dashed through the underbrush; reticulated giraffe; kudu; dozens of dik-dik, a rabbit-size antelope; and an enormous Verreaux's eagle-owl, its pink eyelids clicking open and closed as it stared from a tree limb, watching the body of a dead dik-dik. Every night, hyena yipped and whined nearby, and great disks of dried elephant dung were plas-tered up and down the trail in back of the conservancy buildings, up the hill into the Matthews Range.

Security was strong and effective, as at Lewa. Every day, we were accompanied by at least one trained Namunyak security guard, a lo-cal Samburu man outfitted in uniform, carrying a vintage Kenyan army rifle. The guards also had access to radio equipment to stay in touch with the Kenya Wildlife Service, and in recent years, they had intercepted several attempts to poach elephant. Idi, a tall Samburu man with high cheekbones who went with us on several outings, wore army camouflage fatigues at work, along with platform boots that looked like they hailed from a 1970s blaxploitation movie: local markets in larger Kenyan towns sell the Western world's secondhand clothing. But on Saturdays, Idi appeared in full Samburu regalia, an elaborate beaded headdress and collar, a line of beads strung beneath his lower lip and red cloth tied around his waist. Idi and Saruni, another guard, also interviewed and translated for the herders we came across: they were the only members of our party who spoke Samburu. On one occasion, Idi elicited details about a lion that had recently attacked a livestock *boma*; on another, Saruni helped track a leopard into a ravine. They weren't simply men with guns; they were facilitating the research. What's more, their response to a lion kill that we stumbled across one afternoon convinced me their jobs were changing their attitudes toward predators.

One of the biologists, Sam Andanje, a KWS senior research scientist at Tsavo Conservation Area, had noticed vultures descending near the road. When we pulled over to investigate, the carcass proved to be a young elephant, not an infant, but probably a year old. It looked like an elephant balloon that had been deflated. Walking around the carcass, we could readily tell how the animal had been killed. Unable to negotiate the deep, eroded gullies in the area, the panicked creature must have been cornered by a lion that slashed its trunk, which had a long tear in it, and jumped on its back. A crevice had been opened in the body cavity, where flies teemed around a bloody jumble of flesh and organs.

The view was somehow obscene. As we walked around the reeking corpse, I felt horrified but noticed that the Kenyans were not bothered by such squeamishness. Nduhiu Gitahi, a laboratory technician and graduate student at the University of Nairobi, calmly collected sample hairs and scrapings from the carcass. Idi combed the area on foot, finding lion tracks and scat, an important source of information about the lion's diet. Was it dining on elephant alone? Or was livestock on the menu as well? Andanje marveled at the power of the animal that wrought the destruction.

"It is very important to identify that lion," he said. We were climbing back in the van with the samples. "A lion killing an elephant is not a lion to play around with." Based on the track, he speculated that it might be a solitary male, perhaps responsible for the attack on the *boma* we'd heard about. He decided to return that night, to see if he could catch the lion on film.

Returning near dusk, we stopped at a crossroads where a new green Quonset hut guard post had recently been erected, a gift from Earthwatch. Word had gotten around that we were going back to see if the lion would return, and the van filled with Samburu guards, eager to have a chance to see the predator for themselves. A few disappointed souls could not fit in the van and had to be left behind.

Back at the kill site, we waited over an hour in darkness, the smell of rancid elephant wafting into the vehicle. When the cloud cover parted slightly overhead, allowing more light on the scene, the guard sitting in front of me stiffened, then hissed at me to close the window. The driver switched on the lights. There, in the yellow shaft of headlight, was a startling tableau: three lions had returned to the kill and seemed to be cheerfully worshiping it. Bachelor males, Andanje said. They lacked the luxuriant manes of lions to the south. For reasons as yet unclear, lions in arid climes have less spectacular ruffs, merely a thickening and darkening of the fur around the face.

Nonetheless, they were magnificent. Their heads were enormous, and the animals moved with the curiously regal slow-motion languorousness of lions. One was moving around in the bushes, and another was reclining next to the carcass with his head inside the body cavity, licking the decomposing flesh. Grooming, the lions licked blood off one another's fur. It was impossible to imagine a greater state of contentment, and there seemed almost an intimacy of connection between predator and prey. They were in no hurry. They had all night.

The scene was reminiscent of the "blood-seeking monarchs," the maneless lions whose exploits were chronicled in *The Man-Eaters of Tsavo*, based on the 1898 diaries of J. H. Patterson, a railroad engineer whose job it was to keep thousands of Indian workers laying rail while, night after night, the Tsavo lions picked off another victim from among their ranks. "I could plainly hear them crunching the bones, and the sound of their dreadful purring filled the air," he wrote.[37]

He was particularly offended by the lions' habit of licking the skin off their victims. As we observed these specimens, it was only too easy to imagine their delight in doing so.

Everyone in the van was holding their breath, watching in fascination, the guards most of all. For them, it was a once-in-a-lifetime opportunity to observe a feared yet venerated animal from a position of safety. Killing a lion was once a crucial test of manhood in Samburu culture. Watching Samburu men watching the lions, I could see a transition in progress. If the Samburu came to know and value these lions in the same way that Namibian communities were beginning to value theirs, then a new era of coexistence might be on the horizon. Instead of killing lions, they would be able to milk them for a "lion fund" of their own.

The security guards were not the only members of the community who had become actively involved in conservation work. Anna Maria Lolangwaso, a Samburu teacher at the Gir Gir primary school in the town of Archer's Post, had founded a wildlife club at her school to introduce children to birds, animals, and conservation. "In my class, I have sixty-seven, if they all come," she told me, and she described how hard times were for many local families, whose children walked five or six miles to get to school. "When it rains, everybody feels good," she said. "But life has become very hard because of the drought."[38] She praised conservation donors for sponsoring scholarships and building a girls' dormitory at her school: lions were buying books for her, too.

Not all Samburu were so sanguine. We visited a number of *manyattas* to interview herders about their problems with predators. The researchers worked their way through a long, multipage, multipronged questionnaire with each head of household: Which species were most aggravating? Were attacks on livestock increasing? Were livestock being guarded inside the *bomas* at night and were the fences high enough to keep leopards and lions out? Anything that the scientists could do to reduce conflict might help the Samburu adjust to the new predator-friendly regime.

Universally, everyone we spoke to hated hyenas. "Ay, *minke*—so many!" one woman cried. She complained that hyenas were running into the huts and stealing sleeping mats and clothes. Others suggested that the transition to a conservancy economy—allowing wildlife free rein in exchange for ecotourism revenue—was proving to be rocky. A man named Laisoro, who had lost four goats in one year to leopards, said that he would kill a lion if he caught it attacking his livestock. The *manyatta* across the road was even more hostile to the environmental message. While the headman welcomed us and invited us to sit and talk, another man stood outside the circle of curious children and other onlookers, keeping up a hectic tirade. I later learned that he had been complaining about conservationists who were always coming around asking the same questions and never doing anything to help people. He also said, "Will you go and tell everybody how hungry we are?" They did look hungry, thin and unkempt, and it seemed insulting to ask them only about their problems with wildlife.

The perilous situation of such people raised questions about Kenya's hunting ban. In Namibia, support for conservancies had been secured by permitting limited hunting as wildlife returned: this enabled people to feed their families while tourist businesses slowly began to generate revenue. In addition, the conservancies were allowed to sell lucrative trophy hunting permits, cementing the beneficial relationship between wildlife and communities. But Namibia had a population of two million, Kenya nearly forty million. There were fears, both in Kenya and abroad, that lifting the ban would send the country back to the days of rampant poaching. In 2005, Kenya's president, Mwai Kibaki, announced that it would remain in force. Kenya had been praised for its commitment to the CITES ban on the ivory trade; donors were loath to see a return of the worst poaching days.

But in the absence of hunting, there was little to endear wildlife to some Samburu. There were complaints at several *manyattas* about the Kenya Wildlife Service's failure to follow through on its compensation scheme, to repay herders for livestock lost to wildlife. No one we talked to had been compensated, including Keward (Sammy) Lekalkuli, forty-eight, the headman at a *manyatta* in a remote sub-location of the West Gate Wildlife Conservancy. A tall, authoritative man, despite his startling "Colonoscopies Save Lives" T-shirt, featuring well-fleshed Americans smiling above their clean colons, Lekalkuli headed a household of thirty-six. Since 1990, he had been reporting livestock losses to the KWS, but without response.[39]

Listening to these patiently reiterated litanies, I found myself remembering an anecdote Joseph Kirathe had told me, back at Lewa.[40] A veteran of failed community conservation schemes, Kirathe described a particularly ill-fated and short-lived project he worked on at Tana River, a major river that drains the eastern side of Mt. Kenya, emptying into the Indian Ocean. Designed as a primate monitoring project, it was also a community conservation initiative, offering local people land, health services, and education in exchange for leaving the forest remnants they were gradually cutting and degrading. Kirathe was part of the scientific team that went to look at the forest fragments to assess which species of monkeys were still present. "But in the middle of the night," he said—their first night camped on an island in the river—the locals decided they didn't want them there and untied their boat, a dugout canoe, letting it drift down the river. "We had a terrible night," he said, fearing that they would be attacked. They radioed the KWS, who came and extracted them in the wee hours. "Working with communities is very tricky," he said. "Very tricky."

The problem, in his view, was the sheer amount of money. The project was financed—overfinanced, he thought—with $50 million from the World Bank, among others. He saw it as a cautionary tale demonstrating how smaller-scale investments in conservancies might prove more beneficial than larger ones. Money like that corrupted local leaders, who then "brainwashed the people," he said, convincing them that they could get a cut for themselves. When it was not forthcoming, they mutinied.

The project was eventually canceled. The moral of the failed Tana River community conservation project? To Kirathe, it was simple: conservationists could not expect to helicopter in to communities and thoughtlessly hand out cash. He thought that the conservancies offered a better method, giving the community the tools to identify their own priorities and begin generating their own revenues, which could be reinvested in schools, health clinics, water projects. But even here, he felt, it was a complicated business. "Conservation is about managing people," he said flatly. "It's not about managing wildlife."

Drought

A *moran*, his long braided hair dyed with red ocher, took us to see hyena dens in the rocky outcrops. High up in a pile of rocks was a small abattoir—a scattering of bones, a baboon's skull. He pointed out a slim crevice at the bottom of a giant boulder, a leopard den, festooned with the skeleton of a long-dead camel. Andanje commented on the severe erosion of the cementlike soil, scraped, cracked, and abraded by heavy rains in the past. "The rain is not soaking in," he said. "Look how degraded this area is. The grass disappeared a long time ago." It was a vicious cycle. Trampled and overgrazed, bare soil was then pummeled by rain, which ran off instead of soaking in, carrying away what little topsoil might be left, leaving nothing for grass to grow in, which in turn suppressed antelope and other wild grazers. So predators went for the easiest available prey, livestock.

Drought—and the cycle of heavy rains following drought years, which can be another powerfully destructive force—was everywhere in evidence in the north. The kind of erosion we saw was believed to have contributed to a severe outbreak of anthrax near Wamba in December 2005, which affected equines and, particularly, Grevy's zebra. The anthrax bacillus occurs naturally in soils and is endemic to the region; animals, both domestic and wild, can be exposed to it by disturbances to the soil, including floods. KWS partnered with the Lewa

Wildlife Conservancy to undertake a massive vaccination effort in 2006, inoculating 620 of the endangered zebras, along with some 50,000 head of livestock. It is hoped that their action may reduce the severity of future outbreaks.

The only place where we saw land in excellent condition—thick, lush golden green grass waving over ungrazed hills—was in the immediate area around Sarara Tented Camp, the ecotourism goose that lays the golden eggs in the Namunyak Wildlife Conservancy, the jewel of the Northern Rangelands Trust. We stopped there one afternoon to have a look, but the five luxury double tents, "with open-air bathrooms and private verandas," were fully booked. It was a beautiful site, protected as part of Namunyak's core area for wildlife. But as we drove away, we could see that the rich grassland came to an abrupt end, surrounded by great dry red gullies carved by recent flash floods.

Still, there was ample evidence that the conservancies were succeeding on several fronts: providing security, controlling poaching, and beginning to restore the ecological integrity of the northern rangelands. When I spoke to Ian Craig in 2008, he said that Grevy's zebra had recently returned to the Kalama conservancy for the first time in years, establishing a population of 50 to 150 animals. Previously, they had been crowded into the Samburu national reserve, where they experienced predation pressure from the large lion prides there. "So they've left," he said, "and found a more conducive habitat, and they're thriving. They're doing *really* well."[41] In another surprising development in 2008, Grevy's zebra had moved in significant numbers through the elephant gap in Lewa's northern fence, spreading out into a region largely given over to livestock. If the zebras managed to find their way into the conservancies to the north, they would demonstrate that Lewa's population could help restock the conservancies.

Craig also pointed out that elephants were moving into historical areas where they had not been seen for twenty-five years. "They're not being poached because communities are protecting them," he said. Instead of being crowded into the reserves, they were expanding their range and generating revenue through tourism. Other gains followed from that, one of which was that Kenyan politicians were finally beginning to recognize that wildlife in the NRT represented an economic resource. More communities were banding together: the "South Rift Association of Land Owners Trust" was recently launched, a kind of NRT for the southern stretch of Maasailand.[42]

Drought continues to put pressure on these ventures, particularly newer conservancies that have not yet established a strong revenue flow. Craig acknowledged as much in a Lewa newsletter, writing that "the historical fears" of the Maasai, that they would lose their land to conservation, "came clearly to the fore" during recent episodes of severe drought.[43] Even the Maasai at Il Ngwesi, who had seen their ecotourism business prosper over a decade, began to falter, and some violated the boundaries of the wildlife reserve, bringing cattle to graze on the land that had been set aside. There were confrontations between tribal groups in neighboring conservancies and skirmishes over water.

The drought of tourism was potentially even more troubling, especially in the aftermath of Kenya's postelection violence—which cut tourism so severely that Il Ngwesi had to send most of its staff home—and of the 2008 worldwide economic downturn. Just as in southern Africa, ecotourism can be fickle. While it remains one of the chief drivers of economic growth, tourism is dependent on an ever-expanding economy. Surely the current downturn will come to an end sometime, but the Maasai may not be able to wait that long.

To Ian Craig, the answer to droughts—touristic and climatic—lies in endowing conservancies in perpetuity. Once again, Lewa has taken the initiative and in 2009 embarked on a process to establish its own endowment, a capital fund of several million dollars to support rhino, Grevy's zebra, and other conservation initiatives into the future. It also encouraged the conservancies of the Northern Rangelands Trust—and donor communities supporting them—to do the same. According to Craig, a relatively modest endowment of a million dollars for each, yielding a conservation budget of between $80,000 to $100,000 a year, would do the job.

Improving security and antipoaching initiatives, investment in communities, job creation, migratory corridors, and the prospects for rhinos and Grevy's zebra, Lewa has proved an exceptionally solid model of community conservation. Some critics have questioned the viability of the NRT without Ian Craig to act as a "big brother."[44] But supplying financial advice, technical assistance, equipment, and a platform for radio communications in a region as prone to instability and violence as northern Kenya does not seem like a crutch. It seems like common sense.

Together, Lewa and the Northern Rangelands Trust are well on their way to stitching Kenya's once shattered remnants back together, creating the largest continuous stretch of protected land in the country. They have forged a positive connection between conservation and the Maasai, giving the pastoral community something the Kenyan government has never offered: self-determination, a form of democratic governance, and a market economy of its own. Kinanjui Lesenteria, who once taught Ian Craig the ways of the wild, summed up the benefits of community conservation this way: "When a lion kills your cow, you don't feel as bad because now you're getting something from this wildlife."[45]

CHAPTER 10

THE TIGER MOVING GAME

Royal Rhinos and Community Forests

IN NAMIBIA AND KENYA, CONSERVANCIES WERE INTRODUCING pastoral communities to the fruits of conservation: autonomy, recovering grasslands, ecotourism. Lions were buying schoolbooks, rhinos were putting up health clinics. People were holding on to their traditional cultures while exploring new economic possibilities: Namibian Himba women found themselves collecting *Commiphora* resin and making more money than their husbands while bemused Maasai cattle herders fried eggs for pink-skinned individuals and lived to tell about it. Collectively, communities were rewilding millions of acres, placing huge reconnected areas of their respective countries under conservation, reducing the effects of overgrazing, and allowing threatened species to rebound.

Meanwhile, in the subcontinent of Asia, there was a similar movement under way. While Asia has long been perceived as the most daunting continent for conservation—a place of high population densities, centuries of deforestation, and cultures with little connection to environmental pieties—there were parts of India and Nepal where local people had established a tradition of collectively managing what were called "community forests." Like conservancies, these entities allowed villagers to control their fate by managing their own natural resources, growing native cash crops within the forests and seasonally gathering locally coveted supplies such as grass, reeds, firewood, poles for home construction, and medicinal plants and herbs.

Tropical forests in the region have extraordinary regenerative powers. While temperate forests often take centuries to mature, a function of a cooler climate and seasonal temperature change, the forests of the subcontinent, which play an important role in the local economy, are dominated by fast-growing *Shorea robusta*, or "sal," a valuable hardwood tree that can reach a hundred feet, used to build houses, buildings, and bridges.[1] Beams and decorative woodwork in many ancient temples are constructed from sal wood, and wood-carvers have plied their trade on it for centuries. Sal leaves are sewn together into plates and hats and used to wrap fish or meat. Sal forests often harbor grasslands within them, the product of wide braided rivers and marshy valleys flooded seasonally by monsoon rains. Even severely degraded forests can recover, reaching six to eight meters in height after just a decade, completing a closed canopy overhead that creates a cooler and more humid microenvironment beneath, an ideal habitat for monkeys, birds, bears, and tigers.

The Kingdom of Nepal seemed an unlikely place for the community forest movement to take hold: forests were once the province of the king, who owned numerous forest reserves, the legendary haunt of the greater one-horned rhino, a creature so mythically potent that it was designated a "royal animal."[2] Only the royal family and guests were allowed to hunt it, and poaching was punishable by death. Few people sought to challenge that authority, particularly in the area along the border with India where the jungle forests bred a fatal form of malaria; hunting was safe only during a brief period in the winter, when mosquitoes were absent.

Nepal is a thin, rectangular country, dominated along its northern border by the uninhabitable, unarable, icy Himalayas. Along its southern flank the country was protected from invasion for most of the year by those deadly forests, covering the lowland valleys known as the Terai—a vast, fertile 450-mile sweep of alluvial soil left by rivers draining the Himalayas, overlooked by bands of steep hills—once inhabited solely by the Tharu ethnic group, indigenous people who had developed resistance to the fatal endemic malaria.

But in the mid-1950s, His Majesty's Government of Nepal launched a massive mosquito-eradication program, and malaria was vanquished. Subsequently, the population of the valleys tripled, with a flood of humanity moving down from the hills, where life was harder and agriculture poorer. The influx caused a rapid carving up of the entire Terai, from east to west, and turned it into the rice bowl of Nepal. Until the 1960s, the Terai was thickly populated with elephants, most of which were wiped out when the area was converted to agriculture, leaving a tiny remnant group in the west. Indeed, due to such habitat loss, the Asian species is far more endangered than the African, with only forty to fifty thousand left in the wild. Athough heavily logged and converted to agriculture, the Terai still contained forested pockets that held untold riches in endemic plants, bird species, and three of the most endangered large mammals on earth: Asian elephant, Bengal tiger, and the greater one-horned rhino. In addition, there were two perilously endangered freshwater species, the gharial crocodile and the Gangetic dolphin.

King Mahendra, who assumed the throne in 1955, was a passionate hunter and lover of wildlife, committed to conservation. Alarmed by reports that Nepal's rhino population was dangerously low, a victim of habitat loss and poaching, he ordered the creation of an armed guard for the species, a Rhino Patrol of 130 former military men, a move that would eventually lead to Nepal's unusual decision to defend its parks with the Royal Nepalese Army. But Mahendra's most substantive decision was to institute a Western-style system of national parks. He began with Chitwan, which means "the heart of the jungle" in Nepali, one of his former private hunting reserves.[3] A section of forest, swamps, and grassland between the Rapti River and the Indian border, only twenty-five miles southwest of Kathmandu,

Chitwan measured 210 square miles. Mahendra also earmarked two additional royal hunting preserves in the Terai, west of Chitwan, for the park system: Bardia, a rich section of sal forest along the Karnali River, and Suklaphanta, a swamp and grassland known for its enormous herds of swamp deer. Mahendra died in 1972, but his son and successor, King Birendra, continued his father's conservation support, and the Royal Chitwan National Park was officially established the following year.

Establishment of the park system launched an extraordinary period of conservation activity. The government invited various organizations, including the Smithsonian and the World Wildlife Fund, to conduct the world's first radiotelemetry studies of tigers, the Nepal Tiger Ecology Project, which led to eighteen years of field research on the species and its prey in Chitwan. Over the next few decades, many Nepali students conducted fieldwork and training there, and the long-term tiger study convincingly showed that Nepal's small parks were not large enough to maintain viable populations of the species. That finding encouraged the government to enlarge the boundaries of Chitwan several times and to expand it with two adjacent areas, the Parsa and Valmiki wildlife reserves.

At the same time, the government struggled to develop a forest policy that would control rampant development and logging. By the 1980s, many areas around the national parks had been badly degraded. To deal with these issues, Nepal committed to a nationwide program of community-based conservation. Like Namibia's conservancy law, Nepali policies encouraged villages to take over management of lands once held by the state. Buffer zones were established around the parks, and within them villages were invited to organize community forests, in much the same way as Namibians organized conservancies. Within a legal framework, communities could manage forests and wildlife for conservation in exchange for limited use of resources and tourism revenue. An overall buffer zone committee and elected community forest user group committees would register local inhabitants, write rules on protection and utilization, and arrange for security guards. Those who wanted to use the forest paid a membership fee and contributed to the guards' salaries or volunteered to work as guards themselves. In return, they were allowed to gather certain forest products—leaf litter (used as mulch), grass, leaves, herbs, firewood—in controlled amounts and at certain times. By 2006, there were twelve thousand registered forest user groups across the country, managing 2.1 million acres of forest.[4] Over five million people were involved in the groups, a fifth of the population, and the deforestation rate fell from 2.7 percent a year to 0.7 percent.[5]

In 1995, the Parliament ratified additional bylaws that required 50 percent of revenue generated by Nepal's park entrance fees to be paid to development programs in the buffer zones, a major incentive for communities to support the parks. Prior to this law, few locals living near the parks saw much income from them, because many businesses were foreign-owned. For example, in 1994, tourism at Chitwan, with forty-nine hotels, produced total revenue of $4.5 million, but 61 percent of the hotels were owned by Nepalese or foreigners living outside the area.[6] Only 1 percent of the local population worked in the industry, and little revenue trickled down. But with a share of park fees, communities suddenly found themselves in possession of considerable funds, which could be used to build up their own tourism enterprises as well as local schools, health clinics, and water projects.

Conservation was strong, working to expand connectivity and protection while benefiting the poor. There was a great deal of ground to make up. Nepal was one of the poorest countries in Asia; only Pakistan was poorer. Over 80 percent of Nepal's 25 million people eked out a perilous existence on pennies a day. By the mid-1990s, the Terai's narrow, once densely forested region was home to 6.7 million people, around 30 percent of the country's population. According to one study, over 70 percent could not grow enough food to last their families through the year; over 50 percent were forced to borrow from moneylenders.[7] Yet tiny, impoverished Nepal had set aside nearly 20 percent of its land in a system of protected areas that put richer countries to shame, and its conservation specialists, trained during the era of King Mahendra, were coming into their own.

Hemanta Mishra was one of those specialists, having helped launch Royal Chitwan National Park in the 1970s. A veteran of the radiotelemetry tiger project, he subsequently became Nepal's senior ecologist and was instrumental in creating Sagarmatha (Mt. Everest) National Park—home to highly endangered populations of snow leopard, ibex, and blue sheep—along with nearly a dozen other nature reserves across Nepal.

In the 1980s, Mishra took on his most important conservation project: restoring the greater one-horned rhino to Bardia, where the species had been hunted out. By this point, the rhinos could be found in only two spots on earth: Chitwan and Kaziranga National Park in Assam, India. After a venerable elephant driver had suggested it to him twenty years before, Mishra had become convinced that rhino translocation would provide insurance against catastrophic loss of the species to disease or poaching.

Mishra successfully translocated thirteen rhino to the most accessible area of Bardia in 1986, as a founder population, and the animals thrived, with most of the females becoming pregnant and bearing calves. He laid plans to move twenty-five more, this time to a more remote location within Bardia, the Babai Valley. Bardia was the largest national park in Nepal—377 square miles—crucial for conservation. Shaped like two lobes of a brain, the park lay near the western end of the Terai, close to the Indian border, and was once accessible only during the dry season: there was no road, and during monsoons, flooding rivers cut it off completely. The completion of the King Mahendra Highway in the early 1990s made it possible to transport the rhinos by truck, and Mishra began moving animals into the Babai Valley, a remote section of the park east of the highway that presented perfect undisturbed habitat.

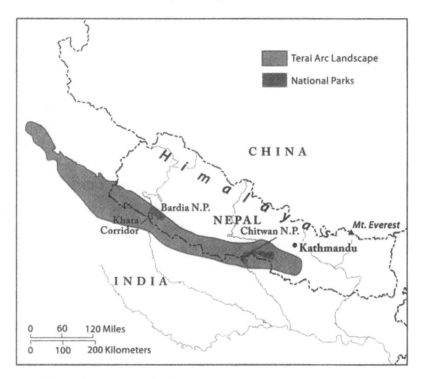

Nepal: The Terai Arc Landscape project aims to reconnect a "green necklace" of protected areas along Nepal's border with India, a last chance for the endangered Bengal tiger, the greater one-horned rhinoceros, and the Asian elephant.

Mishra was joined in this massive translocatory effort by an American biologist, Eric Dinerstein, who had conducted extensive field studies of the rhino in Chitwan and its new population in Bardia. The book detailing his findings, *The Return of the Unicorns: The Natural History and Conservation of the Greater One-Horned Rhinoceros*, is a fascinating compendium of information on this odd creature, its biology, behavior, and importance for the environment. Once widespread throughout Asia, *Rhinoceros unicornis* is slightly smaller than the African species, sheathed in skin creased around the neck and body in thick folds, giving the appearance of plates. Albrecht Dürer's famous woodcut is a mythical interpretation of it: the artist never laid eyes on the animal but had seen sketches of a live specimen given to the king of Portugal in Lisbon in 1515.[8] He depicted a tanklike beast with scaly legs, covered in bizarre armor. The real thing is even more impressive, massively imposing, the pattern of its ribs oddly visible through its heavy hide. Yet, like the African rhinos, the Asian species moves with extraordinary speed and is surprisingly light on its feet.

Nepalis view the rhino with such profound wonder that there seems to be a collective yearning and reverence for every part of the creature. The horn, made of compressed hair fibers, is the most coveted: Hemanta Mishra was called to his father's side during a religious ceremony when he was five and instructed to bring the *khaguto*, a cup made out of rhino horn in which milk and water were offered to appease the souls of ancestors.[9] Everything the rhino produces is valued: Eric Dinerstein saw elephant drivers "risk life and limb" to collect fresh rhino urine, believed to be a powerful antidote to asthma, tuberculosis, and inner-ear infection; he witnessed crowds form around a rhino carcass, waiting for permission to carve it up.[10]

Like elephants, Dinerstein wrote, rhinos are "landscape architects."[11] His four-year study of the interaction between rhinos and plants was triggered when he asked his elephant driver why a common tree always grew in clusters in the middle of the grassland. "Oh, it's the work of *gaida*," the driver said, referring to the rhino, which always returns to the same spots, or "rhino latrines," to defecate, a form of territorial marking.[12] The latrines, full of pooped-out seeds, become nurseries for trees, and Dinerstein examined the rhino's role in the light of the "megafaunal fruit syndrome," a theory that originated to explain the part played by now-extinct megafauna in the evolution of forests in the New World. Dinerstein theorized that the browsing, feeding, and digestive conduct of megafauna like elephant and rhino might well have influenced the structure of Nepal's forests.[13] As in Africa, the tropical forests of the subcontinent evolved in close partnership with these two enormous herbivores, and the relationships between flora and megafauna may well remain essential to their ecological survival.

The rhino translocations were just the beginning. By now the senior scientist at the World Wildlife Fund in the United States, Dinerstein also worked closely with several of Nepal's other notable conservationists and biologists, including one of the first Nepalis to graduate from the system of schools established by Sir Edmund Hillary, Mingma Sherpa, who went on to lead conservation efforts in the Himalayas, and Chandra Gurung, who became the head of the organization's branch in Nepal. Together, they crafted an ambitious, comprehensive large-scale project to restore forested corridors between Chitwan, Bardia, Suklaphanta, and Indian parks and reserves along Nepal's southern and western borders. In 1993, the World Wildlife Fund convinced the Nepali government to back what was now called the Terai Arc Landscape Program, or TAL. Built on the existing system of community forests and buffer zones, taking advantage of the tremendous natural capacity of monsoonal grasslands to recover from intensive use, TAL quickly became the most vaunted landscape-scale project in Asia, possibly in the developing world.

The goal of TAL was to put pieces of the Terai back together again, surround protected areas with restored buffers, eliminate gaps, and reestablish corridors so that tigers, wild elephants, rhinos, and other wildlife could disperse through the landscape. The parks themselves—Dinerstein called them "a green necklace of eleven parks"—included Royal Chitwan National Park, Parsa Wildlife Reserve (on Chitwan's eastern border), Bardia National Park, and several Indian reserves.[14] Shoring up the links between them would create a 19,000-square-mile area managed for wildlife, and it would connect eleven transboundary areas, three Ramsar sites, and two World Heritage Sites.

The corridor design of the Terai Arc works both north-south and east-west, to allow wildlife to disperse during monsoonal flooding in the watershed of the Ganges and other rivers. Although rhinos and wild elephants will happily tramp across agricultural fields—often consuming and destroying them in the process—tigers, Dinerstein wrote, "are averse to crossing gaps in natural habitat that are wider than 5 km."[15] So TAL identified gaps in wilderness where degraded land or land converted to agriculture was wider than five kilometers—a little over three miles—and set out to restore them. Part of that effort involved enticing community forest user groups with a range of benefits—ecotourism, biogas facilities (which convert human waste and manure to cooking fuel), and sustainable livelihood initiatives—as incentives to renew and protect forestland.

The Terai Arc Landscape program was not just a large-scale project to save crucial habitat and improve connectivity. It was overwhelmingly based on the principles of community conservation. Most of its initiatives involved recruiting local people to participate in microcredit schemes, antipoaching patrols, and community forestry. In an e-mail, Mingma Sherpa justified this emphasis: "To me, there is only one solution, that is winning the hearts and minds of the local people to be the custodians of wildlife. The fence and gun methods of protecting parks don't work."[16]

Dinerstein put particular emphasis on "an unanticipated benefit" from the project, pointing out that the returning tigers and rhinos moving into community forests would allow locals to reap the benefits of tourism.[17] One community jumped on the opportunity, he noted: "The enterprising Baghmara forest committee rented domestic elephants, established foot trails, and built treehouses suitable for overnight stays so that adventurous tourists could get to see them. Within weeks of starting operations, these outfits were profitable."[18] Indeed, within the first six months, nearly eight thousand tourists came to the Baghmara forest, generating nearly $200,000 in revenues, which the community used to restore three schools and a health clinic.

There were suspicions and protests early on. When a native tree nursery was established with a $10,000 grant in 1988, and a rosewood tree plantation was set out on a degraded piece of property in a buffer zone near Sauraha, "people were antagonistic," Dinerstein reported.[19] "Some even tried to tear down fences," fearing that these new projects might interfere with cattle grazing. But the locals were quickly won over by the opportunity to harvest thatch grass for roofing and by the increased supply of firewood, from pruning. The plantation eventually became the Baghmara Community Forest, an important expansion of Chitwan's rhino range and a place where the community could bring tourists and harvest plants. These modest beginnings fueled ambitions for TAL, allowing planners to envisage ways to repair critical links between reserves.

During its initial years, the Terai Arc was extraordinarily successful. Rhinos were flourishing in their new habitat at Bardia; tourism was bringing in revenue. Most tourists who traveled to the country to trek in the Himalayas spent a few days visiting Chitwan: In 1990, 36,500 tourists came to the park; by 1997, the number had risen to 96,062. Dinerstein wrote, "We have accomplished what Michael Soulé and Reed Noss call 'rewilding'—returning the land to the creatures that once flourished here."[20] He coined a variation on the term, calling TAL "retigering," defining it as the "largest and most ambitious wildlife and habitat recovery program of its kind in Asia."[21] The project was adopted as the official policy of the Nepalese government and received a coveted imprimatur when the famous naturalist and East Asian wildlife specialist George Schaller called it the "most impressive" effort on the continent.[22]

But then came the war.

The People's War

In 1996, the Communist Party of Nepal, known as the Maoists, declared a "people's war," modeling their movement on Peru's Shining Path, aiming to overthrow the monarchy, the capitalist economy, and the caste system and replace them with a Maoist People's Republic. They were led by Pushpa Kamal Dahal, from Chitwan District, known as "Chairman Prachanda," a nom de guerre meaning "one who kills." For the first few years, the conflict simmered along and the government did not take the threat seriously, assigning the police department rather than the army to handle the insurgency. While the police held sway in Nepal's cities and larger towns, the Maoists recruited heavily in rural areas, particularly in the Terai. They began systematically pressing civil servants and teachers to pay "taxes" and shelter them in schools and public offices; those who refused were threatened and sometimes killed. Throughout the late 1990s, while the government dithered, the Maoist movement gained momentum in the rural south and west.

Then, in 2001, the situation took a turn for the worse when news broke of the most bizarre event in Nepal's history. On June 1, while the royal family hosted a dinner party, the twenty-nine-year-old Crown Prince Dipendra, at odds with his family over his choice of bride and the future of the monarchy, reportedly opened fire with a submachine gun, killing ten, including his mother and father, the queen and king of Nepal and eight others. Dipendra, said to have been drunk, then shot himself; he died three days later. His uncle Gyanendra, who had been out of town during the massacre, assumed the throne. Nepal was swept by rumors. Was Dipendra the murderer, or had the guards shot the family? Had Gyanendra orchestrated a coup to gain power, disturbed at the government's weak response to the Maoists? Eyewitnesses claimed that the prince alone was responsible. The massacre was thought to be the worst killing of royalty since Lenin ordered the assassination of the Romanovs. For the Maoists, it was an opportunity.

Later that year, once peace talks initiated after the massacre fell apart, the rebels launched their most serious attack, killing 186 army and police personnel across the country. From that point on, the war was bitterly fought. An estimated thirteen thousand Nepalis were killed over the decadelong conflict, and hundreds of thousands became refugees in their own country. Human rights organizations cataloged abuses committed by both Maoists and government security forces, including bombings, summary executions, rape, hostage taking, and the conscription of children as combatants.

The civil war was devastating to development as a whole and to conservation in particular. The Maoists targeted the countryside, destroying Department of Forestry offices, ranger posts, training centers, and security camps. After clearing government officials out of the forests, the Maoists used them as safe havens. In the national parks, they killed staff and forced the remaining rangers to flee. In three conservation areas in the Himalayas—Annapurna, Manaslu, and Kanchanjunga—field offices were bombed and conservation leaders killed. In the Terai, where Maoists conscripted the poorest lower-caste ethnic groups to join their ranks, community forests fell under their control, and forest user groups were forced to surrender revenue to the rebels. Nabin Baral, a Nepali ecologist, and Joel Heinen, a colleague at Florida International University assessed the situation in 2006, confirming that the survival of Nepal's community forests was threatened. Long-term implications for conservation, they wrote, were "ominous."[23]

The war destroyed the tourism industry, one of Nepal's chief sources of foreign exchange, which had provided access to a cash economy for the famous Sherpa trekking and mountaineering guides but also funneled revenue to the parks along the southern border. Maoists routinely stopped trekking parties at gunpoint to extort money, boarding buses and holding people for days at a time. While the rebels avoided injuring tourists, they were outspoken in their contempt for Nepalese parks, which were, of course, a legacy of the royal family. On the other side, King Gyanendra, a leader of Nepal's most important national environmental group, the King Mahendra Trust for Nature Conservation, did not endear himself to his people by vastly increasing the royal household budget.

As in civil wars in Africa, the rebels used poaching to pay for arms and feed their troops. Red panda, snow leopard, musk deer, tiger, and rhino were poached inside and outside of the parks. Blue sheep suffered a steep decline in one reserve, falling from 2,200 individuals in 2002 to 563 in 2004.[24] Rhino, easy to locate because of their regular visitation of latrines, were terribly hard hit. In 2000, thirty-three were killed; in 2001, forty-two; in 2002, fifty-five.[25] Parks and reserves lost 70 percent of their guard posts.[26] To raise money, Maoists began gathering and selling rare plants in the Himalayas, pushing species toward local extinction.[27] In addition to poaching, the Maoists also set in motion another wave of migration, with people displaced from mountain areas fleeing to the western Terai, camping in forests, cutting trees, and harvesting flora and fauna to survive.

In 2002, the Maoists issued a threat against foreign-owned tourist businesses, demanding their closure. Tiger Tops, one of Nepal's oldest, most famous companies, which operated lodges in Chitwan and Bardia, rejected the ultimatum, and subsequently a bomb exploded in its Kathmandu office. Two years later, rebels blew up the control tower at the airstrip used by tourists flying to Chitwan, demanding again that Tiger Tops and over a dozen tourist hotels and other businesses close.[28]

Year after year, the Maoists made gains, gradually winning the countryside, closing in on Kathmandu. Gyanendra dissolved the government and declared martial law, but after a series of pro-democracy rallies brought the country to a halt, the king was ousted in 2006 and the Maoists agreed to a ceasefire. Later that year, a peace treaty was signed, with the parties agreeing to place their weapons under UN control.

But if 2006 saw some fragile gains for peace, it was the worst year on record for Nepalese conservation. In May, a press release announced the shocking results of a survey of Bardia National Park: the tigers and rhinos were effectively gone.[29]

During the war years, the World Wildlife Fund had continued to translocate rhinos from Chitwan, and Nepal's population reached a high of 612 in 2000. That year, four were released in Suklaphanta; in 2001, five in Bardia; between 2002 and 2003, twenty more rhinos were moved to Bardia. Of a total of eighty-seven rhinos translocated, seventy-two were released into the Babai Valley. Hemanta Mishra later explained the decision to continue even after the escalation of hostilities, writing, "The political dynamics of Nepal were changing fast. Two years was too long to wait."[30] But once Maoists established a training camp inside Bardia, the Babai Valley became inaccessible to park rangers, several of whom were attacked in 2004.

When a team of forty park staff and wildlife specialists entered the valley on elephants during the ceasefire of 2006, signs of only three rhinos were found. The rest had been slaughtered. Between 1998 and 2001, thirteen tigers were known to frequent the Babai. Only three remained. In other areas of Bardia, only a handful of tigers—out of a population that previously numbered at least twenty-eight—had survived.

It was a cruel blow. The rhino relocation had been hailed around the world, covered by National Public Radio in a "radio expedition" sponsored by National Geographic. Dinerstein had once suggested that "if the protection of rhinos holds, thousands of animals could be flourishing in the Terai Arc reserves within three decades."[31]

Conservationists were still reeling when tragedy struck again. The World Wildlife Fund had decided to take advantage of the relative peace of 2006 to celebrate its handover of a northern conservation area to the local community. Environmentalists from around the world flew in for a ceremony in late September to reward those who had worked on the project since its inception, held in a remote village in far eastern Nepal. After the festivities, visitors boarded a helicopter that crashed shortly after takeoff, in rain and dense fog. All twenty-four on board, including eight conservation advisers and employees, were killed. Among the dead were Mingma Sherpa and Chandra Gurung. A close friend and colleague to many of those who died, Mishra later wrote that Nepal had lost "all the key members of its core parks and forestry leadership, with a combined expertise and experience of two hundred years in conservation."[32]

Goats, Guns, People

In a Kathmandu hotel recently, I came across a Nepalese board game called "Baag Chal," or the Tiger Moving Game. Similar to checkers, the game is played with metal pieces in the shape of tigers and goats: "The goats defend themselves by giving the tigers no open points on which to land. . . . If the goats can encircle the tigers and the tigers have no place to move then the goats win the game."

In Nepal, the goats were winning. TAL's most impressive accomplishment had been the dramatic and expensive translocations, but sadly, most rhinos had not survived the war. Critics suggested that moving animals to the very heart of the Maoist insurgency had been an unwise allocation of resources. They also noted TAL's close association with the monarchy's conservation agenda, a relationship that was bound to count against TAL with a drastic change in government. But whatever the reason, the fact remained that the WWF was losing the Tiger Moving Game. Could the losses have been prevented? Was there any way to protect wildlife during a war? Was it a mistake to rely so heavily on community conservation?

The limits of community conservation were becoming clear. I spoke to Esmond Martin, an American smuggling specialist who traveled frequently to Nepal and whose view was unequivocal. "They have community projects there, and they haven't saved the rhinos," he said.[33] While he was not opposed to such schemes—he lived in Nairobi and sat on the board of the Lewa Wildlife Conservancy—Martin insisted that the chief tool to protect wildlife was improved law enforcement. He pointed to conservation successes in several African and Asian countries. Conservationists, he felt, had "overintellectualized" the problem. He noted that the number of tourists going to Chitwan had been severely affected by the war and that Bardia had lost almost all tourism, cutting park revenues—and their contribution to communities—to almost nothing. Not surprisingly, poachers had stepped into the void, often with the help of villagers.

Martin had interviewed poachers in Nepal and discovered that, while poverty was a motivation, there were also feelings of fear and resentment. "Wildlife *kills* people around Chitwan and Bardia, wildlife *wounds* people, and there are a lot of people who are not sympathetic," he said. I did not go far in Nepal before hearing this confirmed. A guide in Chitwan, Himalaya Keshab Shrehtha, told me his mother-in-law had been trampled to death by a rhino, and he himself bore scars on his mouth from a sloth bear.[34] Negative feelings about wildlife, combined with enormous financial incentives and the opportunities provided by the breakdown of law and order, had overcome the goodwill built up by community conservation during the good years.

As Martin put it, "If law enforcement isn't working, community things are a complete waste of time."

The great-grandson of Henry Phipps, Andrew Carnegie's Pittsburgh steel partner, Martin had a genteel background, but his occupation, as one of the world's foremost experts on the illegal trade in wildlife products, was anything but. His work took him to some of the shadiest corners on earth, and he spent considerable time in the back alleys and markets in Cairo, Khartoum, and Kathmandu, interviewing poachers, carvers, and traders to document the wildlife trade. It was Martin who discovered where rhino horn was going. It was not being used in the preparation of Asian aphrodisiacs, as popularly believed. Most African rhino horn was ending up in Yemen, to be carved by artisans into the handles of ceremonial daggers, called *jambiya*, coveted objects among oil-rich tribes. The discovery led to Yemen's ban on the trade and to a plan encouraging artisans to use water buffalo horn instead, bringing the price of rhino horn down in that country.

In a 2004 report published in *Pachyderm*, Martin described the process of poaching in Chitwan during the war. The poachers, he wrote, were mostly local people, including Tharu villagers: outsiders would soon be spotted. A gang of two to five men with several guns typically entered Chitwan by swimming across the Rapti River. They spent several days hunting a rhino, sleeping in trees or caves. For a single rhino horn, the shooter received the equivalent of $670 to $1,350, a fortune in Nepal; his partners stood to get between $340 and $500. They were paid by a middleman in a nearby town, who then took the horn to Kathmandu to sell it for export, most likely to China: Asian horn—far more valuable even than African—was used in medicinal preparations to treat fever. In 2003, rhino horn was selling for $9,500 to $10,000 a kilogram or ten times what the shooters were paid.[35]

While the army had historically provided security in the parks, the WWF and other private organizations had stepped in during the 1990s to collaborate with the government in creating targeted "Anti-Poaching Units" in Chitwan and Bardia, as part of their community conservation activities.[36] The APUs, consisting of two rangers and thirty game scouts, who patrolled the park and neighboring forests on elephants, also depended on a network of informants, providing information on possible poaching plans. In addition, Chitwan's Buffer Zone Management Committee offered rewards to those who supplied information leading to the capture of poachers.

As hostilities heightened, security faltered. At the beginning of the war, the army had withdrawn from many guard posts in the park in order to consolidate its forces, fearing attack on posts with just a few soldiers. It was a reasonable military strategy but a catastrophe for rhino conservation, leaving large parts of the park with no protection. Disorganized by the war, the forest department, the army, the forest user groups, the police, and the NGOs stopped communicating with one another; the informant network fell apart, contributing to the confidence of poaching gangs; and the park, which had lost so much revenue from the war, stopped paying the APU staff.[37] The result was the rising rhino death toll of 2000–2003.

Yet Martin found that the parties involved were able to suppress poaching effectively—once they had made up their minds to do so—even during the height of the war. In mid-2003, Nepal's parks department finally realized that it had to do something, and a new chief warden was appointed to Chitwan; he changed the patrol strategy, sending out "sweeping operation[s]," large parties of park guards and soldiers patrolling intensively for a week at a time in spots where poaching was common.[38] He improved coordination between the park and the army; payment to informers was resumed with funding from the WWF and other groups, and the intelligence network was bolstered; the parks department visited communities in the buffer zone to reinforce the importance of rhino conservation and its contribution to tourism. The strategy worked, even during some of the worst months of the war. Between July and December 2003, only one rhino was killed in Chitwan and none were killed in Bardia. But the progress was short-lived. Once again, personnel in the park and army were shifted, and the successful strategy was undercut. Once again, rhinos were being killed.

Between 2002 and 2006, 104 rhinos were poached from Chitwan. The mutilated remains of eleven animals were found in the park in late 2006. A warden for the eastern sector of Chitwan told a reporter for the *Daily Telegraph* that his men had been reduced to "data collectors," starved of funding and materials. "Resources have been given in our name," he said, but then they vanished. "We don't even have minimum logistics—no water, no fuel. No training is given."[39] War had wiped out most of the conservation gains of the past thirty years.

Visiting the Terai in 2006, I could see the problems for myself. The war had ended, but the goats were circling the tigers.

Looking for Tigers at Tiger Tops

From the back of an elephant, Chitwan was magical. Ahead, an enormous river valley opened out, reeds and grasses waving about the lobes of the pachyderm's weathered gray head as she pushed through. Behind, a line of forested hills, the celebrated jungles of Chitwan, home to the Bengal tiger, rhino, leopard, and sloth bear. The silence was absolute, broken only by the thrashing of the elephant's trunk and the calling of birds. The horizon was awash in watercolor hues of heat haze, and I was floating, in a lurching kind of way, through a sea of grass, carried on a creature so massive it was impossible to see all of her. She flapped her huge ears, the backs of which presented a startling field of pink covered in gray polka dots, which I learned were a sign of age: the older the elephant, the more dots.

Upon arriving, the only guest at Tiger Tops Tented Camp within the park, I had been led to a set of stairs that ended in open space. Moments later, an elephant backed up to it. I was waved onto the cushioned howdah—a railed platform for passengers, cinched like a saddle around the elephant's stomach—with a vaguely bored air by an attendant, as if an elephant were a jungle subway car and boarding it an everyday occurrence. A brass plate was attached to the top rail of the howdah, inscribed with her name: Dipendra Kali.

Tiger Tops was once one of the premier safari destinations in the world, the most famous in Asia, but it had fallen on hard times during the war. Founded in 1965 by an American company, then acquired by a British businessman, its treehouse-style bungalows built on stilts at the edge of the Rapti River have been visited by everyone from heads of state to rock stars. In the bar of the main lodge are fading photos of Hillary and Chelsea Clinton, Mick Jagger perched atop an elephant (looking less sprightly than usual), Henry Kissinger, Steven Seagal, James Coburn. At its airstrip every January, Tiger Tops still hosts the World Elephant Polo Association Championship to benefit local schools and orphanages; in 2009, England took the trophy from the highly favored Scottish team. According to the rules of the game, which is played with a standard ball but longer sticks, "elephants are swapped at half time to even out any advantage. No elephants may lie down in front of the goalmouth. An elephant may not pick up the ball with its trunk during play. Stepping on the ball is forbidden."

Dipendra Kali, with whom I would develop a fond familiarity over the next few days, even bathing her on one memorable occasion, was forty-five years old. She knew upward of twenty-five commands, spoken by the *phanit*, the elephant driver sitting behind her head, or delivered by the prodding of his bare feet. He urged her on with his toes jammed behind her ears. The *phanit* also carried a metal implement, used on distressing occasions to register his displeasure when Dipendra stopped to have an unauthorized bite to eat or refused to proceed in the direction of his choosing. I sometimes wondered why she didn't snatch it out of his hand and give him a whack with it.

It would have been virtually impossible to travel without her. We were traversing a vast marsh densely covered in grasses twenty feet high. The monsoon season was just ending, and the grass was at its height, the ground soggy. While jeeps made good headway on the road, we were far from any track, and Dipendra sank in up to her knees in spots, pulling her feet out with an audible sucking sound. Had we attempted the journey on foot, we might have been killed by a rhino. The only safe way to approach a rhino is on elephant back: the elephant's smell masks the alarming odor of human. Rhinos detest humans, with good reason.

While Chitwan seemed peaceful, all was not well. I saw no tigers but did see plenty of evidence that people were entering the park illegally, collecting plants and wild animals. Everywhere in the sand were human footprints. My guide and I surprised a barefoot couple carrying a bowlful of snails. Multiply their takings by hundreds, by thousands. As Martin said, tourism in the park was down, and the decline in revenue meant less money for guards and less likelihood that poachers would be discovered.

At Tiger Tops, in Sauraha, a town directly across the Rapti River from Chitwan, it was clear that tourists had vanished, taking the money for the maintenance and expansion of community forests with them. Everywhere, hotels and guesthouses were boarded up and vacant, their signs faded. At the Baghmara community forest user group office, I sat with the longtime chairman of the user group, Bishnu Prasad Aryal, an elderly and avuncular man in the traditional *topi* cap, white shirt, and vest. His family owned the Rhino Lodge, and he said that in better days the forest had used tourist revenues to provide 460 homes with biogas facilities, which spared the need to collect firewood. But times were tough: plans for a new wildlife viewing tower had been frustrated. "We need lots of budget for that," Aryal said.[40] "If there were no political problems, they could have finished it already."

At Bardia, the situation was not much different, although in the surrounding villages, conservation was promoting itself colorfully. Handpainted signs, billboards, and slogans festooned roads and buildings, touting WWF "eco-clubs." "Save the Nature for the Future," read one. A billboard of a blue Gangetic dolphin pleaded, "We Are Endangered—Help Us for Our Survival."

At the best of times, Bardia attracted only a few thousand tourists a year. This was not the best of times, and I again found myself alone at the Tiger Tops Bardia camp. Few tourists were expected in the future. While the park was as beautiful as, if more remote than, Chitwan,

I was surprised to see rice fields in the buffer zone along one border, coming right up to the edge of the forest, poorly protected against incursions from elephant or rhino. Trenches had been dug, and precarious platforms—nests of webbing roofed with thatched grass—slung in the trees as makeshift guard posts. There were a few more substantial wooden watchtowers as well. But a guide showed me muddy patches where elephants had easily crossed: no such low-tech devices could ever keep elephants out.

Hailed as one of TAL's great accomplishments, the Khata corridor—a narrow finger of land oriented roughly north-south, only three kilometers long and one and a half wide, linking eleven community forests—was proof that the program was working. But its success also highlighted the disconnect between conservation goals and community development. Since 2001, the WWF had poured resources into the restoration of the corridor, to bridge a gap and restore wildlife movement between reserves in India and Nepal. Natural regeneration, reforestation, and carefully controlled land use by communities had transformed this degraded strip of forest, partially converted to agriculture, into a migratory corridor used during monsoon months by elephants, tiger, wild boar, and spotted deer moving north out of the flooded areas along the India-Nepal border.

But the Khata corridor was proving so popular with elephants that it was intensifying human-wildlife conflict. Bardia may now have some fifty to sixty elephants, who were, predictably enough, wreaking havoc among nearby villages. An article on a Nepali news site reported on the "titanic struggle."[41] During 2006, the article claimed, elephants had damaged crops, destroyed two huts and part of a ranger's post, knocked down watchtowers, killed one resident, and injured several others. "Every year from August to October we have a miserable time," said Balaram Timilsina, the chairman of a buffer zone user group to the northeast of the lodge. "Herds of elephant raid the adjoining crop fields and we have to take risks in driving the giants away from our rice fields." It remains illegal to kill an elephant in Nepal, and the punishment is severe, a fine of over 100,000 rupees, fifteen years in jail, or both.

Tigers moving up from the south were also taking advantage of the restored Khata and causing no end of trouble for local residents. As we toured WWF projects in and around the corridor, Santosh Nepal, coordinator for Terai programs, and Tilak Dhakal, the TAL project comanager, rarely ventured far from an overwhelming theme: the need to address these conflicts.

Near the corridor, at the modest cement headquarters of the Gauri Mahila community forest user group, the two men led me through a waiting crowd of committee members to greet a local celebrity, Bhadai Tharu, a cheerful round-faced forty-three-year-old man who had lost his left eye to a tiger in January 2003. Santosh Nepal prompted Tharu to tell me the oft-recited tale, introducing him as "a very great person" who had won a conservation award in recognition of his bravery and tolerance.[42] We crowded onto chairs and a bench in the dark room, and those present listened, greatly affected. While he told the story with a practiced air, Bhadai Tharu seemed humbled anew by the dramatic event, patiently allowing a foreigner to view his scars and missing eye.

The father of three, he had been among those who established the forest user group in 1998, becoming its chairman in 2002. On the seventh of January 2003, he opened the community forest for a week to allow people to cut thatch grass and was monitoring their progress. The tiger was hidden in the grass, and when Bhadai Tharu approached it, unknowing, it leapt out, striking him a terrific blow on the head and breaking his eye socket. Santosh Nepal broke in, "It was more than five hundred pounds!" The tiger clawed his arms, and Bhadai Tharu struggled: "I shoved him away. For four to five minutes, I kept fighting."[43] After those terrifying minutes, he was able to run, but his eyeball, Santosh leapt up to demonstrate, "was hanging to that length," holding his hand down to his waist. Friends saw him and ran away in fear. One stood by and said, "Don't run. He might follow you." The room had gone completely still.

An ambulance was called, and Tharu was hospitalized for over a month. Yet he continued to feel, as he said later, that "tigers are ornaments of our forest. We humans have to give them space and learn to coexist."[44] The government awarded him and the group 50,000 rupees (about $750). With half of that money, he bought a small retail shop, since he was no longer able to work in the fields. He remained committed to his ideals. "People need forests," he is quoted as saying on the WWF–Nepal Web site. "If there are forests and jungles, people are happy. And if there are forests, there will be wildlife."[45] The tiger that attacked him still lives nearby, but Bhadai Tharu does not mind. "He is a symbol of conservation here," Santosh Nepal said proudly. It was an impressive tale of tolerance in support of conservation, and I would later hear other examples of courage on the part of the community-based antipoaching units, who, armed only with stout sticks, patrolled the buffer zone around Bardia at risk to themselves.

On a drive through the corridor itself, on a narrow, muddy track between walls of high grass, Santosh Nepal pointed out *machans*, built to "keep an eye on incoming elephants. Rhino also, and deer." The only way to appease the local population for the dangers posed by wildlife, Nepal said, was to offer them something in return. He pointed out a rough cement trough in front of a house: "We support this kind of piglet-feeding trough. This is not our core competency, but we do it." Cement troughs replace the need for wooden ones— one less tree to cut down. "So they feel good about conservation," he said. "But unless we do some things like this . . ." He shrugged.

The implication was clear: here, as in the Northern Rangelands Trust, community conservation outfits had to provide substantial economic support in exchange for cooperation. That afternoon, Tilak Dhakal took me to see more development projects—a juice factory utilizing wild fallen fruits ("rhino apples"), a workshop producing rattan cane furniture. We visited a microfinance bank festooned with a logo reading, "God Loves Originals, Carbon Copies Are Rejected," providing loans of a few dollars to poor or disabled women to buy a piglet or foster a family business making incense sticks or baskets. We dropped by a local school with new TAL-furnished water taps and a home with a biogas toilet.

It was a typical Tharu home, spotlessly swept, a few chicks nestling comfortably inside a rubber sandal. Tall vessels fashioned out of mud, full of rice, divided the space into rooms. Followed by a raft of small children, we inspected the outhouse, remarkably odor-free. Biogas toilets are mounted over a buried anaerobically sealed tank. We peered in the round cement trough where animal waste was deposited and then turned with a crank funneling it into the underground chamber. Methane is produced there and fed by copper pipe to a two-burner stove inside the house. An astonishingly simple yet effective technology, biogas reduces pressure on forests while eliminating household smoke. It frees women from tedious wood gathering, allowing them to pursue education or other ways to earn money. But the facilities are not cheap, and few can afford them.

Continuing farther west, around the city of Dhangadi, after passing through dozens of Maoist "roadblocks"—lines of youth stretching red crepe paper across the road, demanding "donations" for a variety of causes, screaming, "Twenty rupees, twenty rupees!"—I saw more biogas, more community user groups, more feeding troughs. We drove under bright red Maoist banners of the hammer and sickle: there may have been an end to hostilities, but the war for the hearts and minds of the poor raged on. While my World Wildlife Fund hosts assured me that they had a tiger biologist, most of their tasks revolved around tending to myriad forms of community development: building biogas toilets, distributing fuel-efficient stoves, setting up microcredit facilities, supporting health clinics.

Dhan Rai, the doughty WWF economist spearheading efforts there, had recently had a close call with a Maoist bomb, which exploded next to his office while he was away, in Kathmandu. His walls were still pockmarked with holes from the blast. He took me to the outskirts of town, to show me revegetation of formerly degraded fields with young plants, "native species of the Terai," including medicinal herbs and types of asparagus valued in tonics and ayurvedic medicine: the products could eventually be sold to India. Nearby was a tree plantation. "Slowly the people have realized the benefit of conservation," he said.[46] Comparing the area with the Dolpo region, high in the mountains, where everything grows slowly, he noted that the Terai was quick to regenerate: "Here, you can make a big difference in one year's time." Nearby women were pruning and weeding, talking while they worked: their community had successfully undertaken the restoration of a forest in the middle of a civil war, an astonishing accomplishment.

A young assistant to Dhan Rai, Anand Chaudhary, working on a graduate degree in biology, took me to the nearby Ghodaghodi Lakes, choked by invasive plants, the water polluted by an overabundance of pilgrims who descend yearly to observe Hindu marriage rituals and enjoy holy baths. We saw ecoprojects begun and abandoned, including a machine for extracting oil from plants, rusting behind a community forest user group office. Donated by the UN's Development Program, the machine was supposed to have helped local people raise revenue from selling oil. "But they didn't provide proper training," Chaudhary said.[47] "We're trying to find the technician who can train people to use it. Six years it's been here." So far, the community has been able to extract only five liters of oil.

The abandoned hulk, and the baffled community forest president who emerged to ask patiently if we could help him with it, bore eloquent testament to the perils of good intentions. The community development efforts seemed powerfully moving and troubling at the same time. It was impossible to begrudge people who had next to nothing whatever help could be given. But could the slow, incremental progress made by eco-clubs, biogas toilets, and improved stoves act in time to save the tiger and the rhino?

Impressive as TAL's investments in community conservation were, as courageous as staff like Dhan Rai had been, doggedly pursuing conservation in the most difficult circumstances, the program had not yet found the ideal balance between addressing poverty and protecting species. The WWF had successfully opened corridors for wildlife, but while they were at it, something crucial had been neglected: poaching. As a result, most rhinos translocated to Bardia were dead, along with most of the park's tigers; without a change in approach, this was surely the fate that awaited tigers moving in from India. As in Kenya, it felt cruel, pressing Nepalis about this issue when their country was falling apart. At the moment, Nepal was a failed state, its government weak and dysfunctional. Charities and NGOs were offering the few essential services to be had. In the hills, in the Terai, people were starving. According to the *Nepali Times*, the Maoists were planning education "reform," to teach "the life of Mao Tse-Tung" to ten-year-olds, alongside a "general introduction to explosives, grenades, and booby-traps."[48]

Conservation was always finding itself in this position. Wildlife— and the ecosystems that support them—always came last, behind everything of greater importance, which is to say, human importance. The knowledge was there: poaching had been brought to a halt in the second half of 2003, one of the worst periods of the war. But the will to deal with it was not. No matter who was responsible—the government, the parks department, the army, the WWF and other groups— they all understood what a successful antipoaching strategy was. But other matters took precedence, and security took a backseat. There were many things that could have been done, short of arming a private militia—paying rewards, funding antipoaching units, flooding critical areas with patrols. The point, as Martin had put it so eloquently, was protection. Without it, all the plans were worthless—no matter how long-term or large-scale—because there might be nothing left to protect.

That point is irrefutable. Worldwide, there may be as few as fifteen hundred tigers left in the wild.[49] A hundred years ago, there were a hundred thousand. Although there are many in captivity, only wild tigers can teach their young how to hunt and survive. When it comes to big charismatic megafauna like tiger and rhino, vulnerable to poaching, security is everything. Lewa proves it, and also proves that security can be a critical component of community conservation. Three subspecies of tiger have gone extinct in the last century—the Bali, the Javan, and the Caspian. Without stringent protection in public and private reserves, the rest will follow, probably within the next fifty years. I agreed with Mingma Sherpa, that winning hearts and minds was crucial. But as I left Nepal, I also understood that it was not always enough.

In 2007, Nepal's monarchy was abolished, and King Gyanendra—who had worked closely with the WWF for years—moved out of the royal palace in disgrace. The country was reborn as the Federal Democratic Republic of Nepal. Its first elected prime minister, assuming office in 2008, was Prachanda—"the one who kills"—and Maoists won a majority of seats in parliament. While the government frittered away months arguing over ministry appointments, crime escalated. In August 2008, the BBC reported that "a sense of anarchy prevails nationwide. . . . Many Nepalis are in utter despair."[50]

Yet even in the midst of such dejection, Nepal's conservation picture began, yet again, to rebound. Confronted by ever-worsening poaching, the World Wildlife Fund once again bore down on the problem.[51] Anil Manandhar, the new head of the organization in Nepal, set four staff members to study the problem and come up with a plan to protect the rhinos. They analyzed poaching incidents around Chitwan and saw that most animals had been taken along the park's perimeter and in the buffer zones, community forests, and national forests. They could also see that poachers had been watching for opportunities: almost all the attacks took place just after a poaching patrol had ended.

Manandhar used what had worked before. He contacted the government and asked that the army guards once assigned to the park and national forests be remobilized; he encouraged the park to restore abandoned security posts and regular patrolling; and he recruited former army and police members to patrol outside the park on motorcycles and bicycles. Some of these men performed undercover missions, collecting information on local poaching gangs and the market for rhino horn. Six Nepalis were hired and trained to use GPS equipment to track rhinos outside of Chitwan. Once again, an NGO was initiating and organizing what should have been the responsibility of the government, but at least the agencies came on board.

One of the most effective tactics brought to bear in "Operation Unicornis" came about as a direct result of community conservation efforts. The WWF put out the word to its eco-club members, the youth groups active in many villages managing community forests. During an eight-day period, 480 eco-club students from the Chitwan area collected 101,500 signatures on a petition, a roll of white cloth that measured nearly half a mile, in support of the rhino. Presented to the government, the petition inspired a special session of parliament on rhino conservation and poaching issues. A 2006–2011 Greater-One-Horned Rhinoceros Conservation Action Plan was unveiled, dedicating 30 percent of the $1.5 million Chitwan budget to antipoaching patrols—$100,000 a year.[52] In Chitwan, eco-club volunteers used GPS information on rhino movement to assign teams to patrol their location. Some youths began guarding individual rhinos during the night, a technique reminiscent of that used at Lewa. In 2007, the number of rhinos poached in Chitwan fell to one, and the population started moving in the right direction, from a low of 372 in 2005, to 408 in 2008. WWF–Nepal is now planning to expand the program to Bardia. Across the border, India's Kaziranga National Park has expressed interest and may adopt it as well.

Despite the constraints of poverty and unrest, there was still strong support for the rhino, at least among the people living around the parks: Nepal's newspapers carried photographs of the eco-club schoolchildren marching with their enormous banner. Touching as it was, however, that support could not prevent the pendulum from swinging back again. During the first six months of 2009, seven of the country's rhinos were killed. Worldwide, the WWF reported that the previous year had been the worst for rhino poaching in fifteen years. As always, the future of conservation in Nepal is unclear.

During the spectacular celebrations of Tihar—which concludes a monthlong October festival that seems like Thanksgiving, Fourth of July, and Christmas rolled into one, with feasts, religious devotions to Lakshmi, goddess of wealth, fireworks, and boisterous singing late into the night—I saw how Nepalis cherished their animals, crowning startled dogs with garlands of flowers and painting the foreheads of cows with colorful tikas. The people of Nepal have suffered terribly over the past decades, from the corruption of the monarchy, the brutality of the Maoists, the collapse of civil society, and conflicts with wildlife. Yet, when it came to losing the rhino—which may displace the cow as Nepal's national animal—they continued to fight for the sacred *gaida*. Chitwan was still the heart of their jungle.

The Cautionary Tale of Corcovado

The issue of law enforcement in conservation remains perpetually vexed, as the recent history of community-based projects shows. The conservancies in Namibia and Kenya were built on a solid foundation that included community-run security programs. But in Nepal, the picture was murkier: antipoaching patrols functioned well even under the most challenging conditions, but—for whatever reasons—the institutions involved failed to support them consistently. The rhinos and tigers paid the price.

The real value of combining community conservation with law enforcement emerges most clearly in a place that lacks both. Failing to provide such essential services incurs costs that have been particularly conspicuous in an area of irreplaceable biodiversity.

One of the most ecologically important places on earth, Costa Rica's Osa Peninsula is so densely primeval that a fictional island offshore served as the setting for *Jurassic Park*. Located on a major peninsula on the country's southwest coast, comprising an area of a million acres, the Osa is among the most perilously endangered large-scale protected systems on the planet, the last remaining significant block of rain forest on the Pacific coast of Central America.

Osa's richness arises from its geological history: it was itself an island until several million years ago, allowing many species to evolve independently. It contains a great diversity of ecosystems: primary and secondary lowland tropical forests, cloud forests, mangrove swamps, freshwater swamps, beaches, lagoons, and coral reefs. Altogether, 50 percent of the country's biodiversity occurs there—140 species of mammals, 66 freshwater fish, 117 reptiles and amphibians, and perhaps 10,000 species of insects—an amount roughly equivalent to over 2 percent of the world's biodiversity. There are over 700 species of trees, the greatest diversity of all of Central America. Osa also harbors an extraordinary number of endemic species: an estimated 2 to 3 percent of the 4,000–5,000 plant species on the peninsula exist nowhere else. Eighteen of over 375 species of birds are endemic species. In the skies over Osa fly some of the last spectacular, endangered scarlet macaws.

In 1975, the Nature Conservancy, in its first international venture, collaborated with Costa Rica to create the country's first national park, Corcovado, on a third of the peninsula. The group aided the government in acquiring over 86,000 acres, eventually expanded to 100,000. Nevertheless, in creating Corcovado, planners underestimated the needs of its species. As Alvaro Ugalde, a director of the Osa Conservation Area, said in 2004, "Out of ignorance, we created a park that is too small. At the time, we thought it was gigantic. What the years have proven, though, is that Corcovado is very small, . . . especially when we talk about critical species such as the jaguar, peccary, and harpy eagle."[53]

Another long-festering problem lay in the failure of international conservation groups to invest in community development, which served to alienate people living around the park, among the poorest in the country. The density of the peninsula had long isolated it, and only a small indigenous population of Guaymí Indians had settled there before the nineteenth century. But as Costa Rica's population grew, the poor began moving farther from the capital, San José, looking for land to farm. When gold was discovered in Corcovado, thousands flocked there. By 1985, more than a thousand gold miners were trespassing within the park, using destructive placer and pump mining equipment, destroying streams. The miners were evicted in 1985, and many of those who still live around the park hold deep grudges about their treatment.

There were sporadic efforts to involve local people. Some conservation groups tried to engage the population in a sustainable forestry project, in an attempt to create buffer zones around Corcovado. But the program eventually failed due to insufficient funding and an inability to sort out land tenure. That disappointment hardened the attitudes of an increasingly resentful and hostile populace on the park's borders.

One of the results has been a steep increase in poaching in recent years. White-lipped peccary and collared peccary, among the most important prey species of the jaguar, were both drastically affected. A piglike creature covered in dense, short hair, the peccary roams the forest in large herds of fifty to several hundred animals, rooting for tubers. When poaching reduced the number of peccary in Corcovado by 60 percent, jaguars faced starvation and began raiding farms outside the park border, where they were shot. Between 2000 and 2006, lack of prey and illegal shooting cut the number of jaguar in Corcovado from 150 to around 30.[54]

The pillaging might have been stopped by a concentrated application of community development combined with adequate law enforcement. But neither was forthcoming. During the 1990s, the Nature Conservancy included Corcovado in its "Parks in Peril" program. After a three-year effort to improve park management, the group celebrated its "strategy of success," claiming to have installed "an adequate system of protection consisting of 15 trained park rangers."[55] The group also claimed to have promoted local participation and sustainable development, but if so, the results were negligible, since the program preceded the worst period of poaching. The self-administered pat on the back was premature, as the organization failed to ensure consistent security over the long term. In the years since, the Costa Rican parks department has continually struggled to pay its park rangers. In May 2007, a newspaper reported that the Ministry of the

Environment and Energy, which runs the national parks, did not have the money to pay Corcovado rangers that year.[56]

Over the course of several days spent exploring Corcovado, I interviewed a former poacher turned nature guide, Luis Angulo Angulo. Guiding me around the forests and beaches near Sirena Biological Station, a remote camp for park rangers and biologists, Angulo spoke passionately about the history of the park. A small, intense man with dark receding hair, Angulo at fifty had done a little bit of everything. Some years before, he had been shot in the foot by park rangers, who caught him mining in the park. While he had forgiven them, he remained bitter, particularly about the Nature Conservancy's failure to do anything for him or his people. "*Yo soy campesino*," he told me many times.[57] *Campesino* means "peasant" or "laborer," but it carries connotations of political anger, a feeling of being ill-used or downtrodden.

On the wide wraparound porch at the research station, built on stilts to keep it off the sodden ground, Angulo pointed out his village on a faded wall map, showing me where many foreigners, including American movie stars like Woody Harrelson, had acquired valuable land on the peninsula. Above all, he was concerned with basic fairness. Frustrated by the lack of information about conservation fundraising in the area, he knew that the Nature Conservancy and other groups, as well as the government, had raised millions of dollars for conservation in the area. But where had the money gone? He could not see it, nor could other *campesinos*. He saw photographers and tourists arrive, but there was no money for communities coming from such tourism. He knew that photographs of poison dart frogs and jaguars were worth money, but locals weren't getting it.

Angulo's reactions echoed those I had heard from guides in South Africa. While he himself benefited from tourism, he knew that those advantages did not stretch far enough or reach enough people. He had taken a course in parataxonomy, the collection of moths and butterflies for biological research, with the famous conservation biologist

Dan Janzen, and he was convinced of the importance of insects, birds, butterflies. He said he no longer wanted to cut trees or mine for gold. "Because I know," he said, patting the trunk of a vast tree. "Many plants, many insects, many *animales*" made their homes in such trees. Only biologists could have rivaled his ability to identify and explain the utility of the thousands of organisms we encountered.

He was an amazing, preternaturally acute guide. He walked along with his head to one side, listening, crouching, pointing out spider monkeys, coatis, macaws, butterflies, dainty red brocket deer frozen in the underbrush, things I would never have seen without him. Carrying only a pouch and a machete, wearing rubber sandals or walking barefoot, he could identify and speak knowledgeably about scores of species. One afternoon, near the beach, he found fresh tapir tracks and told me to wait for him. He disappeared into dense shrubbery, following the tracks all the way to the tapir, sleeping in a mud wallow with her yearling calf. Returning, he collected me and a few tourists wandering on the beach and led us to the animals, watching proudly as we took our photographs.

But despite his late-found love of conservation, Angulo saw no use for the Nature Conservancy. He accused officials of lining their own pockets and spoke darkly of their expensive homes and big cars. "*Ellos son parasitos*," he spat out. He told me about attending one of their meetings. He asked an official why there was so much money for conservation and none for the people of Osa. He mimed the response: blank face, closed mouth.

He was eloquent on the wildlife conflicts suffered by people living around the park. His mother-in-law had lost all her cassava plants to peccaries, but no one was legally allowed to kill them. His father-in-law was caught killing a tapir that ate his potatoes and had to pay a hefty fine, around $575. "When you are hungry in the stomach," he said, "conservation is impossible."

Luis Angulo Angulo was not the only one saying these things. I asked a Costa Rican biologist, Eduardo Carrillo, why groups like the Nature Conservancy had become so unpopular. Carrillo had spent years tracking peccaries and jaguars through the park, compiling the data that proved poaching had decimated their populations. Carrillo, too, was disappointed with conservation groups.

"You know," he said, "They *must* pay more attention to the people who live in those areas. For example, the Nature Conservancy got money from the Moore Foundation, and they hired park rangers."[58] That was a boon, but the organization's employees behaved in a manner that alienated local people. "They bought cars to work in the peninsula," he said, "big cars with the TNC sign in the door, with air-conditioning. These organizations are not listening to the people, about the problems they have."

Carrillo praised a successful local program that had established an environmental education club for a secondary school in Puerto Jiménez, the town on the peninsula. The club took the students into the park to see the wildlife and got them involved in recycling and making educational materials for other people near Corcovado. "They really believe in conservation right now," he said, "but I can assure you that four years ago they didn't even think about what is important for Corcovado. Maybe the organizations are starting to understand that education is necessary."

Despite those failings, Carrillo was still hopeful about the program to hire new park rangers with the Moore Foundation money. Many of the new rangers were chosen from among the ranks of former poachers, like Luis Angulo Angulo. He told me about training the poachers at a workshop and said one had approached him to say that he now understood why it was important not to kill peccaries. "That touched my heart," Carrillo said. "I thought, we *are* doing something right. Not only chasing people to take them to jail. I feel really proud about that."

The Moore Foundation money may have been keeping the existing, if inadequate, numbers of rangers on the rolls. But even there, to quote Angulo, it was hard to see where the money was going. At Sirena Biological Station, two beleaguered guards trudged miles on their daily rounds, patrolling a vast, remote, and difficult terrain that might have defeated an entire platoon. They complained about lack of money for everything: uniforms, equipment, salaries. There was no secret about their dire situation; it was noted in academic articles and the country's newspapers.[59]

Yet another new plan, the Osa Biological Corridor project, launched by a coalition of local environmental groups, may turn things around. Designed to enlarge and consolidate protected areas in the peninsula by buying private property, the project (now backed by Conservation International and the Nature Conservancy) aims to create a $10 million endowment for Corcovado and to address the elements that have so far been lacking: education and the development of community support. Osa dwellers are apparently becoming more attuned to the value of their natural resources, because there has been growing support for a related program, the "Natural Resources Vigilance Committees," which have recruited people in villages to monitor and report illegal hunting and logging infractions. But with a hundred thousand people living in and around the Osa Peninsula, the vigilance committee members—all fifty-five of them—are facing an uphill battle.[60]

Guns, fences, and what has been called fortress conservation: the community-based movement was dreamed up to put the dark history of parks to rest, placating leftist critics for whom conservation was forever tainted by racially motivated evictions and human rights abuses.[61] At the other end of the spectrum, community projects alienated biologists, whose argument—that conservation and development were aims fundamentally at odds with each other—had merit. Early attempts to wed them had failed. Yet some projects confounded the stereotypes, proving that under the right conditions, guns, fences, and economic development could secure a peaceable kingdom for people and wildlife.

The answers were out there. Namibia found them. Ian Craig found them, and put them to work at Lewa. The Maasai and Samburu recognized and adopted them. There were simple solutions to conservation problems, ready to hand. Conservancies worked, particularly when they granted legal rights and responsibilities; they functioned best when they included law enforcement. But there were also certain situations—wars, conflicts—when community conservation was not enough.

These issues are unresolved. Unfortunately, the conservation community has always been timid about both community relations and law enforcement. Through a fear of alienating their donor base or inciting the wrath of indigenous people, conservation groups have failed to grapple with profound and controversial issues. A refusal to engage honestly with local people about land use and trade treaties sealed the fate of the Mesoamerican Biological Corridor. After moving rhinos into the center of a war zone, the WWF responded to poaching inconsistently. In Corcovado, the Nature Conservancy helped create the park and sporadically supported rangers. Declaring success, it walked away, for a time, from further responsibilities.[62]

Joseph Kirathe was right: conservation is not about managing wildlife. It's about managing people. That may be why putting even simple solutions to work seems to be the trickiest part of a very tricky business.

PART IV

"SUSTAINABLE
CONSERVATION"

CHAPTER 11

RESURRECTION ECOLOGY

From Curtis Prairie to Fresh Kills

REWILDING BEGAN BY WORKING ON CORRIDORS. IT MOVED INTO transboundary planning on a massive scale and made a detour into community conservation, a model for rallying local people to support conservancies and collectively managed forests. But the most surprising progress in rewilding has come in a field so raw, so new, that textbooks are still being written, terms defined, and methods invented. For centuries, we have been Shiva, destroyer of worlds, burning forests, poisoning lakes, emptying oceans. Now we are refashioning ourselves as a creator, recovering wastelands and making them whole.

In the 1930s, when a reporter asked a biologist how long it would take to restore a Wisconsin horse pasture to tallgrass prairie, he said, "Roughly a thousand years."[1] But that horse pasture, consisting of two farms originally plowed in 1836, began to recover within twenty years. The University of Wisconsin's arboretum bought the properties, and work on the sixty-acre site began in 1933, with over two hundred men employed by the Civilian Conservation Corps collecting native seed and seedlings, removing soil from the surface, plowing, reseeding, transplanting sod with native grasses, and mulching with prairie hay. Aldo Leopold, whose restoration efforts in Wisconsin's sand prairie became integral to the project, was a supervisor. Over the decades, important discoveries were made, particularly in the 1950s, when trials with controlled burning revealed that an annual fire regime favored native prairie grasses, controlling shrubs and woody plants.

Curtis Prairie, at seventy-five, is the world's first and oldest experiment in ecological restoration. Subject to continual surveys and analysis, the tallgrass prairie appears fully recovered, with beautiful summer displays of purple aster, goldenrod, and bluestem and Indian grasses. A 2002 survey identified 265 species, 230 of them native. But if the restoration was accomplished faster than expected, it turned out to be neither complete nor perfect. The process can be surprisingly fragile: Curtis Prairie has been subject to repeated invasion by exotic species, including one, reed canary grass, that has colonized sizable areas and proved impervious to efforts to remove it. Destructive storm water runoff from nearby developments, carrying invasive species and fertilizers, has eroded and destabilized soil nutrients. Biologists are currently studying forty-eight isolated plots within the prairie, applying different "cocktails" of native seeds and herbicides, trying to discover what will be effective in eliminating invasives. Joy Zedler, an ecologist overseeing this most recent restoration of a much-restored plot, said recently, "It takes time to identify all the things you don't know. There is no cookbook approach to restoration."[2] Curtis Prairie has taught biologists many things, but the most sobering lesson is this: ecological restoration is never done.

Restoration is the fourth major incarnation of rewilding, but it may become the dominant one, simply because we have already destroyed so much. Indeed, many other rewilding methods incorporate it. Along the European Green Belt, biologists are coaxing wet meadows and frog ponds from the former Death Zone; in the northern rangelands of Kenya, the Samburu are restoring grasslands; in Nepal, the Tharu people are regrowing forests.

But while suffused with hope, these endeavors have also provoked a host of questions: What does it mean to restore a landscape or an ecosystem? Restore it to what? To some prior era of primeval perfection? And once something has been restored, like the Wisconsin horse pasture, how do you keep it that way?

Restoration has also provoked suspicion and skepticism from a broad array of critics. Conservation NGOs felt that it muddied their preservationist message. Deconstructionists accused restorationists of "faking" or "reinventing" nature, acting out paternalistic Western fantasies. Others worried that the field would be sullied by "malicious restoration," the kind of slapdash bulldozing and inadequate replantings that miners leave behind or the monoculture forests planted by logging companies.

In response, biologists began making distinctions. Since restoration ecology emerged as a discipline, they have defined full ecological restoration as nothing less than the reestablishment of a completely functional ecosystem, containing sufficient biodiversity so that it could continue to mature and evolve over time.[3] Rehabilitation, on the other hand, was less ambitious, consisting of beneficial treatment, allowing some ecological recovery but falling short of the full complement of species and processes once present. The most modest category was reclamation, defined as converting land perceived as wastelands (including brownfields, or lands damaged by industry or military activities) to some form of agricultural or recreational productivity.

The methods and tasks involved in restoration are legion: cleaning up pollution, removing invasive species, thinning forests, reestablishing a fire regime or suppressing unnatural fires, removing dams or other artificial water systems, replanting native vegetation. All are organized around a central goal: restoring biological functionality—the dynamic relationships and processes—from the health of the soil to interactions between plants and animals, photosynthesis, recycling of nutrients, and the continuance of the food chain. A functional ecosystem is resilient and self-sustaining, able to withstand stresses caused by periodic drought, severe weather, or fire. The challenges inherent in restoring such ecological processes are so daunting that a journal devoted to the field, *Restoration Ecology*, recently announced a new feature. "Set-backs and Surprises" will be devoted to analyzing projects that prove disappointing, in the hope that others can learn from "negative, equivocal, or unexpected outcomes."[4] Originally, the editors planned to call the section "Spectacular Failures"; they reconsidered when they realized that contributors might recoil from such a label. But the very notion of acknowledging shortcomings in this way suggests the trial-and-error nature of the whole enterprise.

New York City is betting that its staggeringly complex restoration project will not be a contender for the title. Designed to transform the former Fresh Kills landfill on Staten Island—consisting of four gigantic pyramids of trash, 150 million tons of garbage—into the 2,200-acre Freshkills Park, the largest park built in New York in over a hundred years, the scheme has been described by the New York City Department of City Planning as "one of the most ambitious public works projects in the world." The so-called RePark will combine "state of the art ecological restoration" of surrounding wetlands with rehabilitation and reclamation of the more difficult features, including the trash mountains of the former dump.[5] Most closed landfills are capped with soil, but the revegetation and reshaping involved in this project will be far more elaborate. According to the vision of planners, there will be open waterways, sweet-gum swamps, prairies, and wildflower meadows, offering recreation as well as restoration of native biodiversity. Methane vents—capped with silver discs to "reflect the light and color of the sky"—will mark pathways; gray-water ponds will reoxygenate and filter water that can be reused in irrigation. There are plans to remove invasive species, particularly the reed *Phragmites australis*, in favor of native salt marsh grasses. The highest mound, with a heavy layer of soil covering debris from the World Trade Center, will become a "Memorial Forest" of four thousand trees, with victims' families given the option of planting their own.

Rutgers ecologists demonstrated the landfill's capacity for regeneration by planting small outlying areas with native plants in 1989. Whereas most patches of degraded urban habitat languish for years with nothing more than weedy cover—victims of compacted soil and intense heat—their revegetation, featuring a diverse selection of native trees, shrubs, and vines, helped to jump-start the return of greater biodiversity by shading out invasives. As the trees and shrubs grew, they attracted birds—most fruit-eating birds will shun plants under five feet—and they, in turn, brought in seeds of other desirable species.[6]

In 2005, New York State launched an ambitious restoration plan for the larger ecosystem surrounding Freshkills, the Hudson River, with the ultimate aim of making its entire 315-mile length safe for swimming and fishing. The state hoped to enlist two hundred partners, including sixty municipalities, a hundred landowners, and forty businesses and nonprofits, to conserve fifty thousand acres of six key habitats and their representative species.[7] The organizations involved identified four biological goals: revitalizing the signature fisheries of the river and its estuary (sturgeon, striped bass, and American shad), improving water quality and restoring the flow of tributaries, cleaning up contaminants, and conserving upriver habitats for wildlife, including shoreline corridors and unbroken stretches of woodland. Both the Freshkills and the Hudson River plans may take decades to complete, but planners have set a deadline—2016— for the return of the first mature female Atlantic sturgeon to the Hudson estuary.

From Curtis Prairie to the Hudson, it has become apparent that ecological restoration may indeed be the work of centuries, requiring extensive maintenance, recalibration over time, and huge expense. But the benefits are ongoing as well. RePark and the Hudson River project will be costly, but they will create a significant number of jobs and, in the future, produce revenue. RePark, for example, projects an annual future income of $13 million, from wind and solar power generation, entrance fees at its entertainment and recreational venues, and rental of exhibition and conference centers. Aside from the ecosystem services to be regained—revitalized fisheries and agriculture, improved water quality—restoration can augment a sustainable economy.

Worldwide, restoration has the potential to generate millions of jobs, for both skilled and unskilled workers: preparing soil, planting, and tending native species; fighting fires or setting controlled burns; removing invasive species; restoring streams and wetlands; monitoring water quality. People currently despoiling the oceans through illegal fishing and diving can be put to work protecting, repairing, and monitoring coral reefs. Poachers can be retrained as park rangers for restored areas. Vegetation in some projects stores carbon over decades; carbon credits can then be sold to companies looking to offset greenhouse gases.

As RePark suggests, there may be multiple motivations for restoring land. The scientists involved are driven by the intellectual challenge of restoring intricate biological processes as well as by their understanding—and dread—of the consequences of ecosystem failure and biodiversity loss. The general public may be moved by the desire to clean up pollution, create beautiful open spaces and recreational areas, bring back species that have fallen victim to our carelessness, or reconnect with nature. There may be venal considerations as well, issues of real estate value or benefits to other invested parties. When embarking on these expensive long-term projects, planners must carefully consider motivation and bias. All these factors played a role in an early parable of restoration, the resurrection of Lake Washington.

Trade-Offs

On Mercer Island, in the southern half of Lake Washington, that body of water loomed large to everyone, visible out of virtually every window. Growing up there, I was aware that the lake had somehow been saved from ruin, that it had been sick but had recovered. Its salvation seemed incredible but, in fact, required a complex series of trade-offs, exacting a price that is still being paid today. What happened with Lake Washington, which was one of the most expensive, ambitious, and successful restoration projects of its time, suggests that planners ignore underlying motivations—as well as the connections between ecosystems—at their peril.

The second largest in the state, Lake Washington is twenty miles long and, in places, 240 feet deep. The city of Seattle occupies a narrow isthmus of land flanked on the west by the saltwater of Puget Sound and on the east by the freshwater lake, which separates it from the eastern suburbs, now home to the empire of Microsoft. But when this story begins, Bill Gates had yet to be born, and something ominous was happening in the lake, something called *Oscillatoria rubescens*.[8]

A species of blue green algae whose arrival in lake water has been compared by one writer to "the appearance of dead rats in a populous city," *Oscillatoria rubescens* was multiplying rapidly.[9] It heralded eutrophication, the process by which a body of water that has been "enriched"—usually by the dumping of raw or treated sewage, containing a surfeit of phosphates, nitrates, and other nutrients—becomes oxygen-depleted, smothering fish and other life as aquatic plants multiply, die, and decompose.

Scientists had learned about eutrophication during the nineteenth century, when lakes in Western Europe, including Lake Zurich, filled with effluents from burgeoning cities. For Seattle's lake, the summer of 1955 was the turning point: algae began matting across its surface, fouling boat propellers and fishing lines, sticking to swimmers' legs, releasing a terrible stench. Wallis Edmondson, a renowned University of Washington specialist in freshwater systems, whose graduate students were using the lake as their research site, identified the algae by examining under a microscope the water that one student brought back in an old beer bottle.[10] The discovery was big news in the *Seattle Times*—"Lake's Play Use Periled by Pollution."[11] Residents of the city and its suburbs were horrified. From that point on, Lake Washington's beaches, on Mercer Island and around the outer rim of the lake, were often closed.

The sewage was not from Seattle, which had diverted its waste into Puget Sound decades earlier, but from communities on the eastern shore, which were growing rapidly. By that summer, ten plants were dumping twenty million gallons of treated sewage into Lake Washington every day. The treatment did nothing to slow eutrophication; indeed, it speeded up the process, destroying harmful bacteria but making the phosphates and nitrates in the effluent readily available to algae. Visibility in the lake, twelve feet in 1950, dropped to less than three a few years later. At the same time, Puget Sound was also suffering the effects of nearby growth, and raw sewage fouled miles of shoreline near Seattle whenever the city's antiquated drainage system was flooded by rainwater.

The ensuing struggle over how—or whether—to solve the problem bears unmistakable similarities to current debates over how, or whether, to address global warming. A Seattle attorney, James Ellis, who served on a regional planning commission, proposed a comprehensive solution: build a new sewage system routing all sewage from lakeside communities and Seattle into a series of treatment plants that would then discharge treated waste into the sound, the logical end point. The sound was far larger than the lake—over a hundred miles long and nine hundred feet deep—and it was constantly flushed by tides. Because many of the lakeside communities were incorporated cities, it would be necessary to create and fund a regional authority, dubbed the Municipality of Metropolitan Seattle (or "Metro"). A bill to do so passed the Washington state legislature in 1957, but it required a two-thirds majority vote in Seattle and surrounding King County to approve its cost, $2 per household per month.[12]

The political battle over Metro became one of the most divisive in the region's history. The area's progressive voters—upper-middle-class doctors, lawyers, engineers, and scientists, many of whom lived in scenic neighborhoods near the lake—supported the proposal. Conservative residents, including significant numbers of those who made up the workforce of the region's two biggest industries, aircraft production and logging, bitterly opposed the tax hike as a socialist ploy, joining taxpayers' leagues and property-rights groups. Their spokesman, Nicholas Maffeo, announced that if there were a choice between losing Lake Washington and losing "our great American heritage of freedom," he would happily sacrifice the lake.[13] As a solution, he proposed draining and refilling it or using "atomic energy" to clean it. Ads ran in the *Seattle Times* and other area papers depicting the plan as a "Metro Monster," an octopus with the face of Hitler.[14] The plan went down to defeat on its first vote, in 1958. Later that summer, swimming beaches were again closed during a heat wave, and a second vote—this time on a scaled-back version of the plan—passed.

After a few more years of lawsuits, the first of the new diversion and treatment plants finally came on line in 1963, the amount of phosphates dumped into the lake declined, and the water quality improved rapidly and dramatically. Visibility in the lake increased from a couple of feet in the early 1960s to nine feet in 1968, when diversion was completed, and eventually back to twelve feet. Beaches were reopened; fishermen returned; children swam again.

The restoration of Lake Washington was the most expensive pollution-abatement project ever undertaken at the time, and it became a model for the restoration of other lakes around the world, including Lake Zurich and Lake Erie.[15] Metro won an "All-American City" award, and the ecological foresight of Seattle-area residents was lauded in articles in *Harper's*, *Audubon*, and *Smithsonian*. Today, Lake Washington remains one of the cleanest, most well managed bodies of water in the region, and the lake has continued to support high local property values. Bill Gates built his technological dream house on its shores, and another Microsoft billionaire, cofounder Paul Allen, moved to Mercer Island's waterfront. The extraordinary economic growth Seattle experienced during the tech boom would be impossible to imagine if this important natural feature of the city's landscape had been allowed to devolve into a cesspool.

In ways unforeseen, Lake Washington was never quite the same: weirdly, it was better, or at least different. The lake became twice as clear as it had been prior to the pollution, a fact that led scientists to the realization that the balance of power in the food chain had shifted.[16] In the years during and after restoration, Edmondson and other scientists found that the lake's brief bout with eutrophication had altered organisms within it in startling ways, both in distribution and in their very physiology. *Oscillatoria rubescens* declined and the lake was increasingly dominated by creatures belonging to the genus *Daphnia*, tiny crustaceans often called "water fleas," which feasted on algae in place of the other plankton that once filled that role. The three-spine stickleback, a small fish with spines on its back, also adapted to the changes. One scientist documented reverse evolution among the Lake Washington sticklebacks: during the worst of the pollution, the fish lost the genetic trait for spines; visibility was so poor that there was no need for protective armor.[17] When the lake was cleaned up, the stickleback was no longer able to hide in the algae from predatory trout and regained its spines. Work on sticklebacks in the Pacific Northwest has since become one of the hottest areas in evolutionary biology, as scientists continue to probe these rapid manifestations of natural selection.

Wondrous though it was, there was a dark side to the lake's restoration. While the new treatment plants quickly solved the sewage problem in one place, they essentially created it in another, by moving it elsewhere. Decades after Lake Washington recovered, Puget Sound—where all the treated sewage was being dumped, along with other industrial and commercial pollutants—started to collapse.

As creation myths of the area imply, everything in the Pacific Northwest—trees, bears, berries, birds—rests on one extraordinary fish, the salmon. Anadromous species, including sockeye, chinook, and coho salmon, leave their natal freshwater streams to mature in the ocean, returning to spawn and die where they hatched. These species anchor the ecosystems of the North Pacific, renewing nutrients in forests and rivers. Their half-eaten bodies—strewn across beaches and forests by wolves, bears, and eagles fattening up during the fall salmon runs—feed the enormous trees famous on this coastline. Their absence impoverishes it.[18]

When the Lake Washington restoration began, Native Americans in the area—many dependent on subsistence fishing—had already grown alarmed by the decline of local salmon runs. Since the city was founded, numerous streams and rivers in the area, each with its own native salmon run, had been dammed, altered, or straightened. The most abused of these was the Duwamish, a major waterway that flowed through some of the poorest and most industrialized sections of the city, emptying into Seattle's Elliott Bay. Its estuarial tidal flats had once fed hosts of native people, but as the city expanded, the wetlands were filled and paved over. The lake restoration diverted sewage into the Duwamish and thence into treatment plants and out into the sound. After members of the Muckleshoot tribe were arrested in 1963 for fishing near a hatchery on the Duwamish, arguments over native fishing rights and the health of the beleaguered salmon themselves spawned dozens of bitter disputes and court battles. Lake Washington's restoration had come at the expense of the Duwamish—now one of the largest Superfund sites in the country—and ultimately the sound itself, and it worked to the advantage of wealthy neighborhoods over poorer ones. Historian Matthew Klingle, in his environmental history of Seattle, argued forcefully that the cleanup of the lake "tended to benefit those who had power."[19]

Ironically, by the 1980s, just as the city's economic boom made salmon the emblem of its success—decorating the sides of city buses and tossed by vendors in a beloved spectacle at the city's Pike Place Market—local salmon runs had fallen into a catastrophic swoon.[20] Today, with four million people living in the region, the sound hosts less than 10 percent of the salmon that thrived there in numbers so plentiful that they were said to have colored the water. In 1999, nine Pacific Coast salmon and steelhead runs, including those native to metropolitan Seattle, were added to the endangered species list. Shellfish, once so thick on beaches near the city that native peoples left enormous middens along miles of coastline, fell to historic lows, with 25 percent of shellfish beaches closed by 2006 because of contamination by human sewage; most of the remainder were under advisory. High levels of nutrients from treated sewage and storm water runoff were implicated in the decline of salmon and the tainted shellfish. Experts also estimate that an amount of oil equal to that spilled by the *Exxon Valdez* runs into the sound every two years, from concrete, roads, roofs, and every other impervious surface in the region.[21]

The most tragic victim of the salmon's decline and the pollution of Puget Sound has been the orca, the killer whale. The whales occur in all the world's oceans, but there has long been a unique resident population in the southern stretch of the sound. Unlike most other killer whales, which commonly feed on seals and other marine mammals, these relied predominantly on salmon. As a child, in a sailboat in the Strait of Juan de Fuca, the main channel to the ocean, I once saw a local pod surface alongside, their huge black and white bodies slicing silently out of the water, close enough to touch, disappearing as mysteriously as they arrived.

Now the population is critically endangered, primarily because its chief food source, chinook salmon, has declined precipitously. The salmon that remain are contaminated with chemicals, including DDT and PCBs (polychlorinated biphenyls), once widely used in coolants, sealants, and pesticides.[22] Although banned in the late 1970s, PCBs persist in the environment and accumulate in animal fat. Recent toxicological studies revealed that the Puget Sound orcas contain four to five hundred times the amounts of PCBs in human bodies; one scientist has called them "the most contaminated marine mammals in the world."[23] The body of one dead orca, washed up on shore, was found to be so polluted that the corpse was removed by workers in protective clothing and relegated to a hazardous materials disposal site.[24] Contaminants weaken the whales' immune systems and contribute to a low survival rate of offspring. In 2008, seven whales went missing in southern Puget Sound, leaving a diminished total of 83, down from 140 or more in the previous century. Marine biologists fear that the population, stressed by pollution, disease, and starvation, may face extinction within the next twenty years. It will come at a price. In Canada alone, the whale watching industry in the sound is believed to be worth over $100 million a year, more than the entire British Columbian commercial salmon fishery.[25]

The orca is not the only species at risk. Communities along the Duwamish have grown increasingly alarmed over the health risks represented by PCBs and other industrial pollutants dumped over the years that continue to leach into the river. The Environmental Protection Agency has been wrangling with Boeing over pollution dating back to the World War II era. Boeing maintains that the Duwamish can never be restored to a condition that would allow the safe consumption of fish from its waters, but the governor of Washington, Christine Gregoire, insists that "has to be the goal."[26] Amazingly, juvenile and adult salmon still use the waterway.[27] While litigation grinds on, neighborhood activists have organized to begin restoration at several habitat sites along the river.

In response to the overall crisis, the Puget Sound Partnership, a coalition of environmental groups and state agencies created by the governor in 2007, has embarked on one of the most massive sustained ecological restoration projects in history. The effort will take decades and include cleanup of toxins, heavy metals, and other pollutants; shoreline restoration; native plantings; and removal of pilings, debris, and invasive species. Recently, the partnership identified half a billion dollars' worth of "shovel-ready" economic stimulus projects associated with restoring the sound, protecting it in the future, and stopping pollution.[28] The issue is so critical to the area's economy that in 2009 over seventy-five fishing and conservation organizations in the Northwest called on President Obama to create a high-level "salmon director" post at the White House Council on Environmental Quality, responsible for coordinating federal salmon restoration policy.[29]

Big as it was, inspiring as it was, the Lake Washington restoration was simply not big enough, not connected enough, not sophisticated enough. Limited by the knowledge available at the time and influenced by the wealthier communities whose real estate was affected, it did not take into account the fact that the larger ecosystem could itself be overwhelmed by effluents and contaminants. Clearly, when it comes to ecological restoration, underestimating the task at hand—failing to make those connections—is something we cannot afford to repeat.

Shifting Baselines and Pleistocene Rewilding

If you go to the Pacific Northwest today, never having been there before, you will not notice the declining salmon runs, the PCBs in Puget Sound, or the toxic orcas. Like so many other environmental disasters, these critical problems are unrecognizable without prior knowledge or baseline information—essential reference points from the past. On a good day, Seattle looks as gorgeous as it ever has, ferries plying the sparkling water on the sound, bald eagles perched in the evergreens, sea lions bobbing alongside boats. A recent poll in the area found that, while 90 percent of the public wanted Puget Sound protected, over 70 percent thought it was perfectly healthy and not in need of restoration.[30] The inability to recognize the severity of environmental problems, due to a lack of historical knowledge of ecosystems, represents a serious threat to ecological restoration and public support of it.

That phenomenon—the failure to recognize damage already suffered by natural systems—has been dubbed "shifting baselines" by a group of marine biologists and divers who launched a media project and Web site, illustrated with short videos, to describe the problem.[31] Marine scientists were particularly concerned about the issue: aside from specialists and lifelong divers, few people have any sustained experience of marine ecosystems; few people know what underwater habitats look like now, much less what they looked like fifty years ago. Jeremy Jackson does. An oceanographer and professor at UC San Diego, Jackson helped create the Web site as a tool to educate people after decades of watching coral reefs turn to what he despairingly calls "slime."[32] A recent *Shifting Baselines* video about Puget Sound offered one example, noting that in the 1950s there were a thousand pairs of tufted puffins breeding on two small islands in the sound; as of 1995, there were thirteen. "How would you know all these birds were gone if you never knew they were there in the first place?" the narrator asks. "That is the shifted baseline, where we no longer even know what we've done to nature."[33]

Indeed, the fact that there were so many people living on Lake Washington who knew the lake in its once pristine state contributed to support for its cleanup and preservation. Describing the battle to win political support for Metro, the *Smithsonian* article noted that "many people could remember when virtually all the lakes and streams in the Northwest were perfectly clear, and unlike residents of the East and Midwest, they hadn't grown up accepting foul water as an inevitable part of their surroundings."[34]

Even when collecting or re-creating baseline information about an ecosystem is straightforward, the question for restorationists remains: Which baseline to shoot for? In Maryland's Chesapeake Bay, scientists have been bemused by visitors' reaction to a small-scale effort to restore oyster populations, in this case to a 1994 baseline, a relatively poor year for oysters. There was neither the funding nor the resources to aim for a more ambitious baseline, from the 1800s, for example, when countless oysters lined the bay. While those unfamiliar with the history rejoice in this small-scale attempt, seeing it as a sign of recovery, the scientists realize that it represents a fraction of the habitat restoration that must occur before oysters can recover their previous abundance.

If the 1800s is an overambitious baseline, how about thirteen thousand years ago? That was the goal outlined for the ultimate act of guerrilla restoration, in a 2005 commentary in *Nature* titled "Re-wilding North America." Josh Donlan, an evolutionary biologist at Cornell University, and a series of coauthors—including rewilding's founder, Michael Soulé—argued that the extinction of megafauna in North America represented an ongoing threat to ecosystems.[35] They proposed replacing those lost species with imported African relatives, including elephant, lion, and cheetah. The Pleistocene era was the ultimate shifted baseline, with no one alive today fully knowledge-able about species missing for thousands of years.

Crazy as it sounded, there was a scientific rationale behind the plan. One of the coauthors was Paul Martin, a geoscientist at the University of Arizona, who had written an eloquent book, *Twilight of the Mammoths*, about the Ice Age extinctions. Martin pointed out that dozens of species of North American mammals disappeared only ten thousand years ago, including five genera of elephants (mammoth and mastodon among them), two species of bison, four camels, ground sloths, a giant beaver, giant armadillos, ten species of wild horses, a wild cow, and the woodland musk ox.[36] There were once tapirs in California, Kansas, Arizona, and Florida.[37] As these species disappeared, many of their predators also vanished, including the scimitar cat, the saber-toothed cat, the dire wolf, and two kinds of bear. The continent was also patrolled by the American lion and the American cheetah.* All gone.

Martin made an impassioned argument for a program of "resurrection ecology" to bring back a full assemblage of grazers, browsers, and predators in North America.[38] Comparable to the reintroduction of the wolf in Yellowstone, this would be restoration writ large, reenergizing critical ecological mechanisms, including top-down regulation, throughout the continent. He dealt summarily with the inevitable naysayers: "We are obsessively focused on protecting what we have and utterly unaware of what we have lost and therefore what we might restore." He quoted Thoreau's yearning to know "an entire earth" despite the fact that "my ancestors have torn out many of the first leaves and grandest passages, and mutilated it in many places."[39] He called for parks patrolled by African or Asian elephants, to reestablish dynamic relationships between grasslands, forests, and megaherbivores. "Whatever happened to sweep away this rich fauna," he wrote, "we should work to sweep some of it back."[40]

---

* The American cheetah is thought to have been the chief predator of the pronghorn antelope and thus responsible for its remarkable speed.

The 2005 *Nature* paper, the fruit of a meeting between Martin, Soulé, and the other coauthors at one of Ted Turner's New Mexico ranches, was clearly meant to bring that argument to the attention of the world press. It outlined a "bold plan" for the coming century through the immediate establishment of Pleistocene parks.[41] The authors suggested that the stocking of the parks proceed in phases, first with the reintroduction to Big Bend National Park in Texas of the fifty-pound Mexican giant tortoise, or Bolson tortoise, once widely found throughout the Chihuahuan desert. Przewalski's horse, Asian asses, and Bactrian camels were proposed to replace the extinct American species in the prairie grasslands. The second phase would kick off with African cheetahs, Asian and African elephants, and even lions, introduced "on private property."[42] These species would not even need to be imported, the authors pointed out: there were already seventy-seven thousand exotic mammals roaming around on private ranches in Texas. Fencing, they observed in one of their more sober passages, "would be the main economic cost."[43] They acknowledged that their proposals might strike some readers as playing God and noted a few specific objections, including the possibility of disease transmission. But ultimately, they argued, Pleistocene rewilding represented an "optimistic alternative" to the grim projected losses of biodiversity: it could restore North American ecosystems while protecting African and Asian species threatened in their original habitats.

Soulé has always had an elastic imagination, and some of his proposals in earlier papers verged on science fiction. He once explored a futuristic scenario in which "a near total blitzkrieg" of the planet's ecosystems had taken place.[44] But even for Soulé's supporters, Pleistocene rewilding was a step too far, and the article sparked outrage and ridicule. "Loopy" was among the milder rebukes. The WWF's Eric Dinerstein, who had enthusiastically adopted rewilding as his label for the Terai Arc in Nepal, lamented in the *Economist* that it was wrong to allocate scarce funding to outlandish experiments instead of conserving existing ecosystems.[45] Other critics pointed to the nightmarish history of exotic species wreaking havoc outside their native habitats; Daniel Rubenstein, the Grevy's zebra expert, and a group of colleagues objected to the plan on the grounds that America had evolved "for thousands of generations" since the Pleistocene megafauna disappeared. "The consequences of such introductions cannot be predicted," they wrote.[46] Conservationists in Africa were offended by the notion that they were failing to protect and conserve their wildlife. Ian Douglas-Hamilton, a Kenyan elephant expert and the founder of Save the Elephants, called it "a terrible and absurd idea."[47]

The proponents of Pleistocene rewilding defended the science, insisting that North America's Great Plains might perish without the return of keystone species recently driven extinct. One argued that the "African" lion was a "myth" and might well be a genetic match to the lion of North America. But as one commentator put it, "science was never really the issue."[48] The real objection, which neither Martin's book nor the article adequately addressed, was Americans' widespread intolerance of existing predators. If the West cannot handle a few Mexican wolves, how could it ever deal with lions and cheetahs?

Overlooked in the contretemps was that someone had already established such a rewilding park, albeit not in North America. For years, Sergey Zimov, a Russian geophysicist and the director of a scientific research station in a remote northeastern section of Siberia, had been pondering the fact that grassland ecosystems once covered half the earth's land mass, evolving in conjunction with great herbivores like the mammoth, the bison, and wild horses. With much of the current Russian steppe covered in mossy tundra, Zimov speculated that a reintroduction of grazers might restore grasslands there. If the plan worked across the huge and warming expanse of Siberia, it might stabilize the surface soil and keep the temperature of the earth beneath constant, forestalling the expected massive release of carbon from melting permafrost. In a 2005 *Science* article, Zimov noted that the amount of carbon currently sequestered in the "former mammoth ecosystem" of Siberia surpassed that in all the existing rain forests on earth.[49]

In 1989, he launched "Pleistocene Park" on 40,000 acres at his science station. With government approval, he and his partners have reintroduced and protected a suite of species designed to reproduce the Pleistocene grazing regime once managed by mammoths and woolly rhinos: reindeer, moose, wild horses, musk oxen, hares, marmots, and ground squirrels. Predators are present as well, including wolves, brown bears, lynx, wolverines, and foxes. Zimov hopes eventually to reintroduce bison from Canada and to acclimate Siberian tigers, now largely confined to the Russian Far East. While it is too soon to tell if the experiment is working, the past may prove to be a carbon-sequestering prologue.

There is even talk in Pleistocene circles of resurrecting the mammoth, after the astonishing discovery in Siberia of an intact month-old infant, dubbed "Lyuba," who fell into a river or suffocated in mud forty thousand years ago.[50] The most perfectly preserved mammoth ever found, her DNA may, one day, help to resurrect her species.

While wildly imaginative, even romantic, raising mammoths from the dead and importing African wildlife were not the only paths to restoration. In Costa Rica, biologists made use of more prosaic, if ecologically jarring, activities—cattle herding and monoculture planting—to resurrect native forests. Even more startling, these counterintuitive methods worked.

COSTA RICA'S THOUSAND-YEAR VISION

Large-Scale and Long-Term

IN STARK CONTRAST TO ITS CENTRAL AMERICAN NEIGHBORS, Costa Rica made a decision, early on, to invest in its people and environment and to forgo military adventures. The impetus for that choice may have come out of one of the country's first and only such encounters. In 1856, a group of mercenaries sent by William Walker, a soldier of fortune from Tennessee who briefly conquered Nicaragua while championing the conversion of Latin America into a slave state for the American South, entered the northwestern frontier province of Guanacaste, stopping at a house twenty miles from the border. The house was known as the Casona, the ranch house of Hacienda Santa Rosa, one of the oldest ranches in the country, dating from the late 1500s. The Costa Ricans rallied a volunteer force that surprised Walker and his men; twenty-six Yankees were killed, along with nineteen Costa Ricans. The Casona was also the site of a famous 1919 battle, in which Costa Rica fought off an invasion from Nicaragua.

But the country felt no need to repeat its victories: Since 1949, it has had no standing army. What it has instead is the highest standard of living and the best-educated populace in Central America, with a literacy rate of over 90 percent.

In 1966, the government bought four square miles of land at the site of the battle with Walker, including the Casona.[1] The purchase was not entirely encouraging from an ecological point of view: the ranch had been used as cattle pasture since the invasion of the conquistadors. Logged and burned, much of the surrounding region was covered in an invasive grass known as "jaragua." The Pan-American Highway, built in the 1940s, cut through the province as well, further fragmenting habitat. But between the Casona and the Pacific Coast there were patches of original dry tropical forest and secondary growth, preserving a bridge of habitat between the beaches and nearby volcanoes. In 1968, an American forestry official and biologist, Kenton Miller, who, two years earlier, brought graduate students and Peace Corps volunteers to the Casona to study the tourism potential of the area, submitted a report to the government proposing that the country should create a national park by adding more land to the small reserve, expanding it to the coast, and converting the Casona to a historic exhibit.[2]

Miller was particularly keen to protect the dry forest habitat, much of which had already been lost throughout Mexico and Central America. These forests, with an average 10–80 inches of rain per year (compared with 80–160 in the rain forest), have evolved to cope with several months of hot, dry weather every year, and they host a unique combination of deciduous and evergreen trees. Although slightly less biodiverse than rain forests, the dry forest is nonetheless a rich and astonishing ecosystem, with thousands of plants, mammals, and insects, including many endemic ones. What's more, in the Guanacaste region, the surviving remnants of dry forest provided a critical buffer zone around rain forests and cloud forests at higher altitudes. The government took Miller's advice, making Santa Rosa one of Costa Rica's first national parks.

By now, Costa Rica's commitment to conservation is unmatched. In twenty-six national parks, in multiple wildlife refuges, forest reserves, indigenous reserves, monuments, and marine protected areas, the country has placed 25 percent of its land under conservation protection. By comparison, the United States strictly protects less than 6 percent of its land.

Despite struggles in some parks, like Corcovado, there is no doubt that Costa Rica takes its role as steward of the world's most extraordinary ecological riches seriously. Not much larger than Switzerland, but with two coasts—Pacific and Atlantic—volcanoes, highlands, cloud forests, rain forests, and dramatic mountain ranges, it has a wider variety of habitats and greater biodiversity than any other country on earth. No one knows how many species it contains. No one even knows how long it would take to find out. At the very least, there are over 10,000 species of vascular plants, nearly 2,000 trees, over 1,300 species of orchids, 850 birds, 200 species of migrating birds, and 232 mammals, half of them bats. Both the American crocodile and the caiman can be seen in Costa Rica, along with several hundred other reptiles and amphibians—including such legendarily lethal specimens as the boa constrictor, the bushmaster, the fer-de-lance, and the eyelash viper—and a plethora of skinks, geckos, whiptails, and the neon orange and green poison dart and poison arrow frogs. Coastal areas host fish, sharks, whales, and some of the world's most important remaining breeding populations of endangered sea turtles, including the hawksbill, the leatherback, and the olive ridley.

Costa Rica's forests also contain a wealth of insects, over 35,000 species and counting: moths, beetles, wasps, ants, flies, grasshoppers, and 18 percent of the world's butterflies. The long-running "Moths of Costa Rica" project—launched in the 1970s and financed with American tax dollars via the National Science Foundation—has identified some 9,000 moths native to the country. To tally and study this array, scores of the world's field biologists, entomologists, and ornithologists migrate to Costa Rica annually, for training, collecting, and hobnobbing with others of their kind. The country has attracted millions of dollars in research funding and "more WWF projects than any other nation."[3]

Foremost among Costa Rica's superlative parks is the Área de Conservación Guanacaste (ACG), now a complex encompassing the original Santa Rosa National Park and other conservation areas adding up to nearly 300,000 acres of dry forest, cloud forest, and rain forest, along with 106,255 acres of protected Pacific ocean and islands. The ACG amounts to around 2 percent of Costa Rica, its area approaching that of New York City and environs. Dan Janzen, an American biologist instrumental in creating the ACG, has estimated that it may contain 235,000 species, or roughly 2.4 percent of the world's biodiversity. Talking to a reporter from *Popular Science* in 1994, he waved his hand at a volcano in the distance and said, "There are probably more species of animals and plants between us and that cloud bank than in all of the United States."[4] He has defined the role of those who work in the ACG as caretakers of that biodiversity: "We are pro-bono negotiators on behalf of 235,000 species of unknowing and uncaring wee beasties and green lumps."[5]

The transformation of the small Santa Rosa park into a major conservation area is one of the most remarkable success stories in the field of ecological restoration. Passionately recounted in *Green Phoenix: Restoring the Tropical Forests of Guanacaste, Costa Rica,* by William Allen, the story featured scores of Costa Rican politicians, conservation specialists, and park directors in starring roles. President Oscar Arias Sánchez, who won a Nobel Peace Prize in 1987, was one of the heroes, endorsing a debt-for-nature swap, a financial tool enabling conservation groups to purchase part of a country's debt in exchange for its government's setting aside land for protection. Alvaro Ugalde, the park's first superintendent, removed squatters and cattle herds from Santa Rosa and went on to direct the country's Park Service, fighting for the ACG all the while. And then there was Dan Janzen.

Had I encountered him in other circumstances, I might have been tempted to give him my pocket change. One of the most decorated and respected biologists of our time—the Thomas G. and Louise E. DiMaura Professor of Conservation Biology at the University of Pennsylvania, winner of the Crafoord Prize from the Swedish Royal Academy of Sciences, the Kyoto Prize in Basic Science, the Albert Einstein Science Prize, a MacArthur Fellowship—he is also one of the most eccentric. An early student of his has said of him, "'He really

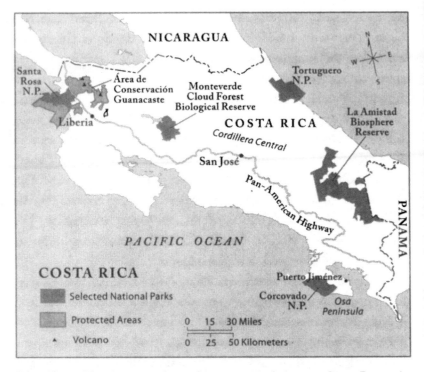

Costa Rica: The divergent fates of two protected areas in Costa Rica—the Área de Conservación Guanacaste in the north and Corcovado National Park in the south—highlight the consequences of different conservation tactics.

looked like a Biblical prophet.'"[6] E. O. Wilson has recalled "buying a 'hungry-looking' Janzen dinner during a visit to Berkeley in the '60s."[7]

When I saw him through the open doorway of the house he lives in at the Santa Rosa administrative area in the Guanacaste Conservation Area—a house so cavelike and approaching collapse into the surrounding forest that it might more accurately be described as a habitat—Janzen was crouched over a laptop computer under the dim light of a hanging bulb, shirtless and sweaty. Rain dripped through the green forest mass above, while a dark brown agouti, a seed-eating rodent, nosed around in the fronds and ferns outside his door. Plastic Baggies full of leaves and pupae drifted overhead, hanging from

clothespins on lines strung through the house. Janzen was
years old, and he was not interested in anyone's pocket char
a few minutes, he told me he needed half a billion dollars.

Janzen has been described as "one of the world's most aggressive
and successful conservation biologists."[8] Of all the great biologists
of this age, and there are many—George Schaller, Michael Soulé,
Paul Ehrlich, Bill Newmark, John Terborgh, Jane Goodall, E. O.
Wilson—Janzen is among the most cunningly creative and vivid advocates for his cause, which is saving tropical nature on a large scale
and for the long term: its species, ecosystems, and processes.

Born in Milwaukee, Wisconsin, in 1939, Janzen—whose father
worked for the U.S. Fish and Wildlife Service—began collecting
caterpillars at the age of nine. After finishing his doctorate at UC
Berkeley in 1965, he published a groundbreaking paper in *Evolution*
about the relationship between ants and acacia trees. In a BBC TV
film, Janzen—pointing to a color slide of a swollen-thorn acacia
in the state of Veracruz, in Mexico—remembered his moment of
epiphany:

> One day I was walking down a road, and a small bush
> more or less like this one, but perhaps a little taller, was
> standing there, and a beetle flew by and landed on one of
> the leaves. The instant that the beetle landed, an ant ran
> up and grabbed the beetle. The beetle jumped off the leaf
> and flew off. And I saw all this happen, but it really didn't
> sink in, and I walked on down the road, and the little
> wheels began to turn and after two or three hours it sud-
> denly hit me that there must be some reason why the ant
> was there. So I turned and walked back to the bush, and
> there were ants walking all over the surface of the plant.
> So I picked up a few insects and threw them on the plant,
> and the ants attacked them. They jumped off, and then
> the immediate question came to mind as to why the ants
> were there and what they were doing there.[9]

Janzen went on to design and carry out ingenious field experiments that themselves became highly influential, studying some five thousand acacias, removing ants, trimming off thorns, observing the consequences of his interference. He discovered that the ants were protecting the swollen-thorn acacias, attacking and stinging other insects or animals that approached, chewing away overshadowing growth, allowing light and nutrients to reach the trees. The ants performed these services while living in the thorns and subsisting on a nectarlike substance from the leaf stalks and the fat and protein produced in the leaflet tips, known as "Beltian bodies," after Thomas Belt, a nineteenth-century English naturalist. Belt and other scientists had noticed aspects of the ant-acacia relationship, but Janzen was the first to document the "fully developed interdependency," a form of "mutualism," or cooperation between species, and of "coevolution," in which two organisms evolve together. Although the famous ant-acacia paper is stiff and proper in its scientific presentation, it reveals a glimpse of Janzen's literary gift. The behavior of the worker ant, he wrote, "is much like that of a hunted squirrel, dodging behind objects and to the far side of branches."[10]

Janzen was well on his way to becoming a much-published and celebrated straight-ahead biologist, studying flora and fauna and writing papers: By 1985, he had published around 250 articles and, feeding his taxonomic urge to catalog, had edited a major tome, *Costa Rican Natural History* (1983), covering geology, climate, soils, plants, reptiles and amphibians, mammals, birds, and insects, writing many of the entries himself. But Janzen had begun to feel uncomfortably aware that the forests he relied on, particularly the scrap of remaining tropical dry forest in Santa Rosa National Park, were disappearing around him. Up until this point, he had never considered himself a conservationist or activist. As he wrote later, "Conservation was 'something' being done by . . . the IUCN, the WWF, TNC, the Government—'those other people.' I studied it, they saved it."[11]

But then several experiences changed his mind about activism. In 1985, Janzen was commissioned by the Costa Rica Park Service, eager to convince international donors of its commitment to responsible management, to write a report about its handling of threats to the tropical rain forest of Corcovado National Park on the Osa Peninsula. This was the period when over a thousand landless poor trespassed into that park to set up mining operations. In discussions with them, Janzen became vividly aware that social conditions outside the parks were bound to affect what went on inside them. He realized that the miners were "by and large hard-working, thoughtful, and respectful people" but with no access to the training they needed to make a lawful living.[12] Moreover, they interpreted the park as "land with no owner" and thus felt that it was reasonable to do something productive with it. Given that prevalent attitude, it was only too easy to predict that what had happened to Corcovado might also happen to Santa Rosa.

Later that year, Janzen and his wife, Winnie Hallwachs, a field biologist who became a crucial partner in the development of the Guanacaste project, traveled to Australia to undertake an assessment of dry tropical habitat there; the government was trying to decide if there was anything left to protect. This trip was even more alarming. Here, they could see that the prospects for long-term viability of socially and ecologically isolated forest fragments were poor. Tropical dry forests in northwestern Australia had been virtually erased by a century of human-caused fires—set by cattle ranchers to ensure a supply of grass—and replaced by "grass plains dotted with fire-resistant eucalyptus trees."[13] They were looking at what could happen in Costa Rica: Santa Rosa, too, was surrounded by cattle ranches and invasive grasslands, constantly under threat of fire. As in Australia, cattlemen had been annually burning the grasslands to keep forests from reinvading and to regenerate fodder for their animals. Australia, Janzen wrote in 1999, "showed us unambiguously that if the anthropogenic fires were not eliminated, very shortly there would be no battered dry forest to conserve in Santa Rosa and no fragments from which to restore the forest."[14]

The moment they returned from Australia, in September 1985, Janzen and Hallwachs turned their hands to writing "an unsolicited strategic plan for the long-term survival of Santa Rosa's dry forest through creating for it the psychological and sociological presence of owners, the 'owners' being both its direct custodians and society near and far." *The Guanacaste National Park: Tropical Ecological and Cultural Restoration* became a universally acknowledged "classic."[15] Inspired by a recognition of the Corcovado miners' underutilized talents and a growing realization that protected areas desperately needed the support of local people, Janzen and Hallwachs decided that what would come to be known as Área de Conservación Guanacaste, needed not only biologists but a cadre of local specialists, people trained to take care of it, people who cared about it because they depended on it. "Thus," Janzen wrote, "we developed our own on-the-job trained experts in fire control, research, police, biological education, restoration/forestry, ecotourism, administration, and maintenance."[16]

The plan called for restoring the forest from existing remnants—allowing wind, birds, and animals to reseed and reforest large areas surrounding the original park—a pioneering idea that would be carried out on an unprecedented scale. Over the years, experimenting as it went, the ACG regrew forest largely by suppressing the unnatural fire regime. In private ranchlands nearby, cattlemen continued to set fires that—driven by wind and heat—could quickly get out of control, threatening areas of the park that were to be restored. In protecting the natural restoration, the park had to stamp those fires out, and to that end, the ACG trained a large corps of fiercely devoted Costa Rican firemen. Their jobs were just one economic benefit that the ACG brought to the area.

What's more, in a brilliant but controversial move that led to conflict with the Costa Rican park officials who, in line with traditional conservation, had battled to drive cattle out of the park—whatever cows don't eat, they trample—the domestic beasts were brought back to act as "biotic mowing machines," grazing pastures slated for restoration. As Janzen wrote, the idea was "to keep fuel loads so low that the nascent fire-control program could manage the occasional fire."[17] Subsequently, as trees grew, cattle were removed. The forests took hold, and the park itself expanded, as ranches and private properties were bought up. Eventually neighboring areas of rain forest and cloud forest were bought, to provide a cushion for some species during the dry season and, potentially, for those that might be "ecologically pushed" out of the drier areas as a result of climate change.[18]

Regrowing dry forest meant suppressing fires, but regrowing areas of tropical rain forest was more difficult. Janzen eventually hit upon the idea of using another conservation villain as a "self-financing tool."[19] While monoculture and plantation logging are generally considered anathema to conservation—monoculture suppresses biodiversity and is vulnerable to pests and diseases—Janzen decided that plantations of a commercial deciduous tree, *Gmelina arborea*, native to India and Southeast Asia, might play a role in fostering rain forest regrowth. While rain forest understory could gain no hold in cleared pastures baked by the sun, the *Gmelina* trees provided necessary shade, allowing it to regrow. The method was analogous to using cattle to suppress the fire load—the dried grasses that become flammable—and, like the cattle, the *Gmelina* could eventually be subtracted, providing a tidy profit. Ever observant, Janzen had noticed that plantation growers often started their crops on old rain forest pastures. "If not weeded," he wrote, "these plantations develop a dense shade-tolerant understory of rainforest shrubs, vines, and tree seedlings, dispersed there by vertebrates. The shade from the *Gmelina* canopy and understory weeds kills the pasture grasses."[20] This method closed the gaps between forested blocks.

These radical experiments did not make Janzen and Hallwachs heroes in the conservation community. "In late 1985 and 1986," Janzen wrote, "I received broad disapproval from international conservation NGOs for expounding a restoration focus. These NGOs were largely surviving on the fund-raising message of 'help us save the tropical (rain) forest now before it is cut, because once cut, it is gone forever.' We were told that the donor public was not sufficiently sophisticated to be able to handle both a conservation and a restoration message."[21] Undeterred, Janzen set out to publicize his work and raise money without the aid of NGOs, presenting his elaborate, colorful stories about tropical species and their complex interactions. Later he would say, "I was told that it's too complicated and a lot of people won't understand it. Bullshit. They can understand it perfectly well."[22]

Janzen went directly to the source, turning out to be a spectacularly successful fund-raiser, perhaps the only biologist in the history of science gifted as a salesman. In the late 1980s, he was able to leverage $3.5 million donated by Sweden into over $24 million by means of a debt-for-nature swap facilitated by the investment bank Salomon Brothers.* In *Green Phoenix*, Allen describes Janzen pitching the idea at the Salomon Brothers offices in the World Trade Center in 1988, convincing competitive, hard-bitten traders to donate their services for free: buy debt, trade it to get better deals, and transfer it to the Central Bank in Costa Rica. By the 1990s, these efforts and additional fundraising had brought over $30 million to the project, much of it invested in an endowment "that provided hundreds of thousands of dollars in interest a year" for the conservation area's operating budget.[23] As a result, there are now over a dozen wasp species named in honor of John Gutfreund, then chairman of Salomon Brothers, and other traders. Ironically, these parasitic wasps were known to prey on a group of spiders known as "money" spiders.[24]

When I met with Janzen in his disintegrating house, I asked what the difference was between conservation in the ACG and elsewhere. He fixed on the management of two key resources, money and time. "In the past three months I've found myself coining a silly phrase," Janzen told me.

> I've found myself saying we need to practice *sustainable conservation*. Now that's ridiculous, it sounds like I'm copying all this jargon about sustainability. But in fact, if you look at everything that's done in the name of conser-

---

* After Arias won the Nobel and Janzen the Crafoord Prize in Sweden, the ACG had received a great deal of press attention in that country; the Swedish Society for the Conservation of Nature helped secure this large donation.

vation, the vast majority of it is not sustainable. That means it's not planned to last for a thousand years. That isn't the goal. It's planned to give somebody a return now, on their ego, on their income, on their watershed, on their sense of good feeling.[25]

I asked how he would plan for a thousand years in Central America. He was emphatic on protecting the viability of large, continuously connected ecosystems over the very long term. "No one's doing conservation that way," he said. "They say: 'There's a gorgeous forest, I don't want to see it cut down,' 'There's a beautiful lake,' 'There's a species going extinct.' It's very one-off. To be crude, it's a one-night stand rather than a marriage."

Conservationists had planned on a large scale, but they had not simultaneously planned over time; many projects took no thought for the future beyond the three-to-five-year granting horizon. Janzen's solution was the simple financial instrument of the endowment. The millions of dollars that had come from Sweden and the Salomon Brothers debt swap had established a permanent endowment for the ACG. "If we had used that money from the debt swap and just spent it, we'd have a much finer park, and it would be dead," Janzen told Allen. "Because when you endow it, you can make mistakes like crazy, and all you lose is that year's interest. Next year, start over. The first of January the check comes in again."[26] Janzen was now looking for a half billion dollars to endow *the entire Costa Rican park system.* Not only was this a significant amount of money, it was money that would be used in a way that would make most conservation donors run for the door.

Indeed, Janzen said, the ACG remains the only endowed national park in the world. (Lewa, for example, is a private reserve.) The financial basis of the park seemed solid, and the project gave every appearance of working, both in terms of conservation and in contributing to the community: its jobs, tourism, and scientific visitors brought over $3 million a year into the region. But endowments had never been popular in conservation: donors preferred to give money in a way that they could control. "That's why this place is still alive," Janzen said, "because we built an endowment for it. If we hadn't endowed it, you wouldn't be sitting here. Think about it for a minute. If you're planning for a thousand years, an endowment is a very obvious mechanism. But most people in conservation want to spend the money now, on whatever piece of it that matters—buildings, jobs, salaries, vehicles. They want to spend the money now. They don't want to save it."

After what had taken place in the Mesoamerican Biological Corridor, with extraordinary amounts of money spent on layers of bureaucracy and consultants, Janzen's impatience with international conservation—its emphasis on short-term projects and pleasing the donor community—seemed sadly justifiable. The genius of Dan Janzen was to discover a way around the problem. While he was at it, he put people to work.

The Parataxonomists

Janzen has always coined new words compulsively. Trolling through his writing, one can see him trying out neologisms, startling metaphors, and strange linguistic coinages, many of which focus on human utilization or reimagination of wilderness. *Gardenification* refers to the "generation of goods and services by a wildland garden."[27] The *green freezer* is the ACG's potential to generate wild seeds, vines, herbs, insects, and pollinators.[28] The *agroscape* is Janzen's name for human-centered landscapes, cities, towns, agricultural fields. Some of these coinages have fallen by the wayside, but there is at least one, coined in both English and Spanish by Janzen and Hallwachs, that has entered the lexicon: *parataxonomist.* Inspired by *paramedic*, parataxonomists were envisioned as on-the-ground foot soldiers in the great struggle to identify and restore the vast riches of the ACG's forests.

The beauty of the parataxonomy plan is that it provides gainful employment and valuable education to those living alongside parks or conservation areas, addressing the needs Janzen first identified in Corcovado. Local people living around the ACG or other conservation areas around Costa Rica are trained in fieldwork and "bioliteracy"— how to collect insect specimens, process, and catalog them, for example. In the ongoing National Science Foundation–funded butterfly, moth, and caterpillar inventory, parataxonomists would not only collect but also rear caterpillars, gathering appropriate plants from the forest to feed them. "Taxonomists need good specimens," Janzen and Hall-wachs wrote, "properly labeled, properly prepared, and in series to portray variation. . . . So our inventory project naturally turned into one of collecting and rearing caterpillars."[29] Beginning in 1989, they trained parataxonomists with a six-month field course, describing the curriculum as a combination of computer and mapping literacy, cool-headed independence, and social graces. Skills taught included how to "drive a car, operate a chain saw, care for and use horses as pack animals, use a computer and a topographic map, use a field guide in a foreign language, manage a budget and petty cash fund, and fathom and tolerate foreigners."[30]

Since the inception of the training program, the parataxonomists have played a role in some astonishing schemes, many originating with Janzen. To take one example, in 1989, the National Biodiversity Institute was launched, a private nonprofit organization that promptly sent what Allen called "a small army of parataxonomists and scientists" all over the country to gather plants, insects, microorganisms, and soil samples; bring them back to a central laboratory; attach a bar code to them; and enter all relevant information into a database.[31] Two years later, the institute signed a million-dollar, two-year deal with Merck, then the largest drug company in the world; the institute would provide Merck with samples of chemical compounds derived from these organisms, which might one day lead to new medicines. If valuable products were developed, the Costa Rican government and the institute would receive a percentage of the royalties, and the government's share would be committed to the country's protected areas.

The deal excited enormous interest in the world of conservation biology. If this "bioprospecting" scheme worked, it might change tropical conservation forever, providing enough funding to protect rain forests across the globe. While the institute never achieved that kind of success—after renewing the agreement twice, Merck decided that the process yielded too little, too slowly—it has inspired other countries to follow suit and become an important scientific institution in its own right, entering into a number of research agreements, working with other botanical institutes, and signing contracts with agricultural, cosmetics, and fragrance firms. It continues to conduct a national inventory of biodiversity, producing detailed volumes on everything from Costa Rica's beetles and bromeliads to its lichens, macrofungi, and tent-roosting bats. And it has inspired parataxonomy programs in other important centers of endemism, including Central Africa and Papua New Guinea.

To see the parataxonomists in action, I drove to the San Gerardo Rearing Barn, a former dairy operation whose emphasis had shifted from cows to caterpillars, about an hour's drive east from the administrative area. Located in intermediate-elevation rain forest, the barn stands on a corridor of land pieced together between Volcán Cacao on the west and Volcán Santa María and the Rincon Rain Forest on the east. It was raining lightly as I climbed and descended the twisting muddy road, crossing rickety plank bridges over raging torrents. There was no answer to my knocking at the front of the first building I came to, but at the back I startled Carolina Cano Cano, a preoccupied Costa Rican parataxonomist. I had interrupted her as she was entering data into a laptop computer. But her ability to "fathom and tolerate foreigners" was well developed, and she listened patiently to my Spanish as I struggled to explain what I was doing there. "Very well," she said, sighing a little, "but do you want to see caterpillars [*gusanos*] or caterpillar wranglers [*gusaneros*]?"[32]

"Both!" I said, and she laughed.

Her *compañeros*, she said, would be returning shortly, but she needed to go out to the forest to gather food for the caterpillars. She seemed astonished when I asked to accompany her but readily agreed, donning a commodious khaki-colored rain poncho and high rubber boots and gathering clippers, a large plastic bag, and a plastic-covered list of plant species she would be looking for. As we walked up the road, she explained that we needed to be very patient. Caterpillars required specific foods, and she needed to take the time to look for them.

After we'd gone about a mile, she paused by the side of the road and then plunged through the vast green curtain into the forest itself, leading me slowly through the sucking mud and dark tangled undergrowth. She moved slowly, looking around her, fingering certain plants. When she had found one of the right species, she clipped off fronds, often reaching high overhead to pull a branch down, then pushed them into the plastic bag.

She told me about a bad fright she'd recently had. "I was in the forest a couple of weeks ago, and a snake lunged at me and was going to bite me," she said. "It was *Terciopelo*," she said, the velvet snake. In English, it's called the fer-de-lance, a feared, aggressive, venomous viper. "I used my bag to fight it off, but now I am really scared, really nervous when I go out." She showed me how her hands shook after she ran from the snake.

Twenty minutes or so later, we made our way through the mud back to the road. She checked out the foliage on the other side of the road and said she would be back "*rápido*," disappearing into the foliage again.

Walking to the barn, she told me she lived in the town I'd passed on the drive there, Quebrada Grande. She had a son, Oscar Daniel. "Now that I do this job," she said, "I'm seeing things all the time. Even when I'm riding on a bus, I see plants. My son says, 'It's your day off!' But I'm constantly looking." I realized that Carolina Cano Cano was part of the solution to the shifting baselines problem: the parataxonomists were creating a local populace that was bioliterate and invested in the diversity around them. In the future, they would be able to recognize potential threats to it.

At San Gerardo, Cano showed me another building with a poured concrete floor, tin roof, and slatted open-air walls, where two of her coworkers, Gloria and Anita, patiently sorted the plants she'd brought. Yes, they said, they had studied with "*el doctor.*" "It is really beautiful work, and I like it," Gloria said. The barn was full of caterpillar and pupa bags, hanging by colored plastic clothespins from lines overhead. On tables underneath stood glass jars that contained parasitoids, some of the thousands of species of parasitic wasps carried by caterpillars or pupae. Individual plastic bags were labeled in black marker; the clotheslines were also labeled by lepidopteran family name: "Sphingidae," read one, for Sphinx moths; "Notodontidae," another, for the family known as prominents.

On the drive back to Santa Rosa, I looped around the volcanoes. Here was the agroscape, what the ACG would have been if Janzen and Hallwachs and Costa Rican politicians and professionals had not worked to save it. There were cleared pastures, shacks, bananas, towns and more towns—El Encanto, Brasilia, Santa Cecilia, Los Inocentes. Everywhere the earth had been forced to bear fruit—pineapples, bananas, oranges. For the oranges, the road had been expensively graded and improved to facilitate their transport to the highway.

By comparison, the Área de Conservación Guanacaste is a storehouse of riches, a vast repository of evolutionary and genetic knowledge, breathtaking in its physical scale and its rewilding ambitions. At first glance, it broke all the rules. It used cattle and nonnative tree plantations in a national park, for its own purposes. When conservation organizations were solemnly lecturing the world about how the rain forest, once lost, was lost forever, the ACG regrew tropical forest, thumbing its nose at the pieties. And the experiment worked.

But most importantly, the people who lived there knew that it worked. Its achievements, after all, were theirs, from the cooks to the director. Every year, the park enters a float in the annual cattlemen's parade in Liberia, the capital city of Guanacaste. Riding on horseback alongside it are the parataxonomists, the program coordinators, and other staff, carrying the flags of Costa Rica, Guanacaste, and the

ACG. That float, Janzen says, is itself a message: "Our sons and daughters go to school with your sons and daughters, we hire your teenagers and you hire our[s]. . . . We are all in this together."[33]

Since the ACG demonstrated what could be accomplished in regrowing tropical dry forest and rain forest, other conservationists around the world have adopted the same techniques. In Indonesia, on the island of Borneo, Willie Smits, a native of the Netherlands and a microbiologist, launched an ecological restoration program with striking similarities to the Costa Rican project. Smits arrived in Borneo in 1981 as a forester but soon went to work as a conservation adviser to the Indonesian government, shocked at the terrifying destruction created by wholesale logging. In 1991, after finding a dying infant orangutan on a trash heap, he founded the Borneo Orangutan Survival Foundation to take in orphans created by the industry. He realized, however, that restoring the primate's habitat was the key: The orangutan could not survive as a species without its forests.

Smits was further motivated by a rash of fires that broke out in 1997 in the wake of uncontrolled logging, which had caused drastic climate change. As in the Amazon—which creates its own weather systems—tropical rain forests in Asia may act as water pumps, pulling in vapor that condenses, generating rainfall.[34] But in Indonesia, the pumps were failing; for months, fires burned hundreds of thousands of hectares on Borneo. A report in *Nature* later revealed that up to 40 percent of the total carbon emissions on earth that year had been produced by those fires.[35]

So in 2001, Smits began buying wasteland in an area he subsequently named Samboja Lestari (for "everlasting forest"), a former palm oil plantation that had been logged and reduced to, as he put it, a "biological desert."[36] Little survived except alang-alang—the jaragua of Indonesia—a fire-prone grass that secretes cyanide. Annual fires completed the devastation, but Smits developed a plan that was striking in its simplicity. "We bought the land, we dealt with the fire . . . we started reforestation by combining agriculture with forestry," he said.[37]

He rallied local Dayak villagers, enlisting them as firefighters. Using his experience in microbiology, he figured that new growth would be sustained by vast amounts of compost, which he concocted with waste from the orangutan sanctuary and manure from livestock, adding sawdust from logging operations. He hired villagers to plant thousands of trees in the newly composted fields, collecting some thirteen hundred species of seed, including some from orangutan feces. He also experimented with planting some commercial trees to provide shade for native seedlings, in much the same way as the *Gmelina* plantations had in Costa Rica.

Throughout the entire process, he engaged local people as partners. Samboja Lestari was located in the poorest district in the province of East Kalimantan, the second largest in Indonesia, and the communities, ravaged by 50 percent unemployment, were apparently delighted to participate in making the project a success. Within four years, the project created jobs for three thousand people, who were further motivated by the chance to grow food—beans, pineapple, ginger, bananas, chocolate, chilies—in parcels of the recovering property. The orangutan sanctuary now buys fruit for its orphans from Samboja Lestari, and the project has also experimented with planting a ring of fire-resistant sugar palms around the growing forest. Tapped twice a day for the sweet liquid produced in the trunk, the palms provided another source of revenue for nearly 650 families. More jobs were created by a tree nursery that grew saplings for the project, and an ecolodge has been built to attract tourists eager to see the recovery of orangutan habitat.

Between 2002 and 2008, over twelve hundred tree species were planted in Samboja Lestari, a thousand seeds and saplings a day. Its accomplishments seem highly promising. By arrangement with NASA and the European Space Agency, Smits has been using satellite photography to monitor atmospheric conditions, the growth of individual trees, and other parameters of the area. So far, he reports, the temperature has fallen by three to five degrees Celsius; air humidity has risen 10 percent; cloud cover has increased over 11 percent; and rainfall—

critically—has risen by 25 percent. Once again, the area is producing its own rain, and life has returned, including 137 species of birds—in the logged area, there were only five—and nine species of primates. In a recent presentation, Smits said of the environmental devastation in Indonesia, "To say it is hopeless is not the right thing to do."[38]

Like Janzen, Smits is something of an eccentric, and the project has not been supported by international conservation organizations, which remain critical of restoration projects in Indonesia, where they hope to preserve what remains of existing forests.[39] They point out that Samboja Lestari's success in reproducing native forest conditions has not been monitored independently. That skepticism brings a fierce debate into focus: no matter how inspirational, restoration may never be able to replace what was lost. Because of the difficulty of re-creating complex soil conditions and the community of microorganisms that contribute to rain forest diversity, restoration—whether intentional, aided by activists like Smits, or naturally occurring on land abandoned by farmers or loggers—has sparked intense battles in the environmental community. In 2005, the United Nations issued a report hailing secondary forest growth in the world's tropic zone—arising from previously logged areas—as a significant conservation gain.[40] As people abandoned rural farms to move into urban areas, seeking jobs and a higher standard of living, an estimated 2.1 billion acres of so-called replacement forest was regrowing, representing an area around the size of the United States. In some regions, areas of secondary forest were increasing by over 4 percent a year. The United Nations, along with a few scientists, suggested that such new forests were "undervalued," particularly in calculations of the carbon sequestration they provided. A senior scientist at the Smithsonian, Joseph Wright, criticized biologists for "acting as if only original forest has conservation value." That, he said, was "just wrong."[41]

Most biologists, however, were horrified by the suggestion that secondary forest be considered an unequivocal boon, arguing that natural regrowth was often limited to a handful of species. Conservation organizations dedicated to saving the last tropical forests in Asia were fearful that the emphasis on new growth might make the world complacent. "Not all forests are equal," one of Wright's colleagues told the *New York Times*.[42]

Eager to counteract the impression that secondary forests might prove an adequate substitute, a group of biologists conducted a 2007 study comparing the diversity found in primary and secondary forests in the Brazilian Amazon.[43] They found that primary forests were essentially irreplaceable: at least a quarter of the species of plants and animals included in the census were never found outside such areas, even though all study sites were located near extensive old growth, offering ready opportunities for recolonization. Their conclusion: we must save primary forests.

Doubtless, a good rule to follow. But in many parts of the world, conservationists are left with few options other than restoration. Heavily logged in recent years, much of Asia is now covered by the same kind of palm oil plantations (for the production of biodiesel) that left Samboja Lestari desolate. Palm oil and other intensive biofuel crops leave behind areas that may host only 15 percent of their original biodiversity. But despite the fact that such restoration may never be perfect, it still offers hope. With ever more ingenious techniques being developed and where there are patches of original growth, there is still a chance. In the soil, in the leaf litter, surviving germplasm—the genetic material found in seeds—still holds the ghosts of a lost biodiversity. In the bush of southwestern Australia, where wheat farmers conducted a valiant, if disastrous, struggle to convert the land to agriculture, a small group of committed ecologists has come to their aid, calling forth the spirit of a once sumptuous land.

CHAPTER 13

REGROWING AUSTRALIA

Extreme Extinction

THE SADDEST ENVIRONMENTAL STORIES YOU'VE EVER HEARD are from Australia. A place of violent extremes, the driest inhabited continent, Australia leads the world in wiping out species, with the highest rate of mammalian extinctions on earth. It seems only fitting that Australia should also have the most inspirational tales of restoration.

Flat and more geologically static than any other land mass, largely untouched by glaciers over 2.8 billion years, Australia has some of the oldest rocks in the world, and its soils are thin and poor. Swept by the ocean's salt winds, baked by the sun, every living thing evolved strange strategies to cope with low levels of nutrients and scarce water.

The continent abounds in bizarrely adaptive flora and fauna. The duck-billed, poison-spurred, web-footed, beaver-tailed amphibious platypus, probably the most primitive mammal to survive to the present day, was so outlandish on first sight that British scientists believed the animal was a hoax, a collection of parts glued together. The platypus is one of three species that make up their own order—the Monotremes—the only egg-laying mammals. The others are two kinds of echidna, found only in Australia and New Guinea. Echidnas look like hedgehogs covered in spines and fur, with a long, hairless anteater-style snout—mouth and nostrils at its tip—that can double as a snorkel when they swim. In addition to koalas, kangaroos, and wallabies, there are carnivorous marsupials: quolls, which look like a cross between weasels and cats, furnished with large white spots all over their bodies; dunnarts, mouselike marsupials with naked tails; phascogales, with extraordinary bottle-brush tails; bandicoots and bilbys, like a cross between kangaroos and rodents. There are possums and gliders of every shape, size, and description, from small to large, from short-haired to luxuriantly furred. Among kangaroos and wallabies, extraordinary specialization has occurred. Gilbert's potoroo eats primarily fungi. Bennett's tree kangaroo, which occurs in only a minuscule area in the north of the continent, climbs trees, leaping from branch to branch, eating leaves and vines.

The arrival of aboriginal people on the continent presaged a wave of extinctions among large mammals.[1] A marsupial lion, a marsupial rhino, a predatory kangaroo, and giant wombats the size of cars all went to the wall. The arrival of Europeans was even more lethal, releasing a plague of introduced species that decimated natives: house cats, pigs, goats, foxes, camels, donkeys, and, infamously, rabbits. Since 1788, these feral animals have wrought, along with the conversion of habitat to agriculture and development, the extinction of sixty-one species of plants, twenty-seven mammals, twenty-three birds, and four frogs, including the crescent nailtail wallaby, toolache wallaby, desert rat-kangaroo, desert bandicoot, lesser stick-nest rat, pig-footed bandicoot, Darling Downs hopping mouse, King Island emu, broad-faced potoroo, eastern hare-wallaby, and the sharp-snouted day frog, the last wild specimen of which was recorded in 1997.[2] The most highly endangered mammal on the planet—the northern hairy-nosed wombat—consists of a population of 115, up from 90 a few years ago, in Epping Forest National Park in central Queensland, near Brisbane. Park conservation officers have had to fence the population completely so that dingoes cannot pick them off.[3]

Perhaps the most pathetic of Australian extinctions was that of the "Tassie tiger." The largest marsupial predator to have survived the earlier wave of losses, the species was extirpated from mainland Australia and New Guinea after the arrival of the dingo but held on in Tasmania. Staring from rare photographs with mute dark eyes, the thylacine looked like a dog with stripes across its hindquarters, although the creature has also been compared to cats, wolves, and hyenas. It hunted wallabies. Australian zoologist Tim Flannery describes its end, hastened by a government bounty program that ran from 1888 to 1912, a pound for a dead adult, ten shillings for a pup:

> The last thylacine to walk the earth was a female kept in Beaumaris Zoo near Hobart. Personnel problems developed at the zoo during 1935–36, which meant that the animals were neglected during the winter. The thylacine was "left exposed both night and day in the open, wire-topped cage, with no access to its sheltered den." September brought extreme and unseasonal weather to Hobart. Night-time temperatures dropped to below zero at the beginning of the month, while a little later they soared above 38 degrees Celsius. On the night of 7 September the stress became too much for the last thylacine and, unattended by her keepers, she closed her eyes on the world for the last time.[4]

Such mythology has grown around the creature—people are so loath to believe that the Tassie tiger is really gone—that sightings continue. While its survival is not a complete impossibility, since large areas of southwestern Tasmania remain roadless and virtually inaccessible, it is probably gone. In 1999, the Australian Museum launched an ambitious project to attempt to clone the species, using DNA extracted from specimens, including an 1866 pup preserved in ethyl alcohol. The museum director has called this effort "the most exciting biological project that's going to occur in this millennium," but others argued the money would be better spent saving species that still exist.[5] Today, the only way to see a thylacine in action is in the pages of David Owen's *Thylacine: The Tragic Tale of the Tasmanian Tiger*. In the bottom corner of each right-hand page, there is a faint image of a thylacine. Flip through the pages, and the Tassie tiger comes back to life.

A Million Acres a Year

Beginning in the 1950s, the Australian government opened up the west to agriculture, releasing a million acres a year to farmers in the War Service Land Settlement Scheme, termed the largest land development program in the Southern Hemisphere. In a modern-day homesteading movement, farmers plowed under an enormous chunk of the continent. But while it took a century to produce the Dust Bowl in the United States, land clearing in the arid southwest of Australia touched off cataclysmic drought, salinity, and bankruptcy after only a decade.

This extraordinary environmental and economic catastrophe was chronicled in a documentary film, *A Million Acres a Year*, first aired in Australia in 2003.[6] With interviews and historical footage, it recounted how the Australian government turned World War II veterans loose, spurred by the million-acre marketing slogan, with no understanding of the forces they were about to unleash. The western Aussie bush was the cheapest land in the country—starting at 1.36 Australian dollars a hectare, with 50 percent off for servicemen—and within six years a half a million acres had been fenced and a quarter million planted with wheat or used to graze sheep. Thousands of hectares were cleared by what one farmer termed "a terrifying machine," a disc plower that tore through native vegetation like a shark. Thousands more acres were burned to remove brush. A local fire captain remembered that burning three hundred acres was considered "a pretty poor day." A farmer recalled the terrible sight of kangaroos "trying to leap out through the flames. . . . It just made me feel sick. . . . It was so unnatural. There was no escape."

There was no escaping the consequences, either. The continent had developed a fragile balance over the course of billions of years of rain, wind, and heat; land clearing destroyed that balance in a single sweep. By 1969, the first drought hit. In 1971–1972, another. In 1981, rains failed again. By 1982 and 1983, the area was well on its way to becoming a dust bowl; 10 percent of the farmland was blowing away. Deep-rooted native plants, adapted to the soils, drying winds, and salinity of the continent, use water efficiently. When they were removed and replaced with thirsty, shallow-rooted wheat, water seeped into saline subsoils, dissolving salts and bringing them to the surface. The new crops altered the climate as well, changing rates of evaporation and minimizing cloud formation. Clearing had destroyed the very climate and soil farmers had hoped to harness.

The government scheme was abandoned. But thousands of farmers were already trapped, unable to sell, unable to scrape by. One farmer said, "You could barely see a hundred meters with the dust and material that was moving around." Another joked, "If you leave your son a farm these days, he's liable to sue you for child abuse." The fallout was so economically, emotionally, and visually dramatic that virtually every farmer interviewed in the film acknowledged that the devastation had been caused by their own actions.

Watching the destructive forces they had unleashed, many of those whose dream had been farming decided to change their way of life. Steve Newbey, a third-generation farmer, made the decision to abandon traditional farming and replant his land with native plants. "I feel I've got a responsibility to repair the damage that my family has done to this land," he said. "Unlike any other business, if you open up a shop somewhere and it fails, well, so what? You can move on and do something else. But if your farm fails and you stuff up the land, that land's stuffed up for everyone for the future, and I don't think we've got the right to do that, and I think real farmers accept that." Newbey had started to see bush returning: "That's a great feeling, you feel like you've really created something. You've got wildlife coming in, blue wrens, quail." An older farmer agreed: "I'd be interested in planting as much back as I could."

It was becoming apparent to those who had held onto scraps of "bush," as native land is referred to in Australia, that the bush itself was the treasure, not the poor farmland they had made of it. The bush was not, as the government had cynically put it, "useless sand plain." It was teeming with endemic biodiversity. In 2000, southwestern Australia would be included in a list of the world's top-twenty-five "biodiversity hotspots," with over eight thousand plant species, 48 percent of which were endemic. Stirling Range National Park, which abutted much damaged farmland, held sixteen hundred plant species; Fitzgerald River National Park to the east, nineteen hundred, seventy-five of which occur nowhere else on earth. In the Australian spring and summer these national parks and the remaining bush around them explode in an exotic display of flowering forms and colors capable of drawing tourists and photographers from all over the world: hakeas, banksias, orchids, dryandras, eucalypts, mountain bells, triggerplant, dwarf kangaroo paw, featherflower, clawflower, velvetbush. Some flowers evolved in conjunction with their own marsupial pollinators: the tiny honey possum inserts its needle nose in blossoms of flowering plants, particularly those in the Proteaceae and Myrtaceae families, feeding entirely on nectar and pollen. Coping with extremes in the climate over millions of years had yielded an extraordinary variety of mechanisms among plants and animals that were only beginning to be understood.

Could this unique ecosystem be regrown? No one knew.

The Puzzle

Gondwana Link is the rewilding response to *A Million Acres a Year.* Its ambition is vast, aiming to protect a zone from the southwestern tip of the continent all the way across the southern peninsula to the

Great Sandy Desert. But it has proceeded by modest steps, seeking to reestablish a physical bridge of native bush between two national parks. That small-scale start, concentrating activities in a single area, has already set Gondwana Link apart from very grand projects like the southern African peace parks, which will take far longer to show results.

Beginning in 2002, several national and regional environmental groups—the Australian Bush Heritage Fund, the Fitzgerald Biosphere Group, Friends of the Fitzgerald River National Park, Greening Australia, and the Wilderness Society—pooled their resources and personnel to create Gondwana Link. The narrator of *A Million Acres a Year*, environmentalist Keith Bradby, coordinated the project, and his team mapped out an overall strategy, focusing purchasing power and expertise on a forty-four-mile-wide swath between the two national parks, Stirling Range and Fitzgerald River, which they took to calling "the Fitz-Stirling." The plan was to buy properties outright, to work with farmers on bush-protection measures, such as fencing and conservation easements, and to encourage "conservation buyers"—environmentally minded people with disposable income—to invest. Degraded properties would be replanted. Intact ones would be protected, creating clusters of existing and restored bush throughout the area.

On paper, the project looks like a partially finished jigsaw puzzle. On the left side of the map, a big green block represents Stirling Range National Park; on the right, a more massive block, extending all the way to the ocean, is Fitzgerald River.[7] In between, on a field of white—cleared land—there are scattered puzzle pieces. Some of the pieces are remaining bushland, never cleared. A big light green piece in the middle represents a proposed nature reserve. The rest—the handful of properties that Gondwana Link has managed to buy and replant—represent one of the most concerted efforts to resurrect nature ever attempted.

I met Keith Bradby at his office, in an old house in Albany, a coastal town south of Perth. Cheerfully obscene, with a mouthful of

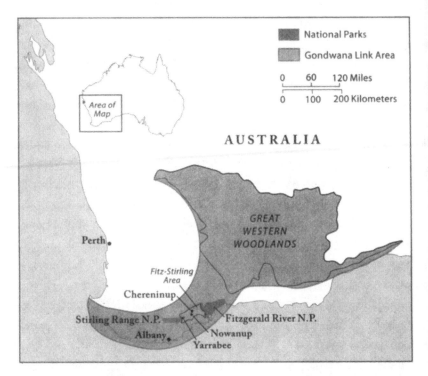

Southwest Australia: In southwestern Australia, Gondwana Link restores native bush between the Stirling Range and Fitzgerald River national parks.

Australian slang, bush boots on his feet, and an Aussie cowboy hat on his head, Bradby pulled out two iconic images of the land. The first was the cover of *Men Against the Earth*, a 1946 novel about a farming family in western Australia, a heroic image of what Bradby called "the old, adversarial approach," a man plowing behind horses, two strangled-looking trees outlined in eerie light as the soil is turned under.[8] The second was a stunning photograph he pulled up on his computer screen.[9] Taken from the air, it showed the Australian landscape northeast of Gondwana Link, a stark line visible down the center. On one side: an expanse of dark green capped by clouds. On the other: straw-dry wheat fields, with barely a cloud in sight. It was absolute, incontrovertible proof of the power of monoculture to commit mayhem.

The line visible in the photograph was the famous No. 1 Rabbit-Proof Fence, the longest fence in the world—1,139 miles—thrown up in 1901 in a desperate bid to keep rabbits out of western Australia, after twenty-four bunnies were loosed on the continent by one Thomas Austin, a British transplant who wanted something familiar to shoot. From the beginning, the rabbit-proof fence was not rabbit-proof. When rabbits were found inside No. 1, numbers 2 and 3 were rapidly erected. They quickly failed: the innate cunning of rabbits.

The fence had inadvertently created one of the world's great climatic experiments, separating bush from agriculture. Native vegetation and crops, such as wheat, have different reflective qualities, known as "albedo."[10] By replacing native bush with wheat, farmers had replaced darker, less reflective plants with lighter, more reflective ones. What's more, they had replaced plants that transpired all year with annual vegetation that only transpired during winter months. Instantly, if unwittingly, they had created the conditions for their failure: the clouds that formed routinely over native bush—cooling and easing evaporation below—literally evaporated. The droughts that followed were man-made. "What we've done is dried it out through clearing it," Bradby said. "Now we're drying it out more, through climate change." Gondwana Link is thus not simply an effort to save and restore native species of plants and animals on a scale that will preserve them into the future. Its scale is a hedge against climate change.

Originally a beekeeper and specialist in native seeds, Bradby is a thirty-year veteran of Australian "landcare," a community and volunteer-based environmental movement that sprang up during the 1980s in response to worries about water and degradation of the land. Landcare is now continent-wide, with activists working on every ecosystem, from coasts and rivers to native bush, restoring native vegetation and removing nonnative species of animals and plants. Large-scale rewilding is the logical extension of landcare, and Bradby gave every sign of making a total, lifelong committment to that enterprise.

Bradby was not focusing on corridors. "Corridor, to me, is an immensely minimalist way of thinking," he said. "A narrow gangway for fauna to scuttle along mainly under cover of darkness." He pointed to the limits of corridors in the Australian setting: "A lot of your fauna will move through your North American settled landscape, but ours will not. Some birds will—but virtually all the critical-weight-range mammals, in terms of anything smaller than a kangaroo, bandicoots, honey possums, potoroos, quokka—won't." He also made it clear that he was pursuing a change in the way conservation was practiced. Planning across a more generous measure of the country-side, he felt, was both a biological imperative and a social one. "If we had $15 million come in tomorrow and could buy land and fence it off and keep everyone out, we wouldn't do it," he said. Like Janzen, Bradby was after something bigger. Something more functional. Something that would bring the whole community along with him.

That was the beauty of Gondwana Link. Its scale was more than generous—embracing everything from people to potoroos—but its methods were consistently, refreshingly doable.

Living in the Link

I visited several pieces of Gondwana Link: Chereninup, revegetated in 2003; Nowanup, site of revegetation and a new meeting center for the region's aboriginal people who have partnered with Gondwana Link; and Yarrabee, a big chunk of land extending out from Stirling Range National Park. As I drove north from Albany, the rapid shifts in landscape were almost bewildering, like switching images on a View-Master reel.

Albany itself, with the only deep water port in the southwest, is a typical rural agrarian Australian town, with its ANZAC (Australia New Zealand Army Corps) monument high on a hill looking out to sea and a freight-loading terminal below. It rains often in town from the coastal weather effect, but a few miles north the rains thin and fields stretch out sere and baked-looking, with patches of bush near riverbeds, which are dry most of the year. The country opens up, and the craggy peaks of the Stirling Range slide into view, a reminder that this was never simply wheat country.

The land is spectacularly strange and varied, from the carpet of native wildflowers near the spooky, foggy summit of Bluff Knoll in the Stirlings to the thick, hot, flat, flyblown bushland of private ranchers. This was the much-vaunted botanical biodiversity of the so-called Fitzgerald Biosphere, where plants evolved in communities based on different soil types, related to the mixed geological mosaic of the bedrock below: marine plain, granite, quartzite, greenstone, red loam, sandy loam, schists.[11] Soils are made up of particles of rocks and minerals mixed with organic matter from decaying plants and animals and are stratified in different layers. Bedrock, beneath layers of subsoil, is the "parent material" of many soil types, contributing to their chemical composition, especially in places like Australia, where plants evolved over hundreds of millions of years, their adaptive processes uninterrupted by periods of glaciation. Part of what makes the plant communities of western Australia so unique and diverse comes from their long evolution out of distinctly different subsoils.

Amanda Keesing and Paula Deegan, two of the Gondwana Link team, drove me out to Chereninup Creek Reserve, where revegetation plantings already reached waist-high. They told me about a memorable day in 2003 when partners, staff, and volunteers had gathered on the rise where we stood to plant thousands of tiny native seedlings and seed of fifty endemic species collected from nearby bush. Several years on, over 80 percent of the seedlings were surviving, even thriving. Seeds in loamy soils had done well, those in clay soils less so, but around 600,000 plants had become established. The shrubs were filling out, erasing the rows in which they'd been planted. "It's wishful thinking that you can get it back to native bush," Keesing said, referring to the more complex mix of species in untouched original growth. "But you can provide different heights and get it as close to native bush as possible."[12]

Later, I had a chance to meet Peter Luscombe, whose Nindethana

Seed Service had provided some of the native seed used in Gondwana
Link projects. His warehouse, outside Albany, is a repository of one
of the most diverse and important collections of Australian flora on
the planet, much of it kept in old ice cream tins. Luscombe himself
comes from a farming clan but as a teenager grew to despise the family business for "destroying everything in the name of making a dollar." Describing the process of revegetation, he noted wryly that it
consists of undoing everything farmers did. "You scrape all the topsoil off," he said. "It's the opposite of what you'd do for normal farming. The topsoil is loaded with introduced annual seeds. It's too rich,
full of trace elements, too wet, too lush." Native plants, he said, "like
to breathe, have a bit of air. So you take all that rubbish off and do a
little rip to get a sterile seedbed. The native plants are so efficient,
they just take off like it's the perfect soil."[13]

And so they did. Animals were already moving into the area.
Deegan led me down a dirt path to point out a big bobtail, an unusual
Australian lizard with a triangular head. When it saw us looking, its
mouth gaped open and a broad blue tongue flopped out like a scarf.
Bobtails mate for life, and their strong pair bonding is unique in the
reptile world. Studies have chronicled the fealty of the creature,
which will stay with a mate killed on a road or fence for days.[14]

Next door to Chereninup, a farmer named Brian Penna had kept
his own substantial native growth, to control salinity. Tall and sun-
burned, he squinted as he said, "Everyone's got degradation issues
here with salt and wind and water. Anyone with a little common
sense sees the advantage of bush." He was helping Gondwana Link
control weeds on its revegetation and watch for fire. With over six
thousand arable acres, Penna had kept fifty-six hundred acres of
bush, much of it along the creek lines. "That's a bigger percentage
of bush than about any farmer I know," he said proudly. "There's a bit
of character around here. The wildflowers are pretty spectacular. I
don't mind a bit of a bush walk."[15]

It was a short drive from Chereninup to Nowanup, but we passed
from one world into another. The landscape went from dry scrubby

bush to grassland, waves of grass kneeling and rising with the wind, running across a valley to come up against dramatic red spongelite cliffs, porous rock formations made of the skeleton of sponges deposited by a warm Eocene sea forty million years ago. There was revegetation at Nowanup, too, but the most striking addition was at the base of the cliff, where a modest Noongar "meeting place" had been constructed in 2007, a collaboration between Gondwana Link and the Noongar people. Under an open thatched roof, round reddish clay plaques were set in a half circle, each with the name of a different Noongar community etched inside the outline of a boomerang: Albany, Warriup, Doubtful Island, Twertup, Borden, Fitzgerald, Ravensthorpe, Gnowangerup, Salt River.

It was a place of great peace, where the wind was both visible and audible. It was without pretensions and far removed from urban or consumer culture. Eugene Eades, a Noongar man who was coordinating "Caring for Country," a program integral to Gondwana Link, called it a "bush university" where elders could gather to experience and remember the old ways and young people could learn about "country," the aboriginal term for the land.[16]

In Australia, aboriginal people were granted separate citizenship status only in 1948, despite the fact that many served their country in both world wars. They were not counted in censuses—and were actually considered part of the country's "fauna"—until 1967, when they were finally granted legal rights and protections in the constitution. Since then, a wave of High Court decisions has established native title without, however, negating nonindigenous rights and titles. Indigenous people may now pursue traditional activities—hunting, fishing, setting up businesses—on land to which they have established native title, but it remains to be seen what additional property rights, if any, will be granted. In 2006, a federal court ruled in favor of the Noongar people's title covering the entire Perth metropolitan region, a decision currently on appeal by the state of Western Australia.

Nowanup was part of Gondwana Link's effort to collaborate with Noongar people in the area and respect their history, culture, and claims on the land. Over the time I was there, it became clear that the project was picking its way through native title controversies with care, trying to learn from the elders and avoid the kind of mistakes that had been made in the Pacific Northwest, where the rich got the benefit of Lake Washington's restoration and the Native Americans got dying salmon runs. Eugene Eades introduced me to one of the elders, a half brother, Aden Eades, at a tea in a tin-roofed farm outbuilding where Gondwana Link was putting on a program about fauna trapping, a survey of local properties to determine what species were found in the belts of bush around streambeds and farm borders. A biologist had brought in a honey possum trapped that morning, and Aden said quietly, "I've never seen a honey possum. And I'm sixty-seven. I'm very pleased as a Noongar that so many people would be interested in these things, and I'm very pleased you invited me here today to have a look at what you're doing. It can only get better. This country's big enough for all of us." He had already told us during a walk in the local bush that he had eaten bobtails as a child. "We ate anything we could find," he said matter-of-factly. "Tail's probably the best part of it."[17]

The honey possum, nervously skittering around at the bottom of a bucket, did not appear so pleased, although it would be released later. Given what Aden Eades's people had suffered, I thought his comments were gracious, but I noticed a coolness between the farmers, who drifted away after they'd drunk their tea, and the Noongar people. This was the Australian version of what was going on all over the world, wherever landowners, conservationists, and native peoples had to grapple with conflicting needs and agendas. I saw the same furrowed brows in the Okavango Delta and sensed the same anxieties in Nepal and in Kenya. Conservation exposes the fault lines between property owners and the landless, between colonizers and indigenous people, between rich and poor. This piece of the jigsaw puzzle is not visible on the map. But it is, perhaps, the most significant piece of all.

At a workshop on "Restoring Connections," those tensions exploded.[18] Eugene Eades was there, along with a dozen or so other Noongar, listening to biologists and other scientists explain projects on restoring wetlands and developing ecotourism products such as quandong jam, made from native fruit gathered by local people. Throughout the day, the Noongar representatives sat quietly, until late in the afternoon, when two archaeologists from the University of Western Australia began describing a project that would examine the entire southwestern landscape for artifacts and other evidence of the ways aboriginal peoples had used the environment. The man sitting next to me, Harley Coyne, a Noongar officer of the Department of Indigenous Affairs, raised his hand and asked about the budget for the project. The answer, 900,000 Australian dollars, was received in heavy silence. People whispered and shifted in their chairs. Coyne demanded to know how the project had been approved, and others raised their voices. One woman shouted that the information the archaeologists were after was "our intellectual property." Coyne grabbed my notebook and wrote, in capital letters, "DUPLICATION!" Eugene burst out, "You'd think from Day 1 you'd put our elders right up top! And give every Noongar a fair and equal opportunity at resource management! They walked that land, told their dreamtime stories, put their fish traps out." He subsided, shaking his head. The scientists, their faces burning, began alternately apologizing and trying to explain that they had attempted, without success, to reach Noongar members to explain the project in advance.

Later, in the parking lot, Coyne was still seething over a budget that, as he'd written in my notebook, seemed to him a duplication of money that had already been spent on similar projects and held little benefit for aboriginal people. "What are we gonna get out of it?" he said. "Fucking $900,000 budget! What a joke. What about education?"[19]

My hosts, Amanda Keesing and Paula Deegan, were chagrined by the confrontation, but it seemed a reasonable airing of grievances. Despite exasperation and recriminations, at least people were talking. In Catron County, the parties would already have progressed to death threats. Gondwana Link was regularly engaged, on a day-to-day basis, in hashing out conflicts and questions, which put it ahead of projects where communication had been, at least initially, a low priority. For all the heated words, everyone seemed committed to conservation. More than any other project I visited, Gondwana Link came closest to reaching Aldo Leopold's "land ethic," bringing the individual into "a community of interdependent parts."[20] The sheer range of people involved in the project was astonishing. Rich and poor, people with private planes and people with barely two sticks to rub together, biologists, seed specialists, orchid lovers, inn owners, taciturn farmers, Noongar people, Australians, Americans—Keith Bradby had recruited them all. It was a rewilding melting pot, a biotic coalition devoted to getting things done.

One of the odd pieces of Australian parlance that any visitor is sure to hear is the ubiquitous "No worries, mate." But now, in Australia, there are worries. On this continent, global warming is not a theory; it has arrived. What happened to the farmers in southwestern Australia is now happening across the country's agricultural sectors. In February 2009, 173 people died in the worst bush fires ever seen; the week before, 200 died in Melbourne during a heat wave and 4,000 flying foxes—large fruit bats that roost in city parks—fell dead out of the trees.[21] Suicide is up. Depression is up. Every major city has enacted water restrictions. Brisbane residents are drinking recycled water, part of a program called "toilet to tap." A farmer in the Murray-Darling Basin near Adelaide, Australia's central wheat and wine belt, told a reporter for the *Los Angeles Times* that he'd come upon a neighbor sitting in his pickup truck weeping helplessly. "It's desperate times," he said.[22]

Gondwana Link's most impressive accomplishment has been to take that despair and channel it into action that can help Australia correct its environmental course. The movement has generated momentum by forging cooperation across the region. Those relationships, nurtured as carefully as if they were revegetation, in frequent get-togethers, consultations, meetings, and even conflicts, have created a community-wide network of protection for different habitat types, riparian areas, and threatened species.

I saw this across the countryside. In Manypeaks, a town just north of Albany, Doug Russell, a politically conservative cattleman who owned three farms and nine hundred head of cattle, had undertaken extensive restoration work, fencing off bush and growing hedgerows of native plants to act as windbreaks. He saw that conserving bush could improve the microclimate on his property, protecting his animals from harsh, drying winds and reducing evaporation at stock ponds. He proudly showed me his own personal protected "karri" forest, an exuberantly messy patch of trees and undergrowth full of native eucalypts, orchids, and vines.[23] Karri and jarrah eucalyptus trees, some of the largest in Australia, can reach two hundred to three hundred feet in moist forests, sheltering blue wisteria and red coral vines, cockatoos and lorikeets, and small marsupials. This was the easternmost stand of karri in all the southwest, and he had placed it under a conservation easement, hosting researchers, bird lovers, and orchid enthusiasts. Like the Malpai ranchers in the American Southwest, he had made the connection between his own economic future and conservation.

In Ongerup, a nearby community, Judy O'Neill, another farmer, took me to see the town's Ecotourism Center under construction, a showcase for the region's flagship bird, the malleefowl, and soon to be a draw for busloads of tourists visiting the nearby parks in the hope of seeing that elusive creature. A few miles away, at a small private protected area known as Gnowangerup ("Place of the Malleefowl"), O'Neill led me through a dense stand of mallee bush to visit a pair tamed by researchers.[24] Whitened mallee roots littered the ground, looking like bleached bones—many eucalypt species grow from a knobby tuber-like underground trunk—but there were also real bones, of introduced foxes, dangling in the shrubbery, a macabre warning to others of their kind.

Waving away clouds of bushflies, O'Neill crept up to a staggeringly large pile of sand, branches, and dirt, the weirdness of it commensurate with the creature. This was no delicate little bowl of twigs but a hill of dirt bigger than a pitcher's mound, bigger than a Barcalounger: rather like crocodiles or turtles, malleefowl entomb their eggs in an insulating earthen chamber, the temperature of which determines the sex of the embryo. The malleefowl itself is no larger than a chicken. The birds live without surface water for months during the summer; they mate for life, with a lifespan of twenty-five to thirty years; they build their incredible mounds over a period of nine or ten months, working to maintain them for three hundred days a year. They lay their eggs within the mound, where they are incubated by the heat of decomposing matter. Then, after all that effort, they abandon it. They lay, and they walk away. Although malleefowl chicks can fly within twenty-four hours after emerging, they barely stand a chance against starvation, snakes, monitor lizards, foxes, and other misfortunes; fewer than 2 percent survive. O'Neill introduced me to "Augusta," a female malleefowl who produced a series of soft, sweet, interrogative sounds, disappointed that we had failed to bring her a treat. For such powerful master builders, malleefowl are modest-looking folk, with beautifully cryptic coloring in bands of beige, taupe, dark brown, and black. Augusta's protected patch of bush was one more chain in the Link.

My next stop was Yarrabee. At 2,300 acres, with 798 acres of bush, the property was the largest yet purchased. It bordered the northeast corner of Stirling Range National Park and extended the reach of the link out from that core area, providing an excellent means by which the flora in the park could gradually expand its range. Replanting Yarrabee's 600 hectares of cleared land would be the largest planting of native vegetation ever in Australia. "This has been our biggest buy," Keith Bradby said wryly. "But then they've all been our biggest buy."[25]

He described Yarrabee as a former "battlers' block," meaning that the previous owners had battled the elements. All too plainly, the elements had won, as evidenced in the abandoned, dilapidated shack where the farmers had set up camp. It was not uncommon, Bradby said, for farmers to live and raise their families in the same metal outbuildings that were commonly used as barns, without insulation, ventilation, or running water. Downhill from the house was more evidence of desperation: salt was appearing in crusty white patches near the edge of the property.

Bradby was anxious about Yarrabee: the initial replanting in 2006 had to be canceled due to drought. So there was little to see at that point, only a few dozen rows of tiny seedlings struggling up through the dry dirt. He was also concerned about the approach of South African love grass, an invasive marching steadily up the road toward the newly acquired property. But the potential of the property for restoration was on stunning display in thick uncleared patches of bush and corridors of woodland where tall myrtle trees swayed in the wind, shedding their bark in long, dangling strips, the wood pale and smooth beneath.

"I don't think we're putting back bush, exactly," he said. "We're controlling salinity, stopping the soil from blowing away." He admitted that there was no way to re-create the sheer complexity of the biodiversity: "We put 120 species in here, and there's three or four *hundred* in the bush. But in Chereninup, reptiles and echidnas are already moving through the land. Most animals can move through it in ten years. Some can't live in it for a hundred years." A sharp contrast to the lightweight promotional rhetoric of more extravagant rewilding initiatives, his concerns were reassuring. He was a realist. Driving past more African love grass, Bradby winced. "We're still in that narrow window of time before restoration problems mount up," he said. "The thing I can't waste is time."

Clearly, time was not being wasted. By 2009, more than 23,200 acres had been protected in the Fitz-Stirling landscape through direct purchase or conservation easements by private owners. Revegetation at Chereninup was head-high, and wildlife found it to their liking. A suite of birds returned, including malleefowl. Black-gloved wallaby turned up, along with echidnas and goannas. Progress continued at Yarrabee, with the entire property planted with native vegetation. The love grass was beaten back. Eighteen local farmers began to plant biodiverse bush on their own land, and neighbors of Gondwana Link properties were actively involved in improving management of soil and water. Across the whole link, over 2,400 acres were planted with 120 native species, and 3,212 additional acres became part of the carbon sequestration and sandalwood program run by Greening Australia.

There was also significant progress in reducing gaps between areas of protected bushland by nearly 40 percent. The largest gap had been reduced from 14 miles to 9. A recently completed purchase of another piece, Monjebup Creek, reduced a gap of 8.5 miles to under 3. With every purchase, every scrap of revegetation, the people behind Gondwana Link were getting closer to restoring the biological and physical connections across a once severed and fragmented landscape.

Meanwhile, at Nowanup, trials continued with native species, including malaleuca broombush (used for brush fencing) and sandalwood, *Santalum spicatum*, which produces nuts valuable both as a food crop and for their oil, used in perfumes and essential oils. Sandalwood grows as a parasite from the roots of other native plants, taking nitrogen from hosts, reducing the need for fertilizer. After it reaches maturity, its wood can be sold as a second cash crop. Its plantings on acacias and other native trees help create biodiverse pockets in cleared land, encouraging the recolonization of birds and other wildlife. Planting sandalwood also sequesters carbon, a potential third benefit for conservationists and commercial growers in countries that are signatories to climate agreements. Gondwana Link plans to utilize these sustainable crops to produce future revenue to support their work.

With enough pieces like Chereninup and Yarrabee, with enough farmers like Russell and Penna, the link through the Fitz-Stirling could well be made in the not-so-distant future. With recent gains there, Gondwana Link began to lay the foundation for their next focus, on the Great Western Woodlands, one of the largest temperate forests left on earth, 16 million hectares of wood and scrubland containing 20 percent of Australia's total plant species.[26] The Great Western

Woodlands represents a major core area in desperate need of improved management and fire control. The area constitutes an important weapon in Australia's fight against further climate change: the woodlands act as a carbon storehouse, holding at least twice the continent's annual greenhouse emissions. To log such forests now, with Australia suffering severe drought and a warming climate, would be catastrophic. Gondwana Link has begun negotiating a memorandum of understanding with native people who have land title in the area and has launched a three-year program to secure its protection and management.

In a recent summary of Gondwana Link's achievements, Bradby identified collaboration and the building of personal and professional relationships as critical. He emphasized starting small but starting smart, "choosing a small number of ecological targets that are achievable." The goal, he wrote, is to "continually consolidate," without spreading resources too thin.[27] Like the Area de Conservación Guanacaste, Gondwana Link had planned on a grand scale but started modestly, raising money, raising awareness, and raising the bar. It did so in a way that built political will first, in every stratum of the community, from farmers recognizing the damage they were doing to the land, to Noongar who wanted to return to it, to a younger generation who wanted no part of farming but cared about environmental restoration. Bradby was not bragging when he said, "People tell us we're the flagship of landscape scale."

Gondwana Link and the projects in Costa Rica and Indonesia are successes rewilding can boast about, knitting up the raveled sleeve of nature in ways visibly real yet miraculous. In a remarkably organic way, the science of restoration and the aims of community conservation coalesced in these projects. Those behind them emerged with their ambitious commitments intact, coming ever closer to the elusive goal of sustainable conservation.

ONLY CONNECT

REWILDING IS ABOUT MAKING CONNECTIONS. FORGING LITERAL connections through corridors. Creating linkages across landscapes and responsible economic relationships between protected areas and people. Forging links between ourselves and the intact ecosystems we need to survive. These connections are not optional: we cannot live without them. Aside from moral or aesthetic considerations, biodiversity is compulsory.

Consider fissures.

On the night of 21 July 2007, while two inches of rain fell on Chandler Heights, one of many satellite subdivisions of the vast desert megacity of Phoenix, Mesa, and Scottsdale, a crack in the earth opened up and swallowed a horse.[1] Citizens of Arizona watched in horror over the ensuing hours of televised struggle, as rescue workers and owners of the eleven-hundred-pound, thirteen-year-old animal, Cash, tried to pull him out of a hole that gaped some fifteen feet deep and ten feet wide. To no avail. After fifteen hours, Cash died of exhaustion.

Like climate change, earth fissures are a little bit of hell on earth, a knotty man-made environmental problem to which we have no answer. Simply put, cracks in the earth appear when the groundwater table has been lowered rapidly within a few decades. In some places in Arizona, the water table has fallen over two hundred feet since 1950, as communities spring up and pump the water out. As the water table drops, sediments shrink and compact, causing the earth to subside. Now found throughout the Southwest, in west Texas, southern Utah, New Mexico, and Nevada, earth fissures have swallowed homes, undermined dams, and destroyed a north Las Vegas subdivision, costing that city $14 million. They are particularly devastating in Arizona, where there are hundreds, possibly a world record. One of them is nine miles long.

Backfilling fissures is pointless: the earth swallows whatever people throw in them. And people are always throwing things in them—cars, tires, pharmaceuticals, solvents, refrigerators, sacks of garbage. Another bad idea, according to geologists: fissures lead straight to aquifers, bypassing the earth's natural filtration system, whereby water seeps through clay, gravel, and rock over time, leaving impurities behind. Although it hasn't happened yet, water specialists in Arizona are worried that crack dumping may eventually lead to contamination of the "stacked" aquifers beneath the state.[2] If that happens, once clean water from the aquifer will have to be treated, at great expense, just like surface water.

Now, consider the prairie dog.

Two animals once played a key role in the water cycle of the shortgrass prairies that swept from north to south, through Montana, Wyoming, Colorado, New Mexico, northern Arizona, and the Texas Panhandle: bison and prairie dogs. Bison created wide, shallow wallows across the prairie, wallows that turned into ponds. According to Alice Outwater, an environmental engineer, buffalo wallows functioned like recharge ponds, funneling rainfall to the water table. Prairie dogs may have been even more significant. Left to their own devices, they create massive underground systems of tunnels, ranging from twenty to over eighty feet long, ten to twelve feet deep. These tunnels once functioned as what hydrologists call "macropores." In *Water: A Natural History*, Outwater describes how the tunnels created a process known as "short-circuit bypass flow," allowing water that would ordinarily be trapped in the root zone of the grass to flow directly to the water table.[3] "The prairie dog population increased the amount of rainfall percolating down to the groundwater—and thence feeding the region's streams and rivers—as surely as the endless wallowing herds of buffalo did," she writes.[4]

Can prairie dogs solve earth fissures? Not by themselves, not anymore, and not without help: the damage to the environment has been too great. But protecting these species could start to repair the harm. Earth fissures are caused by loss of groundwater, and prairie dogs recharge groundwater. There is a connection, if only we can make it.

We have to make it: the Southwest is in serious trouble. Decades of poorly managed grazing leases have obliterated streams and rivers; invasive water-hogging trees crowd the banks of the Rio Grande, sucking out thousands of gallons; firestorms fed by decades of ill-conceived suppression have sterilized thousands of acres. Between 2002 and 2004, millions of acres of native piñon trees throughout the Southwest were killed by drought and the ensuing explosion of bark beetles. Both the drought and the beetle plague were consequences of rapid climate change.

Now that the American West has been systematically dewatered, now that virtually all major western cities are facing water crises and cycles of drought, treating species like prairie dogs as "vermin" is not simply backward but dangerous. Prairie dogs are a keystone species, critical to the biological restoration of prairie ecosystems.[5] They are also well on their way to becoming the next passenger pigeon, with the enthusiastic collusion of the federal government. There has been a drastic decline in their numbers, and the perception that they are common seems impossible to dispel. The naturalist Ernest Thompson Seton estimated that five billion black-tailed prairie dogs—one of five species of the burrowing rodent—populated up to a hundred million hectares of shortgrass prairie in North America in the nineteenth century.[6] Texas alone was thought to contain 800 million, with a single "town" of 37,000 square miles populated by 400 million dogs.[7] Lewis and Clark were fascinated by them, and "by poreing a great quantity of Water in his hole," they managed to catch a live specimen, which they sent to President Thomas Jefferson.[8]

But prairie dogs were loathed as pests and vermin by ranchers who believed they competed for forage with cattle and who feared their horses would break their legs in the holes. The Department of Agriculture declared war on them, and populations were wiped out through poisoning campaigns. By 1960, prairie dogs had declined by 98 percent, and the black-tailed was completely extirpated from Arizona. A photograph taken there in the early 1900s shows a pile of prairie dog carcasses as tall as a man.

Ironically, ranchers were shooting themselves in the foot by eradicating the dogs—another connection missed. Their removal is believed to be responsible for the conversion of six million acres of grassland to desert scrub in Arizona and New Mexico.[9] Mesquite, the woody shrub that virtually destroyed the rich grasslands of west Texas, spread widely after prairie dogs were wiped out: they trim and remove the plant. Studies during the 1970s and 1980s found that the presence of prairie dogs also increases diversity, protein content, and production of plant species, reversing soil compaction caused by cattle. The small creatures' tunneling and clipping of grasses had restored the soil in these arid plains, aerating and mixing it with organic material. "They made loam," Outwater writes, "in a sandy world."[10] Studies indicated that bison and pronghorn antelope grazed preferentially in and around prairie dog towns, which may be true of cattle as well.

But once vermin, always vermin: popular prejudice and local laws have not caught up with science. The prairie dog is still considered an agricultural pest requiring mandatory control in Wyoming, Colorado, Kansas, and South Dakota. Montana, New Mexico, North Dakota, Oklahoma, Texas, and Nebraska have repealed these laws, but landowners may still apply for financial assistance from the government, through the USDA's Wildlife Services Program (formerly Animal Damage Control), which spends over ten million dollars a year to kill a hundred thousand animals, including prairie dogs.[11] On public lands, tens of thousands of acres are poisoned annually, and prairie dogs are exterminated even in our national parks.[12]

The program is counterproductive in another way as well. Prairie dogs hold the only hope of saving one of the country's most endangered species, the black-footed ferret. The ferret is so heavily reliant on prairie dog, its principal food source, that its numbers were reduced to less than twenty in a near-extinction event directly related to prairie dog decline. Required by law to reintroduce the species, the federal government has spent millions to bring the black-footed ferret back—there are currently a few hundred surviving in the wild in several western states and in Mexico—yet the government continues to poison its prey, prairie dogs. The Bush administration refused to add the black-tailed prairie dog to the endangered species list after Colorado congressman Ken Salazar—subsequently appointed as secretary of the interior by President Obama—threatened to sue the U.S. Fish and Wildlife Service if it did so.[13]

Another connection willfully, inexcusably missed.

Even in progressive, environmentally aware Santa Fe, where I live, the connection between prairie dogs, river restoration, and water issues has been surprisingly difficult to establish. Built on the western edge of the Great Plains, Santa Fe grew up right in the middle of a vast prairie dog town, according to historical accounts. Prairie dogs still live here, eking out a precarious existence in parking lots, traffic medians, and waste strips. It's not uncommon to stop at a traffic light and see a prairie dog sitting bolt upright a few feet away, its front paws folded neatly against its chest, its trademark protruding stomach sagging above its toes. Sadly, it is also not uncommon to see the creatures squashed in the road.

But these urban survivors are precious. Santa Fe, along with a few other southwestern cities, holds the last remaining populations of Gunnison's prairie dog, *Cynomys gunnisoni*, one of the most threatened of the five species. Like all other prairie dogs, Gunnison's—more diminutive, shorter, and slimmer than the black-tailed prairie dog—exhibits an array of behaviors that is proving to be extraordinarily complex. A biologist at Northern Arizona University, Con Slobod-chikoff, has been studying communication in colonies of Gunnison's prairie dogs in the Flagstaff area.

A bald man in his sixties, shaped not unlike a prairie dog, Slobod-chikoff travels to cities with prairie dog populations to talk to audiences about the ecological importance of the species. "These are not varmints or mindless robots," he said recently in Santa Fe. And his work on prairie dog language testifies to their astonishing abilities. "Prairie dogs incorporate information about the color, size, and shape of predators, as well as their speed of travel," he said. "They can not only say who the predator is, but they can describe the predator, which is really amazing."[14]

Prairie dogs produce "two kinds of vocalization," according to Slobodchikoff, "social chatter—which we can't decipher, since we don't know what the context is—and alarm calls." His work has focused on the latter, "the Rosetta Stone for deciphering calls' meaning." Using computer analysis of recordings of prairie dog alarm calls, Slobodchikoff has discovered that they employ different alarm calls for different predators: humans, coyotes, dogs, hawks. He and his students record the calls, which can be matched on the computer, and simultaneously videotape the animals' escape responses. When they play the calls back to the animals, they have observed "escape behaviors that are appropriate" to the kind of predator that originally elicited them. Slobodchikoff has published some forty scientific papers on the phenomenon. "The most sophisticated natural animal language so far decoded," he has said of it.[15]

In 2006, Santa Fe—named for St. Francis of Assisi, patron saint of animals—finally started to wake up and make the connection between prairie dogs, water, and drought. Although the town has recently passed an ordinance requiring developers to relocate prairie dog colonies at their own expense, it was almost too late. Prairie dog colonies in town were suffering so severely that one animal was photographed climbing a tree, looking for food. It was a desperate act: Prairie dogs aren't built for climbing. Volunteering for a local group founded to protect prairie dogs from being poisoned in local parks, my husband and I joined a program to feed the animals, hauling past-their-prime vegetables donated by Whole Foods to a park where there was a once prosperous, now starving colony. Although the city had recently invested hundreds of thousands of dollars in planting native grasses and wildflowers in the park, known as Frenchy's, it too had become a victim of drought. The park was overtaken by invasive weeds inedible to prairie dogs, killing the native grasses they needed.

Once a week, we threw handfuls of chopped carrots, celery, lettuce, and corn into holes in the ground, an activity that felt alternately foolish and heartwarming. We were greeted by the dogs' loud "jump-yip" call, followed by their dive down into their "listening rooms," chambers just below the surface, where they monitor activities above.[16] On our way back to our car, we saw prairie dogs popping up across the park, holding a piece of organic produce in their front paws. We learned the difference between "toileting" holes, where dogs push debris and feces out of their burrows, and "spy" holes, which plunge straight down into the ground, furnished with stepping stones.

I have also, on occasion, helped trap prairie dogs around town for relocation: Santa Fe and several towns in Colorado have become the chief suppliers of Gunnison's to reintroduction projects in the Southwest. The volunteers refer to local prairie dog towns by the names of businesses whose properties they've colonized: the "Kinko's prairie dogs," the "Sonic prairie dogs," the "Enterprise prairie dogs," as if they had corporate sponsors. (Of course, the businesses just want to get rid of them.) An experienced volunteer scouts the property, flagging holes with colored flags to identify different "coteries," or families: trapped individuals must be kept with family members lest they attack strangers whose scent they don't recognize. Then traps baited with toothsome morsels—corn kernels, peanuts, sunflower seeds, watermelon, tomatoes, Doritos—are set in loose soil so that the dogs don't feel alien wire underfoot. The trap door is rigged to a rear platform, tripped when the animal steps on it to reach the bait.

On a recent June morning, I set twelve traps at the Santa Fe Rail-yard, an area long colonized by prairie dogs, slated to be developed as a city park. Most of the Railyard dogs had been caught by pumping water through their burrows—a safe and effective method with mature animals—but there were still babies in the holes. Their parents had been caught, and they were home alone. Furtively watching with binoculars from my car, I waited for several hours, trapping nothing but a couple of birds picking up the sunflower seed. At 10:53, the babies popped out, tiny, skinny, starving. They followed the trail of corn kernels laid from the burrow to the trap door, three, then five of them. They managed to extract bait by pulling it out through the sides of the trap, but eventually they lost their caution and entered, eating as they went, practically jumping up and down on the trigger. Nothing happened. They were too light to spring the trap.

By noon, they'd emptied one trap and started on another. Then I hit the jackpot: three of them stood on the trigger at once. Wearing rubber gloves, covering the trap with a towel, I upended it over a pet carrier full of hay. Two fell in but a holdout clung to the stern, like Leonardo DiCaprio in *Titanic*. I blew softly in his face. Horrified, he let go. It was a small personal triumph, a three-dog strike in the rewilding revolution.

The young prairie dogs I trapped would be relocated to one of two sites, the 230,000-acre Sevilleta National Wildlife Refuge to the south, where prairie dogs were eradicated fifty years ago, or Wind River, a private conservation reserve in Mora County, north of Santa Fe. At Sevilleta, three hundred Gunnison's prairie dogs have been temporarily installed in artificial burrows fashioned out of plastic buckets and tubing, capped at the aboveground exits to allow the dogs to become acclimated while keeping predators at bay. Researchers stroll through the colony several times a night to prevent the dogs from being harassed by raptors or badgers. During the first year, 2005–2006, severe weather contributed to a 50 percent mortality rate, but 150 survived. Burrowing owls, which colonize prairie dog burrows, have begun nesting in the area, along with badgers, coyotes, kit foxes, rattlesnakes, and hawks, all natural predators of this keystone species.

Translocating prairie dogs to multiple sites will help to jump-start a "metapopulation," artificially re-creating the means by which one population meets up with another.[17] Metapopulations—different groups that may be separated geographically but occasionally absorb dispersing individuals—strengthen the species, allowing gene exchange. This could be crucial in maintaining Gunnison's, which is susceptible to outbreaks of plague. And Gunnison's prairie dogs, in turn, could help to restore rivers like the Rio Grande, which supply drinking water to Santa Fe and other Southwest cities.

For all the obstacles—especially the stubborn misperception that they are vermin—the fate of these prairie dogs shows that people can make connections. And they are making them, around the country and around the world. From the American Southwest to the anaconda-ridden rivers of Brazil, from the pathways of elephants across southern Africa to the rhino latrines of Nepal, rewilding is approaching a kind of critical mass. We must help it get there.

At the outset, in their first vision of rewilding, Michael Soulé and Reed Noss told us how. "The greatest impediment to rewilding," they wrote, "is an unwillingness to imagine it."[18] In the years since, we have begun to imagine it, making plans, laying foundations, collecting native seed and planting it. Now we have to take it further. Rewilding is a science, but it is also a social movement on a grand scale. To bring ourselves in line with our environment we may need to reengineer our own behavior—our consumption, our habits, our modes of thought. Asking Namibians not to shoot lions and the Tharu to tolerate tigers, rhinos, and elephants, we had better be willing to step up.

We need to rethink the way we practice and finance conservation. We need creative financing—tools to evaluate ecosystem services and methods to extract revenue from rain forests—as has recently been attempted in the Iwokrama, an enormous expanse of tropical forest that forms the core of the Guiana Shield Initiative, an enormous project designed to protect intact forests across six South American countries.

In 2008, the Iwokrama reported revenues of $2.4 million and a profit of close to a million from ecotourism, research, sales of products (honey and natural oils), bioprospecting, and selective logging.[19] The international board of trustees that manages the region is currently negotiating a deal with a private investment firm to sell carbon sequestration services, a project that may become a model for other forests around the world.

We need smarter, more efficient rewilding. Worldwide, conservation spending totals around $10 billion a year, mostly in the developed world, where it is least needed. The developing world, including the planet's entire tropic zone—where biodiversity is concentrated and where it is most endangered—receives less than a billion of that.

We need inspirational leadership and compelling voices: biologists, serious journalists, and people everywhere laboring in the trenches of restoration—from the Maasai to Prince William (who spent part of his gap year working at Lewa), and from the lonely volunteers who guard the last of the northern hairy-nosed wombats in Australia's Epping Forest to the parataxonomists of Costa Rica.

Perhaps most important, we need to reevaluate our place in the world. *Homo sapiens* is not the only species that matters. The naturalist Henry Beston put it eloquently:

> We need another and a wiser and perhaps a more mystical concept of animals. . . . We patronize them for their incompleteness, for their tragic fate of having taken form so far below ourselves. And therein we err, and greatly err. For the animal shall not be measured by man. In a world older and more complete than ours they move finished and complete, gifted with extensions of the senses we have lost or never attained, living by voices we shall never hear. They are not brethren, they are not underlings; they are other nations, caught with ourselves in the net of life and time, fellow prisoners of the splendor and travail of the earth.[20]

Beston wrote that in 1928, in his classic, *The Outermost House*, about his year on a Cape Cod beach. Have we made great progress since then? Every day more evidence arrives suggesting that we are not quite the paragon of creation we would like to think: chimps use tools; dolphins and elephants—even prairie dogs—seem to have a type of language.

Still we resist taking these "other nations"—and our reliance on them—seriously. Moved by the atavistic urge to separate ourselves from whence we came, we trivialize and infantilize wild animals. That is a mistake with consequences. The species we grudgingly share space with are indispensable, from cone snails to killer whales, from hookworms to hippos. They are partners, collaborating in the continuing life of the planet, "so different from each other, and dependent on each other in so complex a manner," as Darwin marveled.[21] Without them, the attainments we take for granted—agriculture, medicine, our tenuous security on this planet—fall away. As a species, it is high time we grew up and put away our childish view of nature, and took up our responsibilities.

The environment must become a top priority. Right now, it is not even on the list. Despite lip service paid to the biodiversity crisis and ecosystem services from the United Nations and the World Bank, despite increasingly dire projections from UN scientists, the governments that agreed to meet the 2010 goal of halting biodiversity loss will fail. They have vacillated, waiting to enact legislation to curb forest loss, erosion, unsustainable agriculture, waiting perhaps until it is too late. Rewilding should look to the success of the movement to confront climate change: the political momentum has at last begun to shift toward economic models that embrace, rather than deny, the need to transform our global energy economy. The biodiversity crisis, which presents us with dangers as profound and costly as climate change does, demands the creation of an equal and corresponding political will to act. The environment *is* the economy. No problem—not poverty, not climate change, not the economic downturn—can be addressed without simultaneously restoring the systems that are life itself.

We are so close. We are edging ever nearer to Aldo Leopold's land ethic, enlarging our communities "to include soils, waters, plants, and animals."[22] We are realizing that conservation is not about managing wildlife as much as it is about managing ourselves—our appetites, expectations, fears, our fundamental avariciousness. If we do not succeed at that, other forces assuredly will. Judging by the pain already caused by climate change and biodiversity loss—flood, famine, fissures—those forces will not be kind.

In a recent interview, E. O. Wilson posed the question, "Isn't it morally wrong to destroy the rest of life?"[23] The answer was self-evident. Such destruction, he said, "should be a horror to people." It is. In the most fundamental ways, people are recognizing the suffering we cause as we destroy species and ecosystems. In northern Kenya, Samburu children are running after tourist vehicles, begging for water: decades of deforestation are drying up the monsoon rains. In the American Southwest, fissures are opening in the earth, swallowing horses, swallowing houses. These horrors must be met with action to restore ecosystems, with a global response commensurate to the problem. But we can start small. Start with a river. Start with a road, making it safe for wildlife. Start with a prairie dog.

Most of us will never lay eyes on that rare night-feeding moth in Madagascar, the Predicta moth. Charles Darwin never did. But he made the connection. Looking at a flower, he imagined the one creature that must exist to pollinate it, to perpetuate it. We must follow his lead. We must look to what is left of our planet. In rewilding, we have dreamed up the ways and means to keep it alive. Now, we must only connect.

NOTES

Introduction: The Predicta Moth

1. William Stolzenburg, *Where the Wild Things Were: Life, Death, and Ecological Wreckage in a Land of Vanishing Predators* (New York: Bloomsbury, 2008), pp. 125–28. Soulé taught at UC San Diego during the late 1960s and 1970s and kept his home there after he moved to the University of Michigan in the 1980s, when he organized work on the canyon studies.
2. Ibid., p. 127. See also Michael E. Soulé et al., "Reconstructed Dynamics of Rapid Extinctions of Chaparral-Requiring Birds in Urban Habitat Islands," *Conservation Biology* 2 (March 1988): 75–92.
3. Soulé et al., "Reconstructed Dynamics." See also Kevin R. Crooks and Michael E. Soulé, "Mesopredator Release and Avifaunal Extinctions," *Nature* 400 (5 Aug. 1999): 563–66.
4. Rachel Carson, *Silent Spring* (Boston: Houghton Mifflin, 1962), p. 2.
5. Ibid.
6. Michael Soulé, Michael Gilpin, William Conway, and Tom Foose, "The Millennium Ark: How Long a Voyage, How Many Staterooms, How Many Passengers?" *Zoo Biology* 5 (1986): 101.
7. Lisa Jones, "Facts Compute, But They Don't Convert," *Sierra*, July–August 2003.
8. Edward O. Wilson, *The Future of Life* (New York: Alfred A. Knopf, 2002), p. 102.
9. International Union for Conservation of Nature, "The Review of the 2008 Red List of Threatened Species," *IUCN Red List of Threatened Species*, 10 Oct. 2008. Web, 15 May 2009.
10. Edward O. Wilson, "On the Future of Conservation Biology," *Conservation Biology* 14 (Feb. 2000): 1.
11. Eric Chivian and Aaron Bernstein, eds., *Sustaining Life: How Human Health Depends on Biodiversity* (New York: Oxford University Press, 2008), pp. 240–46.
12. Ibid., pp. 203–85.
13. Warren E. Leary, "Interior Secretary Questions Law on Endangered Species," *New York Times*, 12 May 1990.
14. Joel Achenbach, "McCain Sees Pork Where Scientists See Success: Candidate Criticizes Ambitious Bear Study," *Washington Post*, 10 March 2008.

15. Andrew Mitchell, "Viewpoint: Time to Invest in Nature's Capital," *BBC News*, 13 Oct. 2008. Web, 14 Oct. 2008.

16. Richard Black, "Nature Loss 'Dwarfs Bank Crisis,'" *BBC News*, 10 Oct. 2008. Web, 10 Oct. 2008.

17. The word "rewilding" was apparently coined by Dave Foreman; it first occurs in Jennifer Foote, "Trying to Take Back the Planet," *Newsweek*, 5 February 1990.

18. Michael Soulé and Reed Noss, "Rewilding and Biodiversity: Complementary Goals for Continental Conservation," *Wild Earth* 8 (Fall 1998): 19–28.

19. Ibid., p. 22.

20. Ibid.

21. David M. Olson and Eric Dinerstein, "The Global 200: A Representation Approach to Conserving the Earth's Most Biologically Valuable Ecoregions," *Conservation Biology* 12 (June 1998): 502–15.

22. Norman Myers, "Threatened Biotas: 'Hot Spots' in Tropical Forests," *Environmentalist* 8 (Sept. 1988): 1–20. See also Russell A. Mittermeier et al., *Hotspots: Earth's Biologically Richest and Most Endangered Terrestrial Ecoregions* (Mexico City: Cemex, 2000).

23. Charles Darwin, *The Various Contrivances by Which Orchids Are Fertilised by Insects* (London: John Murray, 1885), p. 163.

Chapter 1: Rewilding North America

1. "Initiative Inspired by Wandering Wolf," *Connections: Publication of the Yellowstone to Yukon Conservation Initiative* 13 (Winter 2007): 7.

2. Cornelia Dean, "Wandering Wolf Inspires Project," *New York Times*, 23 May 2006.

3. Paul Paquet, telephone interview, 30 July 2007.

4. J. Michael Scott et al., "The Issue of Scale in Selecting and Designing Biological Reserves," in Michael E. Soulé and John Terborgh, eds., *Continental Conservation: Scientific Foundations of Regional Reserve Networks* (Washington: Island Press, 1999), p. 23.

5. Douglas H. Chadwick, *Yellowstone to Yukon* (Washington: National Geographic, 2000), p. 91.

6. Dean, "Wandering Wolf."

7. David Quammen, *The Song of the Dodo: Island Biogeography in an Age of Extinctions* (New York: Scribner, 1996), p. 412.

8. Robert H. MacArthur and Edward O. Wilson, *The Theory of Island Biogeography* (Princeton: Princeton University Press, 1967), p. 3.

9. Ibid.

10. Ibid., p. 4.

11. Ibid., pp. 43–51.

12. Ibid., p. 181.

13. Quammen, *Song of the Dodo*, pp. 428–31.

14. Ibid., p. 11.

15. Jared M. Diamond, "Biogeographic Kinetics: Estimation of Relaxation Time for Avifaunas of Southwest Pacific Islands," *Proceedings of the National Academy of Sciences* 69 (Nov. 1972): 3199–203. See also Quammen, *Song of the Dodo*, pp. 442–43.

16. Quammen, *Song of the Dodo*, p. 444.

17. Ibid., p. 463.

18. Ibid., p. 481.

19. Reed F. Noss and Allen Y. Cooperrider, *Saving Nature's Legacy: Protecting and Restoring Biodiversity* (Washington: Island Press, 1994), p. 132.

20. William D. Newmark, "A Land-Bridge Island Perspective on Mammalian Extinctions in Western North American Parks," *Nature* 325 (29 Jan. 1987): 432. See also, William D. Newmark, "Extinction of Mammal Populations in Western North American National Parks," *Conservation Biology* 9 (June 1995): 512–26.

21. Aldo Leopold, *A Sand County Almanac* (New York: Oxford University Press, 1949), p. 198.

22. James Gleick, "Species Vanishing from Many Parks," *New York Times*, 3 Feb. 1987.

23. Ibid.

24. Rick Ridgeway, "Paths to Survival," *Patagonia*, Winter 2008, p. 1.

25. Dave Foreman, *Rewilding North America: A Vision for Conservation in the 21st Century* (Washington: Island Press, 2004), p. 117.

26. Michael E. Soulé and John Terborgh, eds., *Continental Conservation: Scientific Foundations of Regional Reserve Networks* (Washington: Island Press, 1999), p. 12.

27. William D. Newmark, *Conserving Biodiversity in East African Forests: A Study of the Eastern Arc Mountains* (Berlin: Springer-Verlag, 2002), p. 137.

28. Daniel J. Simberloff et al., "Regional and Continental Restoration," in Soulé and Terborgh, *Continental Conservation*, p. 70.

29. Susan Marynowski, "Paseo Pantera: The Great American Biotic Interchange," *Wild Earth*, special issue (1992): 71.

30. Charles C. Chester, *Conservation across Borders: Biodiversity in an Interdependent World* (Washington: Island Press, 2006), p. 179.

31. Ibid., p. 177.

32. Ibid., p. 171.

33. Paul Paquet, personal interview, 30 July 2007.

34. Tony Clevenger, "Highways through Habitats: The Banff Wildlife Crossings Project," *TR News* 249 (March–April 2007): 15. For more details on the Banff wildlife crossing structures, see "Ten Quick Facts about Highway Wildlife Crossings in the Park," Banff National Park of Canada, n.d. Web, 14 May 2009.

35. Barbara Christensen, "I-90 Safer for Wildlife and People," *Conservation Northwest*, 2 Sept. 2008. Web, 14 May 2009.

36. "Highway Overpass OK'd for Bighorns," *Tri-State Online*, 8 Jan. 2009. Web, 15 May 2009.

37. Paul Beier et al., "Collaborative, Focal Species Approaches to Linkage Conservation: Lessons from California and Arizona," PowerPoint presentation, p. 36.

38. Lance Morgan et al., *Baja California to the Bering Sea* (Montreal: Commission for Environmental Cooperation of North America and the Marine Conservation Biology Institute, 2005).

39. Chester, *Conservation across Borders*, p. 167.

40. William Newmark, personal interview, 11 June 2005. All subsequent quotations from Newmark in this section are from interviews conducted 11–18 June 2005.

41. For details on Berger's pronghorn studies, see "Preserving a Wild West, for Posterity and for Pronghorn," *Wildlife Conservation Society*, n.d. Web, 14 May 2009. See also Cory Hatch, "Biologist: Protect Pronghorn Migration Route," *Jackson Hole News and Guide*, 24 May 2006.

42. Jon P. Beckmann, "Northern Rockies: Using Scat-Detecting Dogs as a Tool to Model Linkage Zone Functionality for a Suite of Species," Society for Conservation Biology, Chattanooga, 15 July 2008, presentation.

43. Stephen E. Ambrose, *Lewis & Clark: Voyage of Discovery* (Washington, D.C.: National Geographic Society, 1998), pp. 154, 162.

Chapter 2: The Problem with Predators

1. Jesse Harlan Alderman, "Idaho Governor Calls for Gray Wolf Kill," Associated Press, 12 Jan. 2007, *CBS News*. Web, 14 May 2009.

2. Kenneth Brower, introd., *A Sand County Almanac with Essays on Conservation*, by Aldo Leopold (New York: Oxford University Press, 2001), p. 9.

3. Leopold, *Sand County Almanac*, pp. 204, 218.

4. Aldo Leopold, *Round River* (New York: Oxford University Press, 1993), pp. 145–46.

5. Ibid., p. 165.

6. Leopold, *Sand County Almanac*, p. 130.

7. Ibid.

8. Ibid., p. 132.

9. See Edward O. Wilson, "The Little Things That Run the World (The Importance and Conservation of Invertebrates)," *Conservation Biology* 1 (Dec. 1987): 344–46, and John Terborgh, "The Big Things That Run the World: A Sequel to E. O. Wilson," *Conservation Biology* 2 (Dec. 1988): 402–3.

10. Terborgh, "Big Things," p. 403.

11. William Stolzenburg, "Ecosystems Unraveling," *Conservation* 9 (Jan.–March 2008): 23. See also Stolzenburg, *Where the Wild Things Were: Life, Death, and Ecological Wreckage in a Land of Vanishing Predators* (New York: Bloomsbury, 2008), p. 94.

12. John Terborgh et al., "Ecological Meltdown in Predator-Free Forest Fragments," *Science* 294 (30 Nov. 2001): 1923–26.

13. William J. Ripple et al., "Trophic Cascades among Wolves, Elk, and Aspen on Yellowstone National Park's Northern Range," *Biological Conservation* 102 (2001): 227.

14. For a discussion of landscape-level change, see Joel Berger and Douglas W. Smith, "Restoring Functionality in Yellowstone with Recovering Carnivores," in Justina C. Ray, Kent H. Redford, Robert S. Steneck, and Joel Berger, eds., *Large Carnivores and the Conservation of Biodiversity* (Washington: Island Press, 2005), pp. 105–7.

15. Warren Cornwall, "Ecological Changes Linked to Wolves," *Seattle Times*, 12 Jan. 2005.

16. William J. Ripple and Robert L. Beschta, "Wolves and the Ecology of Fear: Can Predation Risk Structure Ecosystems?" *BioScience* 54 (Aug. 2004): 755–66.

17. John Terborgh, *Requiem for Nature* (Washington: Island Press, 1999), p. 115.

18. Ibid., p. 117.

19. Cornwall, "Ecological Changes."

20. For a historical overview of the research on top-down regulation, see Ray et al., *Large Carnivores and the Conservation of Biodiversity*, p. 15.

21. Robert G. Anthony et al., "Bald Eagles and Sea Otters in the Aleutian Archipelago: Indirect Effects of Trophic Cascades," *Ecology* 89 (10 Oct. 2008): 2725–35.

22. Stolzenburg, *Where the Wild Things Were*, p. 74.

23. Robert MacArthur, "Fluctuations of Animal Populations and a Measure of Community Stability," *Ecology* 36 (July 1955): 533–36.

24. Julie Cart, "Wolves Protected Again after Excessive Hunting," *Los Angeles Times*, 2 Oct. 2008.

25. Tami Abdollah, "Gray Wolves Regain Endangered-Species Protections," *Los Angeles Times*, 19 July 2008.

26. Defenders of Wildlife, "Statistics on Payments from the Bailey Wildlife Foundation Wolf Compensation Trust," 2008.

27. This program is the Bailey Wildlife Foundation Proactive Carnivore Conservation Fund.

28. Rocky Barker, "Thirteen Years On, Wolves Have Changed Friends, Foes Alike," *Idaho Statesman*, 14 Feb. 2008.

29. Bobbie Holaday, *The Return of the Mexican Gray Wolf* (Tucson: University of Arizona Press, 2003), p. 125.

30. Ibid., p. 161.

31. John Dougherty, "Last Chance for the Lobo: Mexican Wolves Caught in the Crossfire of the Battle over Public Lands," *High Country News*, 24 Dec. 2007.

32. John Duffield, Chris Neher, and David Patterson, "Wolves and People in Yellowstone: Impacts on the Regional Economy," Yellowstone Park Foundation, 2006.

33. Paul C. Paquet et al., "Mexican Wolf Recovery: Three-Year Program Review and Assessment," prepared by the Conservation Breeding Specialist Group for the U.S. Fish and Wildlife Service, Albuquerque, N.M., 2001.

34. Michael J. Robinson, "Wildlife Agency Is 'Collaborating' Gray Wolf to Death," *Albuquerque Journal*, 21 July 2008.

35. Michael Robinson, *Predatory Bureaucracy: The Extermination of Wolves and the Transformation of the West* (Boulder: University Press of Colorado, 2005).

36. *Wildlife Corridors Initiative Report* (Jackson: Western Governors' Association, 2008).

37. Warner Glenn, *Eyes of Fire: Encounter with a Borderlands Jaguar* (El Paso: Printing Corner Press, 1996), p. 8.

38. See Jack L. Childs, "Baboquivari Jaguar," in David E. Brown and Carlos A. Lopez Gonzalez, *Borderland Jaguars: Tigres de la Frontera* (Salt Lake City: University of Utah Press, 2001), pp. 125–28.

39. Peter Warshall, *Northern Jaguar Project*, slide presentation, Santa Fe, N.M., 12 Dec. 2007. Subsequent quotations from Warshall come from this presentation.

40. Craig Miller, telephone interview, 6 March 2008.

41. Jack L. Childs et al., "The Borderlands Jaguar Detection Project: A Report on the Jaguar in Southeastern Arizona," *Wild Cat News*, May 2007. See also Jack L. Childs and Anna Mary Childs, *Ambushed on the Jaguar Trail: Hidden Cameras on the Mexican Border* (Tucson: Rio Nuevo, 2008).

42. Julie Cart, "R.I.P. for Macho B, the Rare Jaguar Euthanized in Arizona," *Los Angeles Times*, 9 March 2009. See also Tony Davis and Tim Steller, "I Baited Jaguar Trap, Research Worker Says," *Arizona Daily Star*, 2 April 2009.

43. B. Poole, "Ruling: Feds Must Rethink Reasons for Not Helping Jaguars," *Tucson Citizen*, 31 March 2009.

44. Daniel Glick, "Leader of the Pack," *Audubon*, March 2006.

Chapter 3: Corridors in Central and South America

1. International Union for Conservation of Nature, *Guidelines for Protected Area Management Categories* (Gland: IUCN, 1994).

2. Pris Weeks and Shalina Mehta, "Managing People and Landscapes: IUCN's Protected Area Categories," *Journal of Human Ecology* 16 (2004): 255.

3. For details on the biodiversity of Central America, see "Mesoamerica," in Russell A. Mittermeier et al., *Hotspots Revisited* (Mexico City: Cemex, 2005).

4. Jocelyn Kaiser, "Bold Corridor Project Confronts Political Reality," *Science* 293 (21 Sept. 2001): 2196–99.

5. Ibid., p. 2197.

6. For the dim view of the MBC taken by academics, see, for example, Liza Grandia, "Between Bolivar and Bureaucracy: The Mesoamerican Biological Corridor," *Conservation and Society* 5 (2007): 478–503.

7. Kenton Miller, Elsa Chang, and Nels Johnson, *Defining Common Ground for the Mesoamerican Biological Corridor* (Washington: World Resources Institute, 2001), p. 31.

8. Ibid., p. 5.
9. Howard Youth, "Chain of Dreams: The Mesoamerican Biological Corridor," *Smithsonian Zoogoer* 34 (2005). The single protected area created by the MBC is the Corazón del CBM between Honduras and Nicaragua. See Alexander López and Alicia Jiménez, "Environmental Conflict and Cooperation: The Mesoamerican Biological Corridor as a Mechanism for Transborder Environmental Cooperation," *United Nations Environment Programme*, Dec. 2007, 22.
10. Wildlife Conservation Society, "WCS Is 'Paving' the Way for Jaguars," n.d. Web, 14 May 2009. See also "Central America Agrees to Jaguar Corridor," *Physorg.com*, 24 May 2006. Web, 14 May 2009.
11. Mel White, "Path of the Jaguar," *National Geographic*, March 2009, pp. 123–33. The first tentative rebranding of Paseo Pantera under a different name, "Paseo Tigre," appeared in the Wildlife Conservation Society's magazine; see Alan Rabinowitz, "Connecting the Dots: Saving the Jaguar Throughout Its Range," *Wildlife Conservation*, Jan.–Feb. 2006, p. 29.
12. Jim Simon, "Dogged Research: Endangered-Species Knowledge May Be Increased by Using Dogs First Trained to Sniff Out Drugs," *Seattle Times*, 25 Aug. 1997.
13. Leandro Silveira, personal interview, 6 July 2005.
14. For more on these initiatives, see Carlos A. Klink and Ricardo B. Machado, "Conservation of the Brazilian Cerrado," *Conservation Biology* 19 (June 2005): 707–13, and Monica B. Harris et al., "Safeguarding the Pantanal Wetlands: Threats and Conservation Initiatives," *Conservation Biology* 19 (June 2005): 714–20.
15. Wagner A. Fischer, Mario B. Ramos-Neto, Leandro Silveira, and Anah T. Jácomo, "Human Transportation Network as Ecological Barrier for Wildlife on Brazilian Pantanal-Cerrado Corridors," *Proceedings of the 2003 International Conference on Ecology and Transportation*, ed. C. L. Irwin, P. Garrett, and K. P. McDermott (Raleigh: Center for Transportation and the Environment, North Carolina State University), pp. 182–94.
16. Anah Tereza A. Jácomo and Leandro Silveira, "Note on *Eunectes Murinus* (Green Anaconda)," *Herpetological Review* 29 (1998): 241.

Chapter 4: Reconnecting the Old World

1. Aldo Leopold, *A Sand County Almanac* (New York: Oxford University Press, 1949), p. 219.
2. Richard Black, "World 'to Fail' on Nature Target," *BBC News*, 13 Oct. 2008. Web, 13 Oct. 2008.
3. Hannah Cleaver, "German Death Strip Starts Fresh Life as a Nature Sanctuary," *Telegraph*, 4 Aug. 2003.
4. International Union for Conservation of Nature, "From Death Zone to Lifeline: Iron Curtain Becomes Green Belt," press release, IUCN, 10 Sept. 2004.
5. Liana Geidezis and Melanie Kreutz, "Green Belt Europe: Nature Knows No Boundaries," *Urbani Izziv* 15 (2004): 136.
6. Stefan Beyer, personal interview, 22 July 2006.
7. Melanie Kreutz, personal interview, 25 July 2006.
8. Geidezis and Kreutz, "Green Belt Europe."
9. Louisa Schaefer, "Germany's 'Problem Bear,' Bruno, Is Dead," *Deutsche Welle*, n.d. Web, 22 Jan. 2007. Allan Hall, "Outcry as Bruno the Bear Shot Dead," *Scotsman.com*, 26 June 2006. Web, 22 Jan. 2007.
10. John D. C. Linnell, Christoph Promberger, et al., "The Linkage between Conservation Strategies for Large Carnivores and Biodiversity: The View from the 'Half-Full' Forests of Europe," in Justina C. Ray, Kent H. Redford, Robert S. Steneck, and Joel

Berger, eds., *Large Carnivores and the Conservation of Biodiversity* (Washington: Island Press, 2005), p. 386.

11. "Hunters Shoot Bruno the Bear," *Italian News General*, 26 June 2006. Web, 23 Jan. 2007.
12. Helmut Steininger, personal interview, 31 July 2006.
13. Alois Lang, personal interview, 2 Aug. 2006. Subsequent quotations from Lang are all from this interview.
14. More about the Prombergers' work can be found in Barbara Promberger, Christoph Promberger, and Jean C. Roche, *Faszination Wolf: Mythos Gefährdung Rückkehr* (Stuttgart: Franckh-Kosmos, 2002).
15. Barbara Promberger, personal interview, 12 Aug. 2006.
16. Chea-hoan Lim, "Official Recognition of DMZ Ecosystems' Importance and Threats to Them," *DMZ Coalition Newsletter* 3 (Aug. 2007). See also Ke Chung Kim, "Preserving Biodiversity in Korea's Demilitarized Zone," *Science* 278 (10 Oct. 1997): 242.

Chapter 5: Peace Parks and Paper Parks

1. Ke Chung Kim and Edward O. Wilson, "The Land That War Protected," *New York Times*, 10 Dec. 2002.
2. "It Began as a Bold Idea," National Park Service, n.d. PDF file.
3. Randy Tanner et al., "The Waterton-Glacier International Peace Park: Conservation amid Border Security," in Saleem H. Ali, ed., *Peace Parks: Conservation and Conflict Resolution* (Cambridge: MIT Press, 2007), pp. 188, 192.
4. For a history of the Condor Corridor peace park, see Saleem H. Ali, ed., *Peace Parks*, pp. 8–10. See also Russell A. Mittermeier et al., *Transboundary Conservation: A New Vision for Protected Areas* (Mexico City: Cemex, 2005), pp. 44–45.
5. For an account of the rapid expansion of international nongovernmental organizations, see Christine MacDonald, *Green, Inc.: An Environmental Insider Reveals How a Good Cause Has Gone Bad* (Guilford: Globe Pequot Press, 2008). The $261 million Moore contribution to Conservation International is described on p. 47.
6. Ibid., p. 21.
7. The first of the series of *Post* articles was David B. Ottaway and Joe Stephens, "Big Green: Inside the Nature Conservancy Nonprofit Land Bank Amasses Billions, Charity Builds Assets on Corporate Partnerships," *Washington Post*, 4 May 2003.
8. See Mark Dowie, *Conservation Refugees: The Hundred-Year Conflict between Global Conservation and Native Peoples* (Cambridge: MIT Press, 2009). See also Mark Dowie, "Conservation Refugees: When Protecting Nature Means Kicking People Out," *Orion*, Nov.–Dec. 2005, and Mark Dowie, "Wrong Path to Conservation in Papua New Guinea," *Nation*, 29 Sept. 2008.
9. Dorothy Zbicz, "Global List of Complexes of Internationally Adjoining Protected Areas," in Trevor Sandwith et al., *Transboundary Protected Areas for Peace and Cooperation* (Gland: IUCN, 2001), appendix 1. See also Mittermeier et al., *Transboundary Conservation*, p. 27.
10. See Kent Biringer and K. C. (Nanda) Cariappa, "The Siachen Peace Park Proposal: Reconfiguring the Kashmir Conflict?" in Ali, *Peace Parks*, pp. 284–85.
11. See Stephan Fuller, "Linking Afghanistan with Its Neighbors through Peace Parks: Challenges and Prospects," in Ali, *Peace Parks*, pp. 300–304.
12. See Michelle L. Stevens, "Iraq and Iran in Ecological Perspective: The Mesopotamian Marshes and the Hawizeh-Azim Peace Park," in Ali, *Peace Parks*, pp. 313–31.
13. See Belinda Sifford and Charles Chester, "Bridging Conservation across *La Frontera*:

An Unfinished Agenda for Peace Parks along the US-Mexico Divide," in Ali, *Peace Parks*, pp. 205–25.

14. See, for example, the extraordinary series of wall posters put out by Korck Publishing, including "Ungulates of Africa," "Primates of Africa," and "Carnivores of Africa," published in conjunction with Jonathan Kingdon, *The Kingdon Pocket Guide to African Mammals* (Princeton: Princeton University Press, 2004).

15. David Western, *In the Dust of Kilimanjaro* (Washington: Island Press, 1997), p. 37.

16. Kent H. Redford, "The Empty Forest," *BioScience* 42 (1992): 412–22.

17. Raymond Bonner, *At the Hand of Man: Peril and Hope for Africa's Wildlife* (New York: Knopf, 1993), p. 278.

18. John F. Oates, *Myth and Reality in the Rain Forest: How Conservation Strategies Are Failing in West Africa* (Berkeley: University of California Press, 1999), pp. 232–33.

19. Ibid., pp. 233–34. See also Arun Agrawal, "Community in Conservation: Beyond Enchantment and Disenchantment," Conservation and Development Forum discussion paper, University of Florida, Gainesville, 1997, p. 13, and Clark Gibson and Stuart Marks, "Transforming Rural Hunters into Conservationists: An Assessment of Community-Based Wildlife Management Programs in Africa," *World Development* 23 (1995): 941–57.

20. Kent H. Redford, "The Ecologically Noble Savage," *Cultural Survival Quarterly* 15 (1991): 46–48.

21. Jonathan S. Adams and Thomas O. McShane, *The Myth of Wild Africa: Conservation without Illusion* (New York: Norton, 1992), p. 247. The position taken by Adams and McShane in this book is puzzling given that they sneer at so-called integrated conservation and development projects (ICDPs)—claiming that "few . . . can claim great success, and some have been catastrophes," on p. 107—and then praise CAMPFIRE as "indispensable" on p. 183. While CAMPFIRE was technically an example of "utilization," both ICDP and utilization projects were part of the same trend to combine conservation and development.

22. Michael Wells and Katrina Brandon, with Lee Hannah, *People and Parks: Linking Protected Area Management with Local Communities* (Washington: World Bank, 1992).

23. George B. Schaller, *The Year of the Gorilla* (Chicago: University of Chicago Press, 1964), p. 7.

24. International Gorilla Conservation Programme, "Influencing Policy" and "Regional Collaboration," *International Gorilla Conservation Programme*, n.d. Web, 30 Jan. 2009.

25. "Gorilla Resurgence: Babies Born in Congo War Zone," *Guardian*, 30 Jan. 2009. Web, 30 Jan. 2009.

26. It should also be noted that the continuing conflicts in the DRC have taken a horrific toll on Virunga National Park rangers: 120 have been killed since 1996. In addition, the Congo government has struggled to pay rangers' salaries, and a number of NGOs have stepped in to fill the gap. See "Wildlife Conservation Society Says Key to Saving Mountain Gorillas Is to Save the Guards That Protect Them," *Wildlife Conservation Society*, 7 Jan. 2009. Web, 28 Jan. 2009.

27. Larry A. Swatuk, "Peace Parks in Southern Africa," PowerPoint presentation delivered at *Parks for Peace or Peace for Parks? Issues in Practice and Policy*, Woodrow Wilson International Center for Scholars, 12 Sept. 2005.

28. As described in William Wolmer, "Transboundary Conservation: The Politics of Ecological Integrity in the Great Limpopo Transfrontier Park," *Journal of Southern African Studies* 29 (March 2003): 261–78. Wolmer gives as his source J. C. Smuts, *Holism and Evolution* (London: Macmillan, 1926).

29. Wolmer, "Transboundary Conservation."

30. Rowan B. Martin, "The Influence of Veterinary Control Fences on Certain Wild Large Mammal Species in the Caprivi, Namibia," presentation at IUCN World Parks Congress, Durban, South Africa, 14–15 Sept. 2003.

31. One of Rupert's obituaries suggested that he had sought "expiation of apartheid guilt" through good works, including his commitment to conservation. See Randolph Vigne, "Anton Rupert," *Independent*, 13 Feb. 2006.

32. Ebbe Dommisse, *Anton Rupert: A Biography* (Cape Town: Tafelberg, 2005), pp. 157–59.

33. Ibid., p. 379.

34. The Peace Parks Foundation Web site featured this quotation on its home page in 2002, as cited in Wolmer, "Transboundary Conservation," n. 21. The formula has been altered over time. For the latest version, which omits the phrase "Dream of an Africa without fences," see *Peace Parks Foundation*. Web, 15 May 2009.

35. John Hanks, "Protected Areas during and after Conflict: The Objectives and Activities of the Peace Parks Foundation," address delivered at Parks for Peace International Conference on Transboundary Protected Areas as a Vehicle for International Cooperation, Cape Town, South Africa, 16–18 Sept. 1997.

36. Dommisse, *Anton Rupert*, p. 387.

37. Robert Mugabe, "Address by the President of the Republic of Zimbabwe, Cde. R. G. Mugabe," 10th Meeting of the Conference of the Parties to CITES, 9 June 1997, p. 18. PDF file.

38. Ibid., my italics.

39. For another, earlier iteration of this philosophy, see David Western, "Ecosystem Conservation and Rural Development: The Case of Amboseli," in David Western and R. Michael Wright, eds., *Natural Connections: Perspectives in Community-Based Conservation* (Washington: Island Press, 1994), p. 28.

40. *An Idea That Binds: How Anton Rupert's Philosophy of Co-existence and Partnership Culminated in Peace Parks*, pamphlet, Peace Parks Foundation, n.d., pp. 24–25.

41. John Hanks outlined these projects in his presentation at the Parks for Peace conference, "Protected Areas during and after Conflict."

Chapter 6: The Great Limpopo

1. Peter Godwin, "Without Borders: Uniting Africa's Wildlife Reserves," *National Geographic*, Sept. 2001, p. 6.

2. Ibid., p. 7.

3. The notion that Kruger or other ecosystems within southern Africa may have exceeded a so-called carrying capacity for elephant is currently a matter of debate. See, for example, Robert J. Scholes et al., *Summary for Policymakers: Elephant Management in South Africa* (Johannesburg: Witwatersrand University Press, 2007), pp. 1–21; and Rudi J. van Aarde and Tim P. Jackson, "Megaparks for Metapopulations: Addressing the Causes of Locally High Elephant Numbers in Southern Africa," *Biological Conservation* 134 (2007): 289–97.

4. G. A. Bradshaw et al., "Concepts: Elephant Breakdown," *Nature* 433 (24 Feb. 2005): 425–26.

5. Charles Siebert, "Are We Driving Elephants Crazy?" *New York Times Magazine*, 8 Oct. 2006, p. 44.

6. Ibid.

7. Bradshaw et al., "Concepts: Elephant Breakdown," p. 426.

8. Sanette Ferreira, "Problems Associated with Tourism Development in Southern Africa: The Case of Transfrontier Conservation Areas," *GeoJournal* 60 (2004): 310.

9. Jané Carruthers, *The Kruger National Park: A Social and Political History* (Pietermaritzburg: University of Natal Press, 1995), p. 113.
10. Rudi van Aarde et al., "Assessment of Seasonal Home-Range Use by Elephants across Southern Africa's Seven Elephant Clusters," Conservation Ecology Research Unit, University of Pretoria, Jan. 2005.
11. Jan Hennop, "Mandela Lets South African Jumbos into Mozambique for 'Superpark,'" Agence France-Presse, 4 Oct. 2001.
12. Anna Spenceley, "Tourism, Local Livelihoods, and the Private Sector in South Africa: Case Studies on the Growing Role of the Private Sector in Natural Resources Management," Sustainable Livelihoods in Southern Africa Research Paper 8, Institute of Development Studies, 2003, p. 96.
13. Marloes van Amerom and Bram Büscher, "Peace Parks in Southern Africa: Bringers of an African Renaissance?" *Journal of Modern African Studies* 43 (2005): 179.
14. Ebbe Dommisse, *Anton Rupert: A Biography* (Cape Town: Tafelberg, 2005), p. 383.
15. "Elephants Escape Cull Project," *Business Day* (South Africa), 4 April 2005. Web, 23 Sept. 2005. See also Leon Marshall, "Elephants Migrate to Mozambique," *Sunday Independent*, 27 March 2005.
16. South African National Parks, "The Great Elephant Indaba: Finding an African Solution to an African Problem," Kruger National Park, 19–21 Oct. 2004. PDF file.
17. Raymond Travers, personal interview, 28 Oct. 2005.
18. "A Park for the People? Great Limpopo Transfrontier Park—Community Consultation in Coutada 16, Mozambique," University of the Witwatersrand Refugee Research Programme, March 2002, p. 9.
19. Andrew Meldrum, "Anton Rupert: South African Tobacco Tycoon Who Promoted Equal Rights during the Apartheid Era," *Guardian*, 23 Jan. 2006.
20. Dommisse, *Anton Rupert*, p. 393.
21. Charles Besançon, "Trade-Offs among Multiple Goals for Transboundary Conservation," IUCN and World Commission on Protected Areas, 2005, PowerPoint, slide 32.
22. Godwin, "Without Borders," p. 21.
23. "Great Limpopo Transfrontier Park: Development of the Buffer Zone of the Limpopo National Park," *Peace Parks Foundation*, n.d. Web, 14 May 2009.
24. "A Park for the People?" p. 7.
25. Stephan Hofstatter, "Southern Africa: Fences Not the Only Barrier for Cross-Border Park," *Africa News*, 13 Aug. 2005. Web, 23 Sept. 2005.
26. S. L. A. Ferreira, "Communities and Transfrontier Parks in the Southern African Development Community: The Case of Limpopo National Park, Mozambique," *South African Geographical Journal* 88 (2006): 170.
27. van Amerom and Büscher, "Peace Parks in Southern Africa," pp. 159–82.
28. "A Park for the People?" p. 11.
29. Ibid., p. 13.
30. Marja Spierenburg, Conrad Steenkamp, and Harry Wels, "Enclosing the Local for the Global Commons: Community Land Rights in the Great Limpopo Transfrontier Conservation Area," *Conservation and Society* 6 (2008): 94.
31. Arrie van Wyck, telephone interview, 21 April 2008.
32. Werner Myburgh, telephone interview, 3 March 2009.
33. Anna Spenceley, "Tourism in the Great Limpopo Transfrontier Park," *Development Southern Africa* 23 (Dec. 2006): 656.
34. Mike Cadman, "Poachers Target Elephants, Rhinos," *Sunday Independent*, 1 April 2007.

35. Hofstatter, "Southern Africa."
36. Werner Myburgh, telephone interview, 3 March 2009.
37. Police Jeroboam Manzini, personal interview, 1 Nov. 2005.
38. According to a press release written by Raymond Travers, Kruger's spokesman, the park's executive director, Bandile Mkhize, signed a memorandum of understanding with community representatives on 16 Feb. 2005, "to encourage economic growth in the area around the park." Mkhize was quoted as saying, "It must be recognised that the KNP [Kruger National Park] is not a development organisation but rather facilitates local economic development without taking sole responsibility for promoting it." Whatever benefits were envisioned by Kruger had not yet reached Welverdiend by later that year.
39. Spenceley, "Tourism, Local Livelihoods, and the Private Sector in South Africa," p. 77.
40. van Amerom and Büscher, "Peace Parks in Southern Africa," p. 175.
41. Julian Matthews, "Can Eco-Tourism Save India's Tigers?" *Telegraph*, 10 Nov. 2008.
42. Spenceley, "Tourism in the Great Limpopo Transfrontier Park," p. 652.
43. See Eleanor Momberg, "Coast-to-Coast Parks Plan for 2010 Tourists," *Sunday Argus*, 24 Sept. 2006, and Leon Marshall, "Transfrontier Game Reserves to Share Out 2010 Benefits," *Sunday Argus*, 17 Feb. 2008.
44. The TFCA Route map was reproduced widely, on the Peace Parks Foundation Web site, in the *Sunday Argus* newspaper articles cited in n. 43, and in a PowerPoint presentation, "Positioning the SADC TFCAs as Africa's Premier International Tourism Destination," at the Meeting of South African Ministers, Johannesburg, 13 June 2005.
45. Sue Segar, "Two Oceans Trek," *Witness*, 6 Nov. 2008.
46. The "Diamond Route" is described on the De Beers Diamond Route Web site. See also "World's First Insect Eco-Tourism Route Announced," De Beers, press release, 7 July 2008.
47. Junie Wadhawan, "Ecotourism—Hope and Reality," *People and the Planet*, 5 Aug. 2008. Web, 15 Feb. 2009.
48. Anna Spenceley and Michael Schoon, "Peace Parks as Social Ecological Systems: Testing Environmental Resilience in Southern Africa," in Saleem H. Ali, ed., *Peace Parks: Conservation and Conflict Resolution* (Cambridge: MIT Press, 2007), p. 95.
49. Leo Braack, telephone interview, 11 April 2008.
50. André Boshoff, "The Baviaanskloof Mega-Reserve," Terrestrial Ecology Research Unit, 2005. PDF file. The Society for Conservation Biology examined the Baviaanskloof Mega-reserve's carbon sequestration claims before selecting the program to offset carbon associated with travel to its annual meetings. See Paul Beier, "SCB's Investment in Carbon Offsets: Frequently Asked Questions about the Baviaanskloof Megareserve Project, South Africa," *Society for Conservation Biology Newsletter* 14 (Nov. 2007): 1, 12–13. See also Benjamin Lester, "Sustainable Science: Greening the Meeting," *Science* 318 (5 Oct. 2007): 36–38.

Chapter 7: The Lubombo Transfrontier

1. Ambitious early plans for this TFCA may have been downscaled after the Southern Africa Development Committee canceled a $2.6 million grant, according to a 2004 World Bank "Implementation Completion Report."
2. Jill Gowans, "Maputaland, The Way It Was," *Citizen*, 25 Feb. 2005.
3. The history of ill-will occasioned in local communities by the creation of the Ndumo reserve is recounted in detail in Anna Spenceley, "Tourism, Local Livelihoods, and the Private Sector in South Africa: Case Studies on the Growing Role of the Private Sector

in Natural Resources Management," Sustainable Livelihoods in Southern Africa Research Paper 8, Institute of Development Studies, 2003, pp. 33–39. For the "fence telegram," see Jennifer L. Jones, "Transboundary Conservation: Development Implications for Communities in KwaZulu-Natal, South Africa," *International Journal of Sustainable Development and World Ecology* 12 (Sept. 2005): 272.

4. Jane Flanagan, "Park Slaughters Hippos to Pacify Evicted Tribe," *Sunday Telegraph* (London), 26 Nov. 2000, p. 39.

5. Roelof J. Kloppers, "Border Crossings: Life in the Mozambique/South Africa Borderland since 1975," diss., University of Pretoria, 2005, p. 156.

6. Ibid., p. 161.

7. Ibid., p. 158.

8. Ibid., p. 159.

9. Roelie Kloppers, personal interview, 19 Oct. 2005. Subsequent quotations from Kloppers are from this interview.

10. "1,800 Jobs Created to Date in TFCA," *Conservation without Borders: Newsletter on Development Projects of Ezemvelo KZN Wildlife in the Lubombo Transfrontier Conservation Area* 18 (Jan. 2007): 1.

11. "Tshanini Produces Top SA Student," *Conservation without Borders* 17 (Oct. 2006): 2.

12. "Update on Tortoise Programme at Tshanini," *Conservation without Borders* 22 (April 2008): 2.

13. "More Than 400 Participate in Training Programme to Date!" *Conservation without Borders* 19 (May 2007): 1.

14. Jones, "Transboundary Conservation," p. 274.

15. "Lion King Brings Regal Cargo Home to Roam," *Cape Argus*, 1 June 2002.

16. Sipho Sibiya, personal interview, 20 Oct. 2005.

17. Thom Mahamba, personal interview, 19 Oct. 2005.

18. "New Community Conservation Area Launched in Northern KZN," *Wildlands Conservation Trust News*, 27 March 2007. Web, 22 Aug. 1007.

19. "Elephant Contraception," Tembe Elephant Park, press release, 27 Feb. 2008.

20. "Elephant Birth Control Practiced in Tembe Elephant Park," *RedOrbit*, 28 Feb. 2008. Web, 15 May 2009.

21. Ibid.

22. "Lubombo: Progress Report, 2007," *Peace Parks Foundation*, 3 May 2007. Web.

23. Werner Myburgh, telephone interview, 3 March 2009. All subsequent quotations from Myburgh are from this interview.

24. Ernest Robbertse, "The Fence," e-mail to author, 1 March 2009.

25. "The Lubombo Route," *The Lubombo Transfrontier Conservation and Resource Area*, n.d. Web, 20 Feb. 2009.

26. "Monitoring Project Now Underway," *Sunday Tribune*, 17 Feb. 2008.

27. Tania Griffin, "Crowning Glory," *Explore South Africa*, Dec. 2007. This article claims that the Royal Jozini is "part of the Trans-Frontier Game Reserve."

28. Ibid.

29. The Peace Parks Foundation Web site posted articles and descriptions confirming the organization's approval of the Royal Jozini. See Suren Naidoo, "Massive Swazi/Jozini Resort," *Mercury Network*, 30 Oct. 2007; "New Partners for Royal Jozini Big 6," *Peace Parks Foundation*, 25 June 2008; and Sue Lewitton, "SA and Swaziland Drop Fences to Form Transfrontier Park," *South Africa Tourism Update*, 1 Oct. 2008.

Chapter 8: Looking for KAZA

1. Russell A. Mittermeier et al., *Transboundary Conservation: A New Vision for Protected Areas* (Mexico City: Cemex, 2005), p. 265.
2. "First Milestone toward the World's Biggest Conservation Area," *Peace Parks Foundation*, 1 July 2008. Web.
3. "The Kavango-Zambezi Transfrontier Conservation Area: A World-Class Destination Now in Development," brochure, Conservation International, n.d.
4. Alison Leslie, telephone interview, 11 May 2007. Subsequent quotations from Leslie are from this interview.
5. "The Fishes of the Okavango Delta Fact Sheet," Harry Oppenheimer Okavango Research Centre, University of Botswana, March 2007. PDF file.
6. Robert S. Steneck and Enric Sala, "Large Marine Carnivores: Trophic Cascades and Top-Down Controls in Coastal Ecosystems Past and Present," in Justina C. Ray et al., eds., *Large Carnivores and the Conservation of Biodiversity* (Washington: Island Press, 2005), p. 112.
7. Details regarding the "Crocodiles of the Okavango" study can be found in Alison J. Leslie, "Expedition Briefing: Crocodiles of the Okavango," *Earthwatch Institute*, 2006. Quotations from Kevin Wallace and Audrey Detoeuf-Boulade from personal interviews, 13–26 Sept. 2006.
8. Wolfgang Saxon, "Richard K. Root, Educator and Infectious Disease Specialist, Is Dead at 68," *New York Times*, 21 March 2006.
9. Alistair Graham, with illustrations by Peter Beard, *Eyelids of Morning: The Mingled Destinies of Crocodiles and Men* (San Francisco: Chronicle Books, 1990), pp. 200–201.
10. "Mischievous Sam the Crocodile," *Botswana Daily News*, 3 Oct. 2005.
11. "Crocodiles Finally Released! Young Crocs Released into Delta," *Ngami Times*, 17 Oct. 2008.
12. Leon Marshall, "Long Walk to Freedom as Elephants Trek to Angola," *Sunday Independent*, 27 Jan. 2008.
13. Linda Pfotenhauer, "Elephants without Borders," *Peolwane*, Sept. 2007. PDF file; Mike Chase, "Interview with Dr. Mike Chase," YouTube video, *Elephants without Borders*, n.d. Web, 15 May 2009.
14. Mike Chase, telephone interview, 24 Aug. 2007. Subsequent quotations are from this interview.
15. Leon Marshall, "Easing the Pressure: Transfrontier Park Will Ease Pressure on the Environment," *Daily News*, 1 March 2007.
16. "Integrated Development Plan for the Zambian Component of the Kavango-Zambezi Transfrontier Conservation Area," *Peace Parks Foundation*, June 2008. PDF file.
17. See "Victoria Falls Could Lose World Heritage Status," *Afrol News*, 14 Nov. 2006, and "Row over Peace Park Erupts," *Botswana Guardian*, 9 Jan. 2007.
18. Joseph J. Schatz, "In Zambia, Battle over Future of Victoria Falls," *Christian Science Monitor*, 21 March 2007.
19. Sibangani Mosojane, telephone interview, 10 July 2007.
20. Mike Chase's documentary, titled *Elephants: Breaking Boundaries*, aired in the United States on Animal Planet, 3 May 2009.
21. Paul Theroux, *My Secret History* (New York: Putnam, 1989), p. 315.

Chapter 9: The Conservancy Movement

1. G. Borrini-Feyerabend, A. Kothari, and G. Oviedo, *Indigenous and Local Communities and Protected Areas: Towards Equity and Enhanced Conservation, Guidance on Policy and*

Practice for Co-managed Protected Areas and Community Conserved Areas (Gland: IUCN, 2004). This IUCN document tiptoes around issues related to the illegal use of natural resources and the trade in wildlife products. It never mentions "poaching," "law enforcement," "illegal hunting," or "bushmeat."

2. "Chapter 11: Principles of State Policy, Article 95, Promotion of the Welfare of the People," *The Constitution of the Republic of Namibia*, 1990. PDF file, p. 46.

3. "More Communities Turn to Conservancies," *Oasys Namibia*, 25 Aug. 2008. Web, 10 March 2009.

4. "Doing African Conservation the Sustainable Way," *Integrated Rural Development and Nature Conservation*, n.d. Web, 5 Feb. 2009.

5. "Conservancy Movement Takes Off in Namibia," *People and Planet*, 11 Oct. 2005. Web, 10 March 2009.

6. "Conservancy Movement Reaps Benefits in Namibia," *USAID Case Study*, n.d. Web, 10 March 2009.

7. Heather Zeppel, *Indigenous Ecotourism* (New York: Oxford University Press, 2007), p. 191.

8. Melissa de Kock, "Exploring the Effectiveness of Community-Based Natural Resource Management in Salambala Conservancy, Namibia," DISS Poverty and Natural Resource Management Seminar, 30 Oct. 2008. PDF file, p. 8.

9. "Updates on the Namibian Commiphora Project," *Milking the Rhino*, n.d. Web, 15 May 2009.

10. "Caprivi Report: 1 July 2007 to 30 June 2008," *Integrated Rural Development and Nature Conservation*, n.d. Web, 15 Feb. 2009. PDF file, p. 11.

11. *Milking the Rhino*, dir. David E. Simpson, Kartemquin Films, 2008.

12. Philip Stander, "CBNRM—Some Comments," e-mail to the author, 14 March 2009.

13. David Western, "Ecosystem Conservation and Rural Development: The Case of Amboseli," in David Western and R. Michael Wright, eds., *Natural Connections: Perspectives in Community-Based Conservation* (Washington: Island Press, 1994), p. 15.

14. Ibid., p. 18.

15. Ibid.

16. Ibid., p. 36.

17. Jerome Monahan, "Cruel Harvest," *TES Magazine*, 21 June 2002.

18. Peter H. Beard, *The End of the Game* (1963; Cologne: Taschen, 2008).

19. For a detailed history of the group ranches in Kenya, see Western and Wright, *Natural Connections*, pp. 30–50.

20. Meera Selva, "Kenya Launches New Poaching Crackdown to Protect Its Wildlife," *Independent*, 23 Sep. 2005. Web, 27 Feb. 2007.

21. Richard Leakey and Virginia Morell, *Wildlife Wars: My Fight to Save Africa's Natural Treasures* (New York: St. Martin's, 2002), pp. 91–92.

22. Bruce Wallace, "Richard Leakey," *Maclean's*, 9 Oct. 1995.

23. A history of Lewa can be found in Edward Paice, *Where Warriors Met: The Story of Lewa Downs, Kenya* (London: Tasker Publications, 1995).

24. *Milking the Rhino*.

25. Ibid.

26. Anna Merz, *Rhino: At the Brink of Extinction* (London: HarperCollins, 1991), p. 158.

27. W. F. Deedes, "Good Fences Make Good Neighbours," *London Telegraph*, 15 Jan. 2004. For more details on security at Lewa, see "Lewa Wildlife Conservancy: Annual Report 2007," *Lewa Wildlife Conservancy*, n.d. Web, 1 June 2008. PDF file, pp. 2–5.

28. *Milking the Rhino*.

29. Ibid.

30. *Community Development*, video, Lewa Web site, n.d.

31. Robert M. Pringle, Truman P. Young, Daniel I. Rubenstein, and Douglas J. McCauley, "Herbivore-Initiated Interaction Cascades and Their Modulation by Productivity in an African Savanna," *Proceedings of the National Academy of Sciences* 104 (2 Jan. 2007): 193.

32. Belinda Low et al., "Partnering with Local Communities to Identify Conservation Priorities for Endangered Grevy's Zebra," *Biological Conservation* (2009).

33. Joseph Kirathe, personal interview, 30 Aug. 2008.

34. "Lewa Wildlife Conservancy: Annual Report 2007," p. 7.

35. For details on the locations, budgets, and donors for conservancies in the Northern Rangelands Trust, see the NRT Web site at <http://www.nrt-kenya.org/home.html>.

36. In one noteworthy scene in the documentary *Milking the Rhino*, Samburu elders of the West Gate Conservancy discuss their concerns about the private tourism operator handling Bedouin Camp.

37. J. H. Patterson, *The Man-Eaters of Tsavo* (1907; Guilford: Lyons Press, 2004), p. 69.

38. Anna Maria Lolangwaso, personal interview, 26 Aug. 2006.

39. Keward Lekalkuli, personal interview, 26 Aug. 2006.

40. Joseph Kirathe, personal interview, 31 Aug. 2006.

41. Ian Craig, telephone interview, 17 June 2008.

42. "The South Rift Association of Land Owners (SORALO) Trust," *SORALO*, n.d. Web, 10 May 2009. PDF file. Another notable example in the south is the Mara Conservancy, founded in 2001 to manage the area around the Masai Mara National Reserve.

43. Ian Craig, "From the Field," *Lewa News* 26 (Oct. 2008): 5.

44. Martha Honey, *Ecotourism and Sustainable Development: Who Owns Paradise?* 2nd ed. (Washington: Island Press, 2008), p. 333.

45. *Milking the Rhino.*

Chapter 10: The Tiger Moving Game

1. Mark Poffenberger, "The Resurgence of Community Forest Management in Eastern India," in David Western and R. Michael Wright, eds., *Natural Connections: Perspectives in Community-Based Conservation* (Washington: Island Press, 1994), p. 60.

2. Hemanta Mishra, with Jim Ottaway Jr., *The Soul of the Rhino: A Nepali Adventure with Kings and Elephant Drivers, Billionaires and Bureaucrats, Shamans and Scientists, and the Indian Rhinoceros* (Guilford: Lyons Press, 2008), p. 45.

3. K. K. Gurung, *Heart of the Jungle: The Wildlife of Chitwan, Nepal* (London: André Deutsch, 1983), p. 1.

4. Nabin Baral and Joel T. Heinen, "The Maoist People's War and Conservation in Nepal," *Politics and the Life Sciences* 24 (18 April 2006): 5.

5. Marty Logan, "Nepal: Community Forests Rise above Obstacles," *Inter Press Service News Agency*, 29 July 2006.

6. Eric Dinerstein, *The Return of the Unicorns: The Natural History and Conservation of the Greater One-Horned Rhinoceros* (New York: Columbia University Press, 2003), p. 187.

7. "Terai Arc Landscape–Nepal Strategic Plan, 2004–2014," His Majesty's Government of Nepal, Ministry of Forests and Soil Conservation, 2004, p. 4.

8. Colin Eisler, *Dürer's Animals* (Washington: Smithsonian Institution Press, 1991), pp. 269–71.

9. Mishra, *Soul of the Rhino*, p. xv.

10. Dinerstein, *Return of the Unicorns*, p. 28.

11. Ibid., pp. 153–77.

12. Ibid., p. 153.
13. Dinerstein, *Return of the Unicorns*, p. 155. The "megafaunal fruit syndrome" was originally described by Daniel Janzen and Paul Martin. See D. H. Janzen and P. Martin, "Neotropical Anachronisms: What the Gomphotheres Ate," *Science* 215 (1 Jan. 1982): 19–27.
14. Eric Dinerstein, *Tigerland and Other Unintended Destinations* (Washington: Island Press, 2005), p. 255.
15. Dinerstein, *Return of the Unicorns*, p. 244.
16. Mingma Sherpa, e-mail to the author, 24 Aug. 2005.
17. Dinerstein, *Tigerland*, p. 257.
18. Ibid.
19. Dinerstein, *Return of the Unicorns*, p. 198.
20. Ibid., p. 254.
21. Ibid., p. 253.
22. Dinerstein, *Tigerland*, pp. 259–60.
23. Baral and Heinen, "The Maoist People's War," p. 5.
24. Ibid., p. 6.
25. Ibid.
26. Ibid.
27. Ibid., p. 7.
28. "35 Nepal Firms May Close after Threats from Maoist Rebels," *Daily Telegraph*, 9 Sept. 2004.
29. Lee Poston, "Alarming Decline in Nepal's Rhinos and Tigers," press release, *WWF*, 31 May 2006. Web, 1 June 2006.
30. Mishra, *Soul of the Rhino*, p. 201.
31. Dinerstein, *Tigerland*, pp. 258–59.
32. Mishra, *Soul of the Rhino*, p. 214.
33. Esmond Martin, personal interview, 6 Sept. 2006.
34. Himalaya Keshab Shrehtha, personal interview, 8 Oct. 2006.
35. Esmond Martin, "Rhino Poaching in Nepal during an Insurgency," *Pachyderm* 36 (Jan.–June 2004): 87–98. Details regarding the 2003 campaign against poaching in Chitwan are all taken from this article.
36. One of the principal organizations involved is a British group, the International Trust for Nature Conservation, founded by the late owner of Tiger Tops, Jim Edwards, which aided the Royal Chitwan National Park in creating its first antipoaching program in 1991 and continues to support it.
37. Information has since emerged confirming that members of Nepal's army were actively involved in poaching during the war. In 2005, the arrest of a notorious rhino poacher, Pemba Lama Yakche, was hailed by conservationists. But in early 2007, Yakche was given an unusually lenient verdict by Chitwan's chief warden, who has quasijudicial authority to hold prisoners and judge offenses; the decision was so suspect, suggesting government complicity, that it touched off a riot at the Chitwan National Park office. Later that year, another significant arrest was made, this time of former airline pilot Ramesh Chandra Pokhrel, who confessed his role in leading a poaching ring active in Chitwan during the war; the ring included army personnel who supplied weapons and ammunition. See "A Nexus Found in the Most Pressing Rhino Poaching Issue," *NepalBizNews.com*, 14 March 2007. Web, 15 May 2009. See also Mishra, *Soul of the Rhino*, pp. 215–16.
38. Martin, "Rhino Poaching in Nepal during an Insurgency," p. 94.
39. Thomas Bell, "Let Down by Its Protectors, Pursued for a Myth, the Asian Rhino Could Soon Be Extinct," *Daily Telegraph*, 3 January 2007.

40. Bishnu Prasad Aryal, personal interview, 10 Oct. 2006. Rajesh Aryal served as translator.

41. Basant Subba, "Growing Human-Elephant Conflict," *Rising Nepal*, 8 Dec. 2006.

42. Santosh Nepal, personal interview, 19 Oct. 2006. Subsequent quotations from Santosh Nepal are from this interview.

43. Bhadai Tharu, personal interview, 19 Oct. 2006.

44. "Bhadai Tharu: Chairman of the Gauri Mahila Community Forest User Group," *Tiger Watch* 7 (Fall 2004): 5.

45. "Threats in Nepal: Human-Animal Conflicts," *WWF*, 26 Oct. 2006. Web, 15 May 2009.

46. Dhan Rai, personal interview, 23 Oct. 2006.

47. Anand Chaudhary, personal interview, 24 Oct. 2006.·

48. "Little Maoists," *Nepali Times*, 23–29 March 2007.

49. "Tiger Numbers Could Be Doubled in South Asia," *ScienceDaily*, 6 Nov. 2007. Web, 19 May 2009. There is some irony in the fact that the information in this piece, confirming the dire situation of tiger populations throughout Asia, was presented as a hopeful opportunity. The source was the Wildlife Conservation Society, citing a study examining the potential capacity of existing reserves. While lauding that capacity, the organization did not explain how poaching would be addressed. Since then, the WCS has announced yet another new "Global Tiger Initiative," with $2.8 million in funding from the World Bank and the Global Environment Facility, promising to work with China and Vietnam to attack the illegal wildlife trade. See "Tigers Get a Stimulus Plan," *ScienceDaily*, 27 Feb. 2009. Web, 19 May 2009.

50. Charles Havilland, "Prachanda: The Challenges Ahead," *BBC News*, 15 Aug. 2008. Web, 18 Aug. 2008. On the peril to wildlife caused by the decline in security, see also Navin Singh Khadka, "No Peace Dividend for Nepal's Wildlife," *BBC News*, 5 Feb. 2007. Web, 24 March 2009.

51. Details of Operation Unicornis can be found in Tom Dillon, "Answering the Call of the Rhino: Spring Report 2008," *WWF*, n.d. Web, 31 March 2009.

52. "The Greater One-Horned Rhinoceros Conservation Action Plan for Nepal (2006–2011)," Government of Nepal Ministry of Forests and Soil Conservation, Department of National Parks and Wildlife Conservation, 2006, p. 29.

53. Katiana Murillo, "On the Record," interview with Alvaro Ugalde, posted on *Eco-Index: Connecting Conservationists across the Americas*, Jan. 2004. Web, 20 May 2007.

54. Eduardo Carrillo, "Tracking the Elusive Jaguar," *Natural History* 116 (May 2007): 33.

55. "Parks in Peril: Corcovado National Park," *The Nature Conservancy*, n.d. Web, 11 May 2009. The inadequacy of fifteen park rangers patrolling Corcovado was made clear in 2004, when the Moore Foundation donated $8 million, in part to address the problem; Costa Rica's Environment and Energy Ministry subsequently hired a staff of sixty-seven to manage Corcovado, including fifty-four rangers. Ironically, the Nature Conservancy was selected to manage the donation. Much of the grant went to purchase further properties around the Osa Peninsula, and by 2007 the money was exhausted. See Steven J. Barry, "Foundation Donates $8 Million to Stop Poaching in Southern Zone," *Tico Times*, 12 Nov. 2004.

56. Dave Sherwood, "Osa Park Guards Next Endangered Species as Donated Funds for Their Salaries Run Out," *Tico Times*, 25 May 2007.

57. Luis Angulo Angulo, personal interviews conducted 2–6 June 2007.

58. Eduardo Carrillo, telephone interview, 7 Aug. 2007.

59. See, for example, Thomas T. Ankersen, Kevin E. Regan, and Steven A. Mack, "Towards a Bioregional Approach to Tropical Forest Conservation: Costa Rica's Greater

Osa Bioregion," *Futures* 38 (2006): 421. The authors note: "There are too few park guards, many park staff lack uniforms and field equipment, and some guards suffer health problems from the difficult working and living conditions." See also "Costa Rica: Wildlife in Danger," *Global Exchange*, 18 March 2005. Web, 21 May 2007.

60. Ankersen et al., "Towards a Bioregional Approach," p. 426.

61. See Dan Brockington, *Fortress Conservation: The Preservation of the Mkomazi Game Reserve, Tanzania* (Oxford: James Currey, 2002).

62. In 2007 and again in 2009, I attempted to obtain the Nature Conservancy's response to these charges. Although Zdenka Piskulich, director of TNC's Costa Rica Country Program, initially agreed to speak with me, her office failed to return subsequent calls or answer e-mails.

Chapter 11: Resurrection Ecology

1. Cory Ritterbusch, "Curtis Prairie Restoration," *Prairie Works*, 28 Aug. 2007. Web, 5 Oct. 2008.

2. Paroma Basu, "Embattled Curtis Prairie a Test Bed for New Restoration Techniques," *University of Wisconsin–Madison News*, 6 July 2005.

3. Andre Clewell, John Rieger, and John Munro, "Guidelines for Developing and Managing Ecological Restoration Projects," 2nd ed., *Society for Ecological Restoration International*, Dec. 2005.

4. Richard Hobbs, "Looking for the Silver Lining: Making the Most of Failure," *Restoration Ecology* 17 (Jan. 2009): 2.

5. "Freshkills Park," *New York City Department of City Planning*, n.d. Web, 15 April 2009. See also "RePark: Recycle, Recollect, Recreate," PDF file.

6. George R. Robinson and Steven N. Handel, "Forest Restoration on a Closed Landfill: Rapid Addition of New Species by Bird Dispersal," *Conservation Biology* 7 (June 1993): 271–78.

7. "New York State Offers Hudson River Restoration Plan," *Environment News Service*, 24 Dec. 2005. Web, 15 April 2009.

8. Bill Gates was born in October 1955, after the discovery of algae in the lake.

9. Daniel Jack Chasan, "The Seattle Area Wouldn't Allow Death of Its Lake," *Smithsonian* 2 (July 1971): 6.

10. Ibid., pp. 6–7.

11. "Lake's Play Use Periled by Pollution," *Seattle Times*, 11 July 1955, as cited in Matthew Klingle, *Emerald City: An Environmental History of Seattle* (New Haven: Yale University Press, 2007), p. 212.

12. Klingle claims the fee was initially $2 per house per year (*Emerald City*, p. 220) while another source, "Control of Eutrophication in Lake Washington" (see n. 15, below), describes it as $2 per month. I am inclined to believe the higher figure, because of the bitterness of the subsequent debate over Metro.

13. Ibid., p. 214.

14. Ibid. One of the anti-Metro ads is reproduced on p. 215.

15. "Control of Eutrophication in Lake Washington," *Ecological Knowledge and Environmental Problem-Solving: Concepts and Case Studies* (Washington: National Academies Press, 1986), p. 306.

16. Ibid., p. 309.

17. Sandi Doughton, "Darwin's Fishes: The Threespine Stickleback of the Pacific Northwest," *Seattle Times*, 15 Feb. 2009.

18. For the importance of salmon to the Pacific Northwest ecosystem, see Klingle, *Emerald City*, pp. 17–18.

19. Ibid., p. 229.

20. Ibid., p. 3.

21. Hedrick Smith, senior producer and correspondent, *Poisoned Waters, Frontline*, PBS, 21 April 2009.

22. Rachel Ehrenberg, "Pacific Northwest Salmon Poisoning Killer Whales," *ScienceNews*, 23 Jan. 2009. Web, 10 March 2009.

23. M. L. Lyke, "Killer Whales Are Full of Toxic Chemicals, New Study Finds," *Seattle Post-Intelligencer*, 25 Oct. 1999.

24. *Shifting Baselines in the Sound*, video, *Shifting Baselines*, 6 April 2008. Web, 14 April 2009.

25. "Killer Whale Backgrounder," *Sierra Legal Defense Fund, Western Canada Wilderness Committee and the Georgia Strait Alliance*, 23 May 2007. PDF file, p. 1.

26. Smith, *Poisoned Waters*.

27. "Habitat Restoration," *Duwamish River Cleanup Coalition Superfund Focus Sheet*, Spring 2004, n.p.

28. "Puget Sound 2009–2010 'Shovel-Ready' Economic Stimulus Projects," *PugetSound-Partnership*, 3 March 2009. Web, 15 April 2009.

29. "Save Our Wild Salmon," press release, *Save Our Wild Salmon Coalition*, 9 March 2009.

30. Rachel La Corte, "New Agency Tackling Puget Sound Cleanup," *USA Today*, 26 Dec. 2007.

31. See the Web site, *Shifting Baselines: Common Sense for the Oceans*, at <http://www.shiftingbaselines.org>.

32. Jeremy Jackson, "Oceans on the Precipice: Scripps Scientist Warns of Mass Extinctions and 'Rise of Slime,'" *Scripps News*, Scripps Institution of Oceanography, 13 Aug. 2008.

33. *Shifting Baselines in the Sound*, video, *Shifting Baselines*, 6 April 2008. Web, 14 April 2009.

34. Chasan, "Seattle Area," p. 10.

35. Josh Donlan et al., "Re-wilding North America," *Nature* 436 (18 Aug. 2005): 913–14.

36. Paul S. Martin, *Twilight of the Mammoths: Ice Age Extinctions and the Rewilding of America* (Berkeley: University of California Press, 2005), pp. 30–40.

37. Ibid., p. 38.

38. Ibid., p. 200.

39. Ibid., pp. vii, 56.

40. Ibid., p. 211.

41. Donlan, "Re-wilding," p. 913.

42. Ibid., p. 914.

43. Ibid.

44. Michael Soulé, Michael Gilpin, William Conway, and Tom Foose, "The Millennium Ark: How Long a Voyage, How Many Staterooms, How Many Passengers?" *Zoo Biology* 5 (1986): 102–3.

45. "Back to the Future," *Economist*, 20 Aug. 2005, p. 62.

46. Dustin Rubenstein, Daniel Rubenstein, Paul Sherman, and Tim Caro, "Rewilding Rebuttal," letter, *Scientific American*, Oct. 2007, p. 12. See also Dustin R. Rubenstein, Daniel I. Rubenstein, Paul W. Sherman, and Thomas A. Gavin, "Pleistocene Park:

Does Re-wilding North America Represent Sound Conservation for the 21st Century?" *Biological Conservation* 132 (Oct. 2006): 232–38.

47. Bogonko Bosire, "African Conservationists Denounce Proposal for Giant US Wildlife Park," Agence France-Presse, 18 Aug. 2005.

48. William Stolzenburg, "Where the Wild Things Were," *Conservation in Practice* 7 (Jan.–March 2006): 34.

49. Sergey A. Zimov, "Pleistocene Park: Return of the Mammoth's Ecosystem," *Science* 308 (6 May 2005): 798.

50. Tom Mueller, "Ice Baby," *National Geographic*, May 2009, pp. 30–55.

Chapter 12: Costa Rica's Thousand-Year Vision

1. For the history of the Casona, as for much else about Dan Janzen and the evolution of the Area de Conservación Guanacaste, I am indebted to William Allen's wonderfully complete *Green Phoenix: Restoring the Tropical Forests of Guanacaste, Costa Rica* (New York: Oxford University Press, 2001), pp. 55–56.

2. Ibid., pp. 58–59.

3. Laura Tingley, "Costa Rica—Test Case for the NeoTropics," *BioScience* 36 (1986): 300.

4. Robert Langreth, "The World according to Dan Janzen," *Popular Science*, Dec. 1994, p. 80.

5. Daniel H. Janzen, "Costa Rica's Area de Conservación Guanacaste: A Long March to Survival through Non-Damaging Biodevelopment," *Biodiversity* 1 (2000): 18.

6. Allen, *Green Phoenix*, p. 33.

7. Ibid., p. 32.

8. Ibid., p. 31.

9. Ibid., p. 28.

10. Daniel H. Janzen, "Coevolution of Mutualism between Ants and Acacias in Central America," *Evolution* 20 (Sept. 1966): 264.

11. Janzen, "Costa Rica's Area de Conservación Guanacaste," p. 9.

12. Allen, *Green Phoenix*, p. 43.

13. Janzen, "Costa Rica's Area de Conservación Guanacaste," p. 10.

14. Ibid.

15. Allen, *Green Phoenix*, p. 47.

16. Janzen, "Costa Rica's Area de Conservación Guanacaste," p. 11.

17. Ibid., pp. 10, 11.

18. Allen, *Green Phoenix*, p. 255.

19. Janzen, "Costa Rica's Area de Conservación Guanacaste," p. 9.

20. Ibid., p. 15.

21. Ibid., p. 11.

22. Allen, *Green Phoenix*, p. 102.

23. Ibid., p. 192.

24. Ibid., p. 193.

25. Daniel Janzen, personal interviews, 29–30 May 2007. Unless otherwise noted, the quotations from Janzen that follow are from these interviews.

26. Allen, *Green Phoenix*, p. 278.

27. Daniel Janzen, "Gardenification of Tropical Conserved Wildlands: Multitasking, Multicropping, and Multiusers," *Proceedings of the National Academy of Sciences, USA* 96 (May 1999): 5987. See also Janzen, "Gardenification of Wildland Nature and the Human Footprint," *Science* 279 (27 Feb. 1998): 1312–13.

28. Janzen, "Gardenification of Tropical Conserved Wildlands," p. 5988.

29. Jeffrey C. Miller, Daniel H. Janzen, and Winifred Hallwachs, *100 Butterflies and Moths: Portraits from the Tropical Forests of Costa Rica* (Cambridge: Harvard University Press, 2007), p. 232.
30. Allen, *Green Phoenix*, p. 216.
31. Ibid., p. 224.
32. Carolina Cano Cano, personal interview, 31 May 2007. The quotations that follow are taken from this interview.
33. Daniel H. Janzen, "Lumpy Integration of Tropical Wild Biodiversity with Its Society," in *A New Century of Biology*, ed. W. John Kress and Gary W. Barrett (Washington: Smithsonian Institution Press, 2001), p. 143.
34. Douglas Sheil and Daniel Murdiyarso, "How Forests Attract Rain: An Examination of a New Hypothesis," *BioScience* 59 (April 2009): 341–47.
35. Susan E. Page et al., "The Amount of Carbon Released from Peat and Forest Fires in Indonesia during 1997," *Nature* 420 (7 Nov. 2002): 61–65.
36. Jane Braxton Little, "Regrowing Borneo's Rainforest—Tree by Tree," *Scientific American*. Web, 7 Jan. 2009.
37. Willie Smits, "Willie Smits Restores a Rainforest," presentation delivered at TED Conference (Technology, Entertainment, Design), Feb. 2009. Web, March 2009. See also "The Fairy Tale of Samboja Lestari," in Gerd Schuster, Willie Smits, and Jay Ullal, *Thinkers of the Jungle* (Konigswinter, Germany: H. F. Ullmann, 2008), pp. 300–317.
38. Ibid.
39. Little, "Regrowing Borneo's Rainforest."
40. "State of the World's Forests Report: 2007," *United Nations Food and Agriculture Organization*, as cited in Elisabeth Rosenthal, "New Jungles Prompt a Debate on Rain Forests," *New York Times*, 29 Jan. 2009.
41. Rosenthal, "New Jungles Prompt a Debate on Rain Forests."
42. Ibid.
43. J. Barlow et al., "Quantifying the Biodiversity Value of Tropical Primary, Secondary, and Plantation Forests," *Proceedings of the National Academy of Sciences, USA* 104 (20 Nov. 2007): 18555–60.

Chapter 13: Regrowing Australia

1. See Tim Flannery, *The Future Eaters: An Ecological History of the Australasian Lands and People* (New York: Grove Press, 1994).
2. Michael Parfit, "A Harsh Awakening: Australia," *National Geographic*, July 2000, pp. 2–31. See also the accompanying double map supplement, which contains figures on Australian extinctions.
3. Ken Eastwood, "Saving the Northern Hairy-Nosed Wombat," *Australia Geographic*, Oct.–Dec. 2003, pp. 72–83. See also a great classic of natural history, James Woodford, *The Secret Life of Wombats* (Melbourne: Text Publishing, 2001), pp. 130–51. I rejoice to add that Australian conservation officials have recently announced plans to establish a second colony of northern hairy-nosed wombats at a private nature refuge.
4. Tim Flannery, *A Gap in Nature: Discovering the World's Extinct Animals*, illus. Peter Schouten (New York: Atlantic Monthly Press, 2001), p. 146.
5. David Owen, *Thylacine: The Tragic Tale of the Tasmanian Tiger* (Crows Nest: Allen and Unwin, 2003), p. 182.
6. *A Million Acres a Year: A Devastating Examination of an Ecological Disaster*, dir. Frank Rijavec, writer, Keith Bradby, Snakewood Films, 2002.

7. Amanda Keesing, "FitzStirling Private Conservation Reserves as of September 2006," Gondwana Link Coordination Unit, 6 Sept. 2006.

8. Keith Bradby, personal interview, 3 Nov. 2006.

9. T. J. Lyons, "Clouds Prefer Native Vegetation," *Meteorology and Atmospheric Physics* (2002).

10. Ibid.

11. Nathan McQuoid, *Reasons for Richness: The Nature of the Fitzgerald Biosphere Flora*, comp. Amanda Keesing, Gondwana Link Coordination Unit, 2006.

12. Amanda Keesing, personal interview, 4 Nov. 2006.

13. Peter Luscombe, personal interview, 6 Nov. 2006.

14. Phillip Mayes, "Mates for Life: Pair Bonding in the Bobtail Skink," *Environmental Issues in Western Australia* (2006): 112–13.

15. Brian Penna, personal interview, 7 Nov. 2006.

16. Eugene Eades, personal interview, 4 Nov. 2006; see also Eades, "Unique Noongar 'Bush University' Promotes Conservation and Healing," press release materials developed for Nowanup Meeting Place Launch, *Greening Australia*, 9 May 2007.

17. Aden Eades, personal interview, 4 Nov. 2006.

18. "Cultural and Natural Heritage Management of the Pallinup River Catchment: Information Workshop," *Restoring Connections between People and Land in Southwestern Australia*, 9 Nov. 2006.

19. Hatley Coyne, personal interview, 9 November 2006.

20. Aldo Leopold, *A Sand County Almanac* (New York: Oxford University Press, 1949), p. 203.

21. Julie Cart, "What Will Global Warming Look Like? Scientists Point to Australia," *Los Angeles Times*, 9 April 2009.

22. Ibid.

23. Doug Russell, personal interview, 7 Nov. 2006.

24. Judy O'Neill, personal interview, 7 Nov. 2006.

25. Keith Bradby, personal interview, 8 Nov. 2006.

26. See Alexander Watson, Simon Judd, James Watson, Anya Lam, and David Mackenzie, "The Extraordinary Nature of the Great Western Woodlands," *The Wilderness Society*, June 2008.

27. Keith Bradby, "Gondwana Link: A Landscape Scale Restoration Project in South-west WA," *Gondwana Link*, 2009, p. 5. PDF file.

Conclusion: Only Connect

1. Mimi Diaz, "Chandler Heights Fissure Swallows Horse," *Arizona Geological Survey*, 24 July 2007. PDF file, pp. 1–3.

2. Sam Stoker, "The Problem with Fissures: As Arizonans Pump Evermore Water out of the Ground, Deadly Cracks Continue to Develop," *Tucson Weekly*, 27 Sept. 2007.

3. Alice Outwater, *Water: A Natural History* (New York: Basic Books, 1997), p. 76.

4. Ibid., p. 77.

5. The classification of the various species of prairie dog as keystone is accepted by a number of experts but not all. For a discussion of the issue, see Brian Miller, Richard P. Reading, and Steve Forrest, *Prairie Night: Black-Footed Ferrets and the Recovery of Endangered Species* (Washington: Smithsonian Institution Press, 1996), pp. 155–62, and C. N. Slobodchikoff, Bianca S. Perla, and Jennifer L. Verdolin, *Prairie Dogs: Communication and Community in an Animal Society* (Cambridge: Harvard University Press, 2009), pp. 136–38. For the argument questioning keystone status, see Paul Stapp, "A

Reevaluation of the Role of Prairie Dogs in Great Plains Grasslands," *Conservation Biology* 12 (Dec. 1998): 1253–59.

6. William Stolzenburg, "Understanding the Underdog," *Nature Conservancy*, Fall 2004, p. 28.

7. C. Hart Merriam, *The Prairie Dog of the Great Plains* (Washington: Yearbook of the United States Department of Agriculture, 1902), pp. 257–70.

8. *The Journals of Lewis and Clark*, ed. Frank Bergon (New York: Penguin, 1989), p. 42.

9. "Restoring the Prairie Dog Ecosystem of the Great Plains," *Predator Conservation Alliance*, Nov. 2001, p. 12.

10. Outwater, *Water*, p. 74.

11. See William Stolzenburg, "Us or Them?" *Conservation* 7 (Oct.–Dec. 2006): 16. For a detailed history of government poisoning campaigns in the United States, see Slobodchikoff et al., *Prairie Dogs*, pp. 178–81.

12. Ibid.

13. The Mexican (*Cynomys mexicanus*) and Utah (*Cynomys parvidens*) prairie dogs are currently listed as endangered; the black-tailed (*Cynomys ludovicianus*), white-tailed (*Cynomys leucurus*), and Gunnison's are not. In 2007, a federal court ordered the U.S. Fish and Wildlife Service to issue a new determination on Endangered Species Act protection of the Gunnison's, after it was revealed that the refusal to list the species—a decision issued the previous year by the deputy assistant interior secretary, Julie MacDonald—had been the result of political interference. MacDonald, who had no training in biology, subsequently resigned amid a congressional investigation into political manipulation of the endangered species program.

14. Con Slobodchikoff, "The Language of Prairie Dogs," presentation sponsored by Wild Earth Guardians and People for Native Ecosystems, Santa Fe, N.M., 25 Aug. 2005.

15. For more on prairie dog language, see C. N. Slobodchikoff et al., *Prairie Dogs*, esp. ch. 4, "Communication."

16. Russell A. Graves, *The Prairie Dog: Sentinel of the Plains* (Lubbock: Texas Tech University Press, 2001), p. 49.

17. Outwater, *Water*, p. 185.

18. Michael Soulé and Reed Noss, "Rewilding and Biodiversity: Complementary Goals for Continental Conservation," *Wild Earth* 8 (Fall 1998): 25.

19. "Growing on Trees: A Profitable Rainforest," *Economist*, 18 May 2009.

20. Henry Beston, *The Outermost House: A Year of Life on the Great Beach of Cape Cod* (New York: Henry Holt, 1949), pp. 24–25.

21. Charles Darwin, *On the Origin of Species: The Illustrated Edition* (New York: Sterling, 2008), p. 513.

22. Aldo Leopold, *A Sand County Almanac* (New York: Oxford University Press, 1949), p. 204.

23. E. O. Wilson, interview, *Bill Moyers Journal*, PBS, 6 July 2007.

ACKNOWLEDGMENTS

Since I first heard the spooky and evocative phrase *demographic winter* I have wanted to know more. In answering my questions, many people were extraordinarily generous with their time, expertise, and hospitality, and I am indebted to them all.

As influences, I turned often to David Quammen's *Song of the Dodo* for its wonderfully accessible explanation of the theory of island biogeography and history of the intellectual battles of conservation biology. Dave Foreman's talks, books, and Web site on rewilding—a word he coined—and a personal interview on the history of the Wildlands Project proved rich funds of information. The remarkable Michael Soulé answered questions on several occasions, supplied hard-to-find articles, and engaged in a thoughtful discussion on the future of conservation.

In North America, I thank Paul Paquet, John Oates, Gary Tabor, Joel Berger, and David Johns, as well as Sam Wasser and Carly Vynne at the University of Washington for an introduction to their incredible scat-detector dog program. Denise Saccone, now with Santa Fe's Prairie Dog Alliance, taught me the fine art of trapping prairie dogs. At the Sky Island Alliance, David Hodges and Kim Vacariu brought me up to speed on conservation in the Southwest, and Janice Przybyl, leader of Sky Island's peerless focal species program, tutored me in the art of tracking. Sergio Avila invited me to Mexico and drew his vision of jaguar corridors in the dust; may that vision become a reality. It was well worth struggling up and down Idaho hillsides after Bill Newmark, whose complex and nuanced discussions of corridors and conservation were a revelation.

For additional interviews, I thank Karen Baragona and Eric Dinerstein at the WWF, Hall Healy of the DMZ Forum, Patrick Bergin at the African Wildlife Foundation, Steve McCormick and Joy Grant at the Nature Conservancy, and Mark Chandler and Blue Magruder at Earthwatch. Nancy Baron, friend to journalists, was a fount of contacts at the 2006 Society for Conservation Biology conference.

On my travels—no matter how bemused by my purpose—conservationists and scientists helped me at every turn. In Brazil, Leandro Silveira and Anah Tereza de Almeida Jácomo were always gracious, despite occasionally slamming on the brakes for giant armadillos. I have never enjoyed midnight drives as much as those along the back roads of Emas with Cyntia Kashivakura and Mariana Furtado. Thanks as well to Andrew Terry at the IUCN, Michael Cramer in Berlin, Stefan Beyer in Mitwitz, and Melanie Kreutz in Nuremberg. Alois Lang's crash course in the dark history of European borders made me understand just how transformative the Green Belt might be. In Romania, Christoph and Barbara Promberger provided an entrée to the fascinating rural countryside, wildlife, and farming culture of Sinca Noua.

It was in Africa that I came to appreciate the gravity of the social issues that conservation is struggling with and the sacrifices it asks. Jennifer Jones took the time to discuss the social scientist's view. At Tembe Elephant Park in South Africa, Ernest Robbertse, Thom Mahamba, and Sipho Sibiya shone a light on life in rural KwaZulu Natal, and Roelie Kloppers shared his wealth of knowledge. I would also like to thank Willem van Riet and Werner Myburgh at the Peace Parks Foundation, Leo Braack at Conservation International, and John Hanks. Mike Stephens in Kruger and Jonathan Braack and Jimmy Ndubane at neighboring Ngala, one of South Africa's gorgeous private reserves, effortlessly produced an ark of African wildlife, from millipedes to elephants. In Welverdiend, Lotus Khoza, Isaac Tembe, and James Currie of the Africa Foundation introduced me to the other side of Kruger.

Many thanks to Samuel Andanje, William Ogara, Karl Ammann, Apollo Otieno, and Ian Craig for offering their views on Kenyan conservation. I am particularly indebted to my Wamba roommate, Anna Maria Lolangwaso, for a glimpse of Samburu life. In Nairobi, Esmond Martin dropped whatever he was doing when I rang and invited me to his home to discuss one of the most contentious aspects of conservation. His candor and plainspokenness were welcome gifts, as were a pile of monographs on the ivory and rhino horn trade. Driving around on the dirt roads of Lewa must always be a pleasure but was especially so with Joseph Kirathe, whose insights stayed with me.

In Botswana, Kevin Wallace and Audrey Detoeuf-Boulade made freezing nights grappling with reptiles on the Okavango Delta positively festive. I thank also Alison Leslie, as well as Sibangani Mosojane and Gaseitsiwe Masunga in Botswana's Department of Wildlife and National Parks.

In Nepal, Steve Webster offered invaluable travel advice and introductions as the war was winding down and a haven at Shivapuri Heights. My thanks to Jas Bahardur Rai and Arjun Thakuri, Jitu Chaudhary at Tiger Tops in Chitwan, Anita Thapa and T. B. at Tiger Tops Bardia, and Hari Shankar Aryal at Rhino Lodge. Despite the fact that I arrived during the holiday season, several WWF employees went to great lengths, at a most difficult moment for that organization, to ensure that I was able to see the Khata corridor and their community conservation work around Bardia and Dhangadi. I remain awed by the dedication of those I met there, including Tilak Dhakal, Santosh Nepal, Dhan Rai, Lakshmi Chaudhary, and Anand Chaudhary. Many thanks to Trishna Gurung at the WWF office.

Australia's southwest should be as famous for its unpretentious hospitality as for its remarkable accomplishments in conservation. Amanda Keesing and Paula Deegan ferried me everywhere, taking me into the Stirling Range, farmers' kitchens, and the karri-tingle forest, and I learned much from our far-ranging conversations. I am deeply grateful to Keith Bradby for discussions of the Gondwana Link philosophy, dinners, and for a personal tour of Yarrabee. I benefited from talks with Peter Luscombe, Angela Saunders, Jack Mercer, Doug Russell, Brian Penna, Judy O'Neill, and Simon Judd. I particularly wish to thank Eugene Eades for introducing me to his Noongar elders—including Aden Eades and Averil Dean—and for sharing the heartfelt story of his people.

Costa Rica and its people were a revelation—warm, open-hearted, fantastically committed to conservation. I thank Alejandro Masis, subdirector of the ACG, Róger Blanco, its associate director of research, and Carolina Cano Cano at the Area de Conservación Guanacaste. The masterful Luis Angulo Angulo laid bare the facts about Corcovado as he knew them, and I also thank Eduardo Carrillo for his forthrightness. I can add little to what has already been said about Dan Janzen and Winnie Hallwachs, except to say that they are two of the greatest gladiators conservation has ever known.

This book owes everything to its editor, the incomparable Sara Bershtel, a paleontologist of prose. Presented with miscellaneous bones and assorted body parts, she expertly assembled the beast, ferreting out its elusive argument, attaching essential structural components and connective tissue, discarding the odd bits that didn't belong. She took extraordinary pains to improve this immeasurably, and I cannot thank her enough. At Metropolitan Books she is surrounded by fine professionals, including Megan Quirk and Jason Ng, who kept everything on track. My thanks to Roslyn Schloss, whose copyediting smoothed out many errors and infelicities, conferring a wonderful clarity and polish. I'm grateful to two agents: Nina Ryan, who helped refine the proposal, and, later, Cynthia Cannell, who provided words of encouragement at a distressing moment.

Urging me on were Harbour Fraser Hodder, Rita and Doug Swan, Jeff Dreiblatt, William Walker, and Sheryl Whitney. Along with Mark Wheatley, the wickedly cheerful Colin Wheatley and his sister, Sarah Wheatley—staunch defender of wildlife since she could walk—exhibited such irrepressible enthusiasm for this book, I knew I had to finish it. Katie Wheatley kindly translated materials from Spanish, and Christopher Fraser and Linda Warner inspired the section on Puget Sound—but above and beyond those contributions, they are my essential support.

Since our earliest forays into the wilderness, startling bears, beavers, and frequently ourselves, Hal Espen has been my naturalist companion-in-arms. He went on many trips with me and tolerated a four-month stint when I went off by myself. Cautioning, challenging, advising, editing, fact-checking, he read countless versions and buoyed me through waves of turbulence. I hardly know how to thank him, but he has all my admiration, respect, and love. I hope that, in these pages, something of his wise counsel and joy in wombats and lorikeets shines through.

INDEX

CPSIA information can be obtained
at www.ICGtesting.com
Printed in the USA
LVOW10s1758250817
546384LV00002B/208/P